D1807072

THE LOGIC OF INNOVATION

This is a towering work of imagination and insight, elegantly written and compellingly argued. If all scholarship was like this our intellectual world would be infinitely enriched. Reading intellectual property law through the looking glass proves an effective device for understanding its maze of rabbit holes, where they lead us and why it matters.

Fiona Macmillan, Birkbeck University of London, UK

Johanna Gibson is Herchel Smith Professor of Intellectual Property Law and Director of the Queen Mary Intellectual Property Research Institute (QMIPRI), Queen Mary University of London, where she researches and teaches in intellectual property law. Gibson brings expertise in literature, art history, critical and cultural theory, and law, and consults regularly to industry, the profession, and UK and European government institutions. Gibson is widely published, including the following Ashgate monographs, *Intellectual Property, Medicine and Health* (2009), *Creating Selves: Intellectual Property and the Narration of Culture* (2006), *Community Resources: Intellectual Property, International Trade and the Protection of Traditional Knowledge* (2005), as well as being the editor of *Patenting Lives: Life Patents, Culture and Development* (2008). Before moving to academia, the author was in commercial practice in intellectual property, media and competition law at the Melbourne, Australia office of a top-tier international law firm.

Intellectual Property, Theory, Culture

Series Editor: Johanna Gibson, Herchel Smith Professor of
Intellectual Property Law, Queen Mary University of London, UK

This series presents theoretical and cultural examinations of intellectual property laws, developments, and policy. Volumes in the series may be identified by their innovative and critical analyses and their original contributions to international debate. Interdisciplinary in approach, the series will be of interest to intellectual property experts and stakeholders, policy advisors, and NGOs, as well as students and researchers in the very critical areas of intellectual property law, anthropology, and cultural studies.

Also in the series:

Food Security, Biological Diversity and Intellectual Property Rights
Muriel Lightbourne
ISBN 978-0-7546-7611-9

Intellectual Property, Medicine and Health
Current Debates
Johanna Gibson
ISBN 978-0-7546-7218-0

Patenting Lives
Life Patents, Culture and Development
Edited by Johanna Gibson
ISBN 978-0-7546-7104-6

The Logic of Innovation
Intellectual Property, and What the User Found There

JOHANNA GIBSON
Queen Mary University of London, UK

ASHGATE

© Johanna Gibson 2014

All rights reserved. No part of this publication may be reproduced, stored in a retrieval system or transmitted in any form or by any means, electronic, mechanical, photocopying, recording or otherwise without the prior permission of the publisher.

Johanna Gibson has asserted her right under the Copyright, Designs and Patents Act, 1988, to be identified as the author of this work.

Published by
Ashgate Publishing Limited
Wey Court East
Union Road
Farnham
Surrey, GU9 7PT
England

Ashgate Publishing Company
110 Cherry Street
Suite 3-1
Burlington, VT 05401-3818
USA

www.ashgate.com

British Library Cataloguing in Publication Data
A catalogue record for this book is available from the British Library

The Library of Congress has cataloged the printed edition as follows:
Gibson, Johanna.
  The logic of innovation : intellectual property, and what the user found there / by Johanna Gibson.
     pages cm.—(Intellectual property, theory, culture)
  Includes bibliographical references and index.
  ISBN 978-1-4094-5417-5 (hardback)—ISBN 978-1-4094-5418-2 (ebook)—
ISBN 978-1-4094-7338-1 (epub) 1. Intellectual property—Philosophy. I. Title.
  K1401.G539 2014
  346.04'8—dc23

                                                                 2013039374

ISBN 9781409454175 (hbk)
ISBN 9781409454182 (ebk – PDF)
ISBN 9781409473381 (ebk – ePUB)

MIX
Paper from
responsible sources
FSC
www.fsc.org FSC® C013985

Printed in the United Kingdom by Henry Ling Limited,
at the Dorset Press, Dorchester, DT1 1HD

Contents

The Logic of Innovation

Acknowledgments

This book has been perhaps one of the more personal journeys that I have made into the analysis of intellectual property. The 'heritage' of this book is long and long-winded, intertwining the many facets and many figures of my academic life and debt – from literature, cultural theory, art history through to law. I have felt a genuine affective response in reading, the kind of experience that is part of the 'value' of the intellectual enterprise. I write because I read. The magnificence of *Alice* as the universe in which all these passions come together is both enticing and reassuring. It just makes sense.

My journey here has not been about defining an audience or addressing a solution, as such, and for this I take reassurance from Lyotard (and so might we all): 'The instability of criteria, even in fashions, comes from this experimental situation. It also makes for the fact that today the majority of people who write interesting things, write without knowing to whom they are speaking. That is part of the workings of this society, and it is very good. There is no need to cry about it.'[1] I long for experimentation, I long to wonder.

And just like the 'tangled tale' of intellectual property, I too became almost unravelled during the process, knitted together by an impossibly patient family and some invaluable friends – Phillip who read and reread and kept me in constant coffees, my mother, Dawn, who has always encouraged and indulged and led by example, and my brother, John, who despairs of my eccentricity but inevitably fuels it nonetheless. And of course, during the entire process, I have been kept constant company by my own Wonderland of intriguing creatures – Leo the Bedlington Terrier and Finnegan the Otterhound, joined at the end of the writing process by a new intrusion, Huw the Bedlington Terrier. *Huw's there?*

And I have a wonderful group of colleagues who have made the wondering not only stimulating but also tremendous fun. Indeed, writing, like a particular genre of crisis, lets you know who your real friends are! I am grateful to a number of people, the perfect counterpoint to the isolation of writing with their indulgence of my absences, cancellations and self-imposed incarceration. Some very treasured individuals have made every difficulty considerably less – Dorit Samuel and Sheldon Halpern have been incomparably supportive friends and colleagues for some years now, from whom I have always received unconditional and limitless understanding, guidance and wisdom; Fiona Macmillan for her ceaseless wisdom on copyright and culture, her renegade humour and her intellectual stimulation;

---

1   J-F Lyotard in J-F Lyotard & J-L Thébaud, *Just Gaming*, W Godzich (trans), U of Minnesota P, Minneapolis, [1979]/1985: 9.

The Logic of Innovation

Susy Frankel for her tremendous insight on treaty interpretation and international economic law, and for her equally satisfying insights into academic life; my very dear friend and colleague Jaime Stapleton, confidante and co-conspirator, one of the first people I met in London while musing over a digital art exhibit at the ICA, for always intriguing and provoking me and for having a ticket to all the best parties; Florian Koempel, for his kindness, support and grounded wisdom; Malcolm Langley for a shared love of archives and rare texts; Gaetano Dimita for the good humour, gaming and all the coffees; the wonderful, 'political' Peter Petrov, wise and wistful (please come back to London soon!); and John Frow, who honestly started me on all of this fantastic journey. And to all at Ashgate, especially the indefatigable and undefeatable Alison Kirk, I am enormously grateful, working with you all is not just professional and rewarding, it is a distinct pleasure.

Importantly, true friends shine through with incredible patience and understanding. I am so grateful to Amanda for her good humour, kindness and friendship, to Ella for keeping me updated on the outside world, to the irrepressible Lottie and her irreverent outlook on life, and to Amy for regularly matching my 'artistic' temperament! A special mention must be made of one particularly discerning literary critic, Alfred (Mr Mutt), who thought so much of a late draft that he embraced the Wonderland references literally … and ate the first page. To Caroline, Lisa, Kipper and Yuri, and all in the crew, thanks for keeping me distracted in the occasional sunshine. To Johnson, thank you for the music! And the last word will go to good friend Jo, for the comedy, venting, recipes and Billy Bragg, and for a special moment of insight she offered me during a frenzied Eurovision evening of online fact checking, 'has the Internet taken away wondering?' I wonder.

Of course, I have an enormous gratitude to the genius of my interlocutor, Lewis Carroll, who, together with the curiosity of Alice, has provided me with the most magical vehicle for reimagining the discourse of innovation. What an enormous joy and infinite intrigue their contribution to my questions has been. And to the White Rabbit, thank you for leading me to wonder.

Acknowledgments always fail as an accounting exercise and of course, this is just the beginning. Ambiguity intended.

> Lastly, she pictured to herself how this same little sister of hers would, in the after-time, be herself a grown woman; and how she would keep, through all her riper years, the simple and loving heart of her childhood: and how she would gather about her other little children, and make *their* eyes bright and eager with many a strange tale, perhaps even the dream of Wonderland of long ago.[2]

I hope at the very least this strange tale will make some eyes a little bright and eager, with perhaps even the dream of Wonderland.

Johanna Gibson

---

2   *Through the Looking-Glass*, 103.

Works of Lewis Carroll

Alice Adventures in Wonderland, in L Carroll, *Alice in Wonderland*, Wordsworth, Ware, Hertfordshire, [1865]/1992: 1–103 (*Wonderland*).

'The Hunting of the Snark', in L Carroll, *Alice in Wonderland*, Wordsworth, Ware, Hertfordshire, [1876]/1992: 229–52 (*Snark*).

'Preface to the Eighty-sixth Thousand of the 6/- Edition of *Alice's Adventures in Wonderland*', in L Carroll, *Alice's Adventures in Wonderland* and *Through the Looking-Glass and What Alice Found There*, H Haughton (ed), Penguin, London, [1896]/1998 (*Preface*).

Sylvie and Bruno, in L Carroll, *Lewis Carroll: The Complete Works*, CRW, London, [1889]/2005: 101–80 (*Sylvie and Bruno*).

Sylvie and Bruno Concluded, in L Carroll, *Lewis Carroll: The Complete Works*, CRW, London, [1893]/2005: 181–264 (*Sylvie and Bruno Concluded*).

'A Tangled Tale', in *Alice in Wonderland*, Wordsworth, Ware, Hertfordshire, [1885]/1992: 253–95 (*Tangled Tale*).

Through the Looking-Glass and What Alice Found There, in L Carroll, *Alice in Wonderland*, Wordsworth, Ware, Hertfordshire, [1872]/1992: 105–98 (*Through the Looking-Glass*).

'What the Tortoise Said to Achilles', in L Carroll, *Lewis Carroll: The Complete Works*, CRW, London, [1895]/2005: 455–56 (*Tortoise to Achilles*).

Preamble

For authors, if they are great, are more like doctors than patients. We mean that they are themselves astonishing diagnosticians or symptomatologists. There is always a great deal of art involved in the grouping of symptoms, in the organization of a *table* where a particular symptom is dissociated from another, juxtaposed to a third, and forms the new figure of a disorder or illness. Clinicians who are able to renew a symptomatological table produce a work of art; conversely, artists are clinicians, not with respect to their own case, nor even with respect to a case in general; rather, they are clinicians of civilization.[1]

The philosopher's treatment of a question is like the treatment of an illness.[2]

It is often repeated to the point of indifference that the intellectual property system is broken, unsound, infirm, languishing. The repetition of this refrain simply loses meaning, becomes disconsolate nonsense. It has become too easy to engage in shorthand rhetoric that has become so removed from its original object that it has deteriorated into unreflective and polarizing trumpery. Rather than merely pointing to a problem in a process of meaningless amphigory, it is essential to engage with the explanation of the question itself.

Nevertheless, there persists an undeniable disquiet among all participants in this arena, whether identified variously as producers, consumers, rights-holders. Whatever the role, these are all users of the same system and it is a system which, in the age of intellectual work and resources, has become ubiquitous in making sense (and value) of all types of production. In other words, mechanical and industrial production is similarly being understood through its subordination to the prominence of the intangible and to the logic of intellectual property.

Intellectual labour, effort and participation have been overwhelmed by a transformation in the social, and the nature of social life as itself a sphere of production. The aggregated, associational and public nature of production in social media may indeed be understood as giving rise to what will be explained here as 'familiar production',[3] but such production is itself merely indexical of a revolution in community integration, mutual productivity, privacy and personal integrity. The digital is not simply a technological environment; it is, much more importantly, a shift in the way in which people not only communicate but also

---

1   G Deleuze, *The Logic of Sense*, M Lester & C Stivale (trans), CV Boundas (ed), Columbia UP, New York, [1969]/1990: 237.

2   L Wittgenstein, *Philosophical Investigations*, GEM Anscombe (trans), 3rd ed, Blackwell, Oxford, [1953]/1995: §255.

3   See the more detailed introduction to this concept in 'Use'.

account for that communication. It is analogical, it is personal, it is a production of the familiar.

It is therefore essential and judicious to examine not merely a supposed problem, but what indeed it means to describe the system as broken. If the system is to be pathologized in this way, then it is necessary to invoke the social context for the problem in its very resolution, the very 'life' of the system, the anticipation, the wages and the losses.

> 'What is literature?'; literature as historical institution with its conventions, rules, etc., but also this institution of fiction which gives *in principle* the power to say everything, to break free of the rules, etc., to displace them, and thereby to institute, to invent and even to suspect the traditional difference between nature and institution, nature and conventional law, nature and history. Here we should ask juridical and political questions.[4]

It is not possible to rework or reform intellectual property within the conventional parameters provided by the dominant discourse and debates – the conditions for innovation and creativity – as there is no real consensus on the nature of the very object itself. Within the discourse of law, business, economics, it is not possible for it to reflect upon itself if the tools of interrogation are also the very object of inquiry. One cannot treat oneself. Therefore, the literary clinician is engaged as witness to the dynamic, in this talking cure: 'an evaluation of symptoms might be achieved only through a *novel*.'[5] A therapeutic approach literally emphasizes the use and activity in the system, as distinct from the culpability or otherwise of its rules and its players.

The interlocutory undertaken here engages exemplary reconfiguration of the discourse on intellectual property through the literary lens of wonder, and the intrigue of curiosity: 'A main cause of philosophical disease – a one-sided diet: one nourishes one's thinking with only one kind of example.'[6] Arguably, diversification of examples and 'imaginary scenarios'[7] will be crucial to legal and economic analysis that is both elucidating and demystifying: 'For each one of these sentences I can imagine circumstances that turn it into a move in one of our language-games, and by that it loses everything that is philosophically astonishing.'[8] By negotiating the calculable through chance, the judgment

---

4   J Derrida, *Acts of Literature*, D Attridge (ed), Routledge, New York, 1992: 37.

5   Deleuze, *The Logic of Sense*, 237.

6   Wittgenstein, *Philosophical Investigations*, §593.

7   Imaginary scenarios are deployed by Wittgenstein throughout his later work in order to interrogate possible conditions and relations. See further discussion in D Cerbone, 'Don't Look but Think: Imaginary Scenarios in Wittgenstein's Later Philosophy', 37 *Inquiry* 1994: 159–83.

8   L Wittgenstein, *On Certainty*, GEM Anscombe & GH von Wright (eds), D Paul & GEM Anscombe (trans), Blackwell, Malden MA, [1969]/1975: §622.

through indecision, the dictate through silence, measured clarity emerges in the literary shadows:

> The magic of *Alice in Wonderland:* of drying out by reading the driest thing there is.
>
> With the magical healing of an illness, one *directs* the illness to leave the patient.
>
> After the description of any such magical treatment, one always wants to say: If the illness doesn't understand *that*, I don't know *how* one should tell it to leave.
>
> Nothing is so difficult as doing justice to the facts.[9]

Alice is our clinician. Her dialogue with herself, her 'inward speech', resonates with the same kind of grammatical investigation that propels the philosopher's dialogue; a dialectical, diagnostic, clinical examination of the riddles of innovation and the malaise of intellectual property, the driest thing there is. This inquiry into innovation and property is traced through the wonder of Alice's journey. Literature is the cartography of the treacherous terrain of the legal landscape, and the mapping of an otherwise ghostly digital domain. Through the curiosity of Alice the built environment emerges from the shadows of the digital; like throwing silver dust onto a light beam, the magic of the literary restores the measure of the law. Relationships in intellectual property are somewhat indisposed, and there is an acquired immunity to the prescriptive discourses that have become entrenched in this field. Literature is not only a 'treasury of information', but also an encyclopedia of transformation. Alice will take us on the journey towards the political will to get well … the dialogical intensity of familiar production:

> "It's hardly fair," muttered Hugh, "to give us such a jumble as this to work out!"
>
> "Fair?" Clara echoed bitterly. "Well!"
>
> And to all my readers I can but repeat the last words of gentle Clara:
> FARE-WELL![10]

And so on to the journey of innovation. Fare well.

---

9   L Wittgenstein, *Philosophical Occasions: 1912–1951*, JC Klagge & A Nordmann (eds), Hackett, Indianapolis, 1993: 129.

10   *Tangled Tale*, 295.

To those teachers, past and present, who led me to wonder

USE

Use

The law is made. The law is mad, is madness; but madness is not the predicate of law.[1]

There Goes Innovation

I would like to say: 'I must *begin* with the distinction between sense and nonsense. Nothing is possible prior to that. I can't give it a foundation.'[2]

Innovation.
What is innovation? What is the meaning of innovation?

What is the meaning of a word?
Let us attack this question by asking, first, what is an explanation of the meaning of a word; what does the explanation of a word look like?
The way this question helps us is analogous to the way the question 'how do we measure a length' helps us to understand the problem 'what is length?'[3]

On the one hand, what does it mean if the book, this questioning, this inquiry (in terms of both investigation and trial), is to commence with an attempt at defining innovation? Is this to explain the meaning of innovation? Or on the other hand, is meaning and understanding arrived at through identifying the very process itself of defining innovation? That is (and perhaps more usefully, as it were), the preoccupation of this work is rather the wider (cultural, social, economic, political) process of attempting to arrive at the meaning of defining innovation. More fundamentally, the 'meaning' of innovation is not found in the object but in the mythology, the wonder, the folklore. It is thus the contemporary fixation on innovation (rather than the identification), together with the values and mythology attached to dominant beliefs and ideas of innovation, that fascinate, intrigue and ultimately are understood to illuminate the meaning of innovation, not only for intellectual property discourse but also for a broader understanding of what it means to be creative, to be adventurous, to be innovative. Therefore, the task of this inquiry is not to define innovation, but rather, to use it: 'The importance of

---

1   Derrida, *Acts of Literature*, 251.
2   L Wittgenstein, *Philosophical Grammar*, R Rhees (ed), A Kenny (trans), Blackwell, Malden MA, [1974]/1980: 81.
3   L Wittgenstein, *The Blue and Brown Books*, 2nd ed, Harper & Row, New York, [1958]/1965: 1.

this examination lies in this, that it applies to the relation between learning the meaning of the word and making use of the word. Or, more generally, that it shows the different possible relations between a rule given and its application.'[4] In other words, 'Let the use *teach* you the meaning.'[5]

If therefore the meaning of innovation is achieved not by reference to terminology, but rather by engaging with practice, what is the significance of this for any attempt to speak around the topic of innovation? It is useful to commence with economic discourse and the attempt to quantify and measure innovation, to provide the 'evidence' of the value of innovation. But at the outset, that representation is itself guarded and reserved. As a purposeful narrative, economic discourse defines innovation through 'hindsight', as it were, that is, through the gathering of 'evidence' of new products, methods, markets and the like. Joseph Schumpeter, somewhat the 'signature' of innovation, emphasized its operation (but not its definition) throughout his career. Innovation is felt by its imprint, not by its face. Similarly, in a very real way, the value of the intellectual property is in its infringement, not in its actual materials.

Schumpeter is credited with introducing the process of technological change into economic theory. The so-called Schumpeterian trilogy of technological change provides that innovation is one of three stages in the narrative: invention, innovation and diffusion.[6] That is, and rather intriguingly, innovation is a kind of 'entrepreneurialism' of the familiar, as distinct from creating the new. Innovation is not the creativity of the product, but the imagination of the market; that is, the interpretation of the technology in its use.[7] In this model, innovation identifies the products after the fact of invention, creating the 'need' through the entrepreneur: 'Sometimes innovation is so conditioned, whereas the corresponding invention occurred independently of any practical need.'[8] The 'innovative' stage in Schumpeter's model is therefore use itself.[9] Invention merely generates innovation

---

4   Wittgenstein, *Blue and Brown Books*, 11.

5   Wittgenstein, *Philosophical Investigations*, 212e.

6   See further the discussion of the Schumpeterian trilogy in ES Anderson, *Joseph A Schumpeter: A Theory of Social and Economic Evolution*, Palgrave Macmillan, Basingstoke, 2011: 141.

7   See the discussion in JA Schumpeter, *Business Cycles: A Theoretical, Historical and Statistical Analysis of the Capitalist Process*, McGraw-Hill, New York, 1939: 100–107.

8   Schumpeter, *Business Cycles*, 81 fn 25.

9   This raises questions immediately in relation to social media. While social media platforms are considered in more detail in Chapter 2 and throughout, it is worth introducing briefly here the relationship between the user and production in social media platforms. While social media may provide extensive platforms in which to produce within the social and digital space, there is also the risk of a kind of alienation from creativity in the conventional personal sense and in the collateral lack of privacy. Intriguingly, increased use and participation is in some ways a protection from the loss of privacy, in that it exhausts the private and maximizes use. Use is not only paramount, but also 'private'. The innovation in social media participation is not an end in itself but a tool to further production.

in so far as it provides the circumstances or resources for innovation: 'the output of this process, innovation *per se*, does not occur until the changes being generated either come to the market or are used for the first time.'[10] *Use is paramount.*

Schumpeter's analysis is thus of particular intellectual and theoretical significance for introducing innovation as an instrument of economic growth, as distinct from an end in itself, where technical change thus becomes a key variable in margin analysis.[11] This application of a product (and need) after the creative event, interrupts the traditional linear, causal model of research and development (R&D), which traces a question through to a solution; that is, the traditional R&D model of basic research, applied research and development.[12]

While it should be noted that the invention-innovation-diffusion trilogy has been suggested as mapping neatly onto the more systematic (and possibly linear) R&D model,[13] this supposed parity is perhaps open to dispute. Arguably the potential of the specific event of innovation in 'creating' the technological process through product might disrupt the linear model more significantly than that would suggest:[14] 'As soon as it is divorced from invention, innovation is readily seen to be a distinct internal factor of change.'[15] This absorptive capacity of the innovation process is crucial to contemporary technological development. In so far as innovation is a desirable value, it is value not in a product, but in change itself; it is the difference and production in repetition.[16] It is therefore at once

---

10   P Stoneman, *Soft Innovation: Economics, Product Aesthetics and the Creative Industries*, Oxford UP, Oxford, 2010: 2–3.

11   See, for instance, the discussion in M Blaug, *Economic Theory in Retrospect*, Cambridge: Cambridge UP, 1983: 462–63. For a more general discussion of Schumpeter's influence and work in the fields of sociology and economics, see R Swedberg, *Joseph A Schumpeter: His Life and Work*, Polity P, Cambridge, 1991 and R Swedberg, *Schumpeter: A Biography*, Princeton UP, Princeton, 1995.

12   A definition of R&D is provided in the OECD *Frascati Manual*, 6th ed, OECD, Paris, 2002: 30. See further a discussion of the impact of a linear market-driven model of R&D in relation to medicines and access to scientific research in J Gibson, *Intellectual Property, Medicine and Health: Current Debates*, Ashgate, Farnham, 2009. This relationship between linear models of innovation and directionless innovation is considered throughout later chapters in the present discussion.

13   P Stoneman, *The Handbook of Economics of Innovation and Technological Change*, Blackwell, Oxford, 1995. See further Bettina Peters who describes the Schumpeterian trilogy as linear in *Innovation and Firm Performance*, Physica-Verlag, Heidelberg, 2008: 21–22.

14   See further Schumpeter's own refutation of linear dependence and causality throughout *Business Cycles*.

15   Schumpeter, *Business Cycles*, 82.

16   Deleuze's early work in the theory of difference is concerned with the problem and object of representation in relation to a stable identity. This problem of representation is central to the narrative of intellectual property and innovation and to the analogous scenarios devised for the digital in the context of an attachment to an original, and in the construction of digital 'originals' and authenticity. See further G Deleuze, *Difference and Repetition*,

incomparable, indefinable and unrepresentable, recognized only by its effects. Change is fundamental to production and to capital: 'Capitalist reality is first and last a process of change. In appraising the performance of competitive enterprise, the question whether it would or would not tend to maximize production in a perfectly equilibrated stationary condition of the economic process is hence almost, though not quite, irrelevant.'[17]

While change is a fundamental value of capitalism, nowhere does this suggest any emphasis on a predictable trajectory of 'innovation', nor on a subjective 'value' of that change; that is, the innovation narrative does not presuppose the solution provided by the invention (in a simple sense, its use, its value and meaning as through its application). This is illustrated very well through the practice of rule-based, computer-generated music. In this case, the 'invention' of the music might be calibrated, however the outcome, the interpretation, is unpredictable. Brian Eno explains this in interview with art curator and critic Hans Ulrich Obrist: 'Each piece of music is a little machine, in a way, for producing. It is like each set of rules is a single kind of genome, and then each individual performance is one of that species. They are related, but they are not identical ... The initial material can be quite small, but the number of reconfigurations is huge.'[18] In other words, the remarkable value lies not so much in the original material (the original work, original identity, original presentation), but rather in the difference that is proliferated through reconfiguration and repetition,[19] through use. The unique quality of repetition is not through imitation but through the exposure of difference: 'Art does not imitate.'[20] This critical aspect of Eno's explanation highlights much

---

P Patton (trans), Columbia UP, New York, [1968]/1994. The difference in repetition and the dilemma of representation as a moral and imperative form is considered throughout the present discussion, but see in particular the discussion of original, copies and simulacra in Chapter 9.

17   JA Schumpeter, *Capitalism, Socialism and Democracy*, R Swedberg (intro), Routledge, London, [1943]/1994: 77 n 1.

18   Brian Eno, in interview with Hans Ulrich Obrist, in HU Obrist, *Interviews: Volume 1*, Charta, Milan/Fonazione Pitti Immagine Discovery, Florence, 2003: 214–19, at 216.

19   This individuality through repetition is critical to the work of Gilles Deleuze and 'difference-in-itself' as examined in *Difference and Repetition*; in particular, see chapter 1.

20   In *Difference and Repetition*, Deleuze explains, 'Perhaps the highest object of art is to bring into play simultaneously all these repetitions, with their differences in kind and rhythm, their respective displacements and disguises, their divergences and decentrings; to embed them in one another and to envelop one or the other in illusions the "effect" of which varies in each case. Art does not imitate, above all because it repeats; it repeats all the repetitions, by virtue of an internal power (an imitation is a copy, but art is simulation, it reverses copies into simulacra). Even the most mechanical, the most banal, the most habitual and the most stereotyped repetition finds a place in works of art, it is always displaced in relation to other repetitions, and it is subject to the condition that a difference may be extracted from it for these other repetitions. For there is no other aesthetic problem than

of the substantial concern and challenge in the age of virtual reproduction and digital innovation. In his response, Obrist observes what he calls the 'uncertainty gap', which artfully illustrates the significant difference between invention and innovation (use): 'this uncertainty gap between the imagination of how the piece should be or should look and the interpretation, the various interpretations that can be given following the instructions'.[21]

The question therefore appears to us (like the Cheshire Cat?) as to the extent to which 'innovation', in the broadest sense, indicates the technology or indeed the use: 'how much you think the work consists of the process of making it and how important it is for you to make something that sits separate from you and your explanations of it'.[22] If it comes down to the relationship of use, perhaps this means anything can be rendered innovative, just as anything can be rendered art: 'Can one then "paint anything" now? Yes. But this perhaps is still to insist too much on painting. One should rather say: let everyone join the game of images, and start to play.'[23] A notion of innovation must therefore refrain from placing too much emphasis on the 'industry' and system of invention, of authorship, of design. Innovation means participating in the game. *Start to play.*

This relationship between the 'personal' technology (or creativity) and the unpredictable use (or 'market' in the loosest sense) is perhaps key to the institutionalization of relevant and effective models for the dissemination and remuneration of creativity in a digital environment. The site of value is perhaps no longer in the objectified artefact of creativity, but rather in the hyper-proliferative experience of use. That 'use' is manifest through the game itself, including the games of intellectual property, whether in terms of infringement, coordination (for instance, in cross-licensing and pools) and so on. The critical challenge, therefore, is in the intelligible reckoning of experience within a commercial digital environment.

Returning to the discourse on innovation, more recently innovation has been appropriated to become a general umbrella term for technological change,[24] with the attending implications of economic growth and development. Therefore, although innovation continues to be emphasized in the discourse of invention, patents and industry, its application has been extended quite dramatically (and perhaps also

---

that of the insertion of art into everyday life. The more our daily life appears standardised, stereotyped and subject to an accelerated reproduction of objects of consumption, the more art must be injected into it in order to extract from it that little difference which plays simultaneously between other levels of repetition' (293).

21   Obrist, *Interviews: Volume 1*, 216.

22   Brian Eno, in interview with Hans Ulrich Obrist, in Obrist, *Interviews: Volume 1*, 217.

23   M Foucault, 'Photogenic Painting', in G Deleuze, M Foucault & G Fromanger, *Le Peinture Photogénique*, A Rifkin (intro), S Wilson (ed), Black Dog, London, 1999: 83, at 103.

24   Stoneman, *Soft Innovation*, 3 and 20–21.

strategically) to encompass more traditionally 'creative' pursuits. Indeed, the use of 'innovation', 'industry' and similar language in relation to creativity and artistic life is arguably inextricable from attempts to 'professionalize' and 'industrialize' creative pursuits within a wider economic and political agenda. Therefore, perhaps in part due to its import in policy language of economic growth, the term 'innovation' has been introduced in the more 'industrial' language surrounding creativity and should be understood to signal the full range of practices of creativity, inventiveness and imagination. Thus, in contemporary analyses of an innovation culture and society, notions of creativity, change, difference and contradiction must inform a much broader and meaningful notion of innovation. This is the wider experience of innovation invoked here in the present work.

Unusable Incentives

Along with the discourse on innovation is the collateral interest in incentives. Incentives raise particular questions with respect to use, including whether the 'social' interest in incentives is related to specific benefits or, more realistically and practically, to maintaining markets without regard to specific innovations. Notwithstanding any ambiguity in incentives to innovate and benefits to the public, almost invariably when the topic of incentives arises it is suggested that there is a causal relationship between intellectual property and the incentive to innovate.[25] However, the assumption of this causal relationship is problematic. Despite numerous attempts it is perhaps impossible to prove (although simple to abbreviate and suggest) and so is also easy to refute if innovation and access to knowledge are approached from a precautionary perspective.

An emphasis on use is clearly significant in this respect in that this not only foregrounds the user-consumer but also does not leave the discourse on rights vulnerable to arguments that the rationale for intellectual property (namely, as an incentive to innovation) is a specious and therefore refutable one. From the perspective of use (and thus access to use), recommendations are made based not on arguments of incentives to innovate, but instead on arguments of incentives to provide access to use.[26] In other words, an emphasis on use provides a more

---

25   For instance, see M Landes & R Posner, *The Economic Structure of Intellectual Property Law*, Belknap-Harvard UP, Cambridge MA, 2003: 13–14; L Marshall, *Understanding Copyright Law*, 5th ed, LexisNexis, New Providence NJ, 2010: 21–22; A Roughton et al., *The Modern Law of Patents*, 2nd ed, LexisNexis, London, 2010: 1049.

26   Indeed, in relation to historical origins of copyright in Britain, Ronan Deazley notes, 'Copyright, in eighteenth century Britain, was never simply concerned with the bookseller or the author. What emerges from a close study of the movement of the law during this period is that copyright, with both the passing of the *Statute of Anne* and the factual decision of *Donaldson*, was primarily defined and justified in the interests of society and not the individual. A statutory phenomenon, copyright was fundamentally concerned

realistic and tangible characterization of incentives as directed towards ways in which to facilitate transactions in access and use, and so to protect and promote the relationship between different users of the system (consumers, producers). In this respect, for instance, discourse on limitations, exceptions and other mechanisms for reform are concerned with benefits both to industry and to consumers. The dominant assumption of incentives as directly related to the motivation to innovate is arguably not only inaccurate but also damaging to the legitimacy of the law and actual validity of the debate concerning greater efficiency and effectiveness of the relationships and transactions themselves.

So, while there has been extensive consideration of the relationship between intellectual property and creative/innovative endeavour in terms of incentives, facilitation and remuneration with respect to the activity of innovation itself, it is perhaps crucial to consider intellectual property instead as a system of representation, through dialogue and language that concerns the meaning of innovation through use. However, in the dominant discourse on innovation and incentives, the language of intellectual property is not only organized by creativity but also organizes the construction and representation of creative intent. In this respect, the intellectual property framework is cast as a self-referential system, redressing all innovation within its (economic) paradigm. The very paradox of the representative function of the system is the impossibility of its own reflection: 'A *picture* held us captive. And we could not get outside it, for it lay in our language and language seemed to repeat it to us inexorably.'[27] In this way, the proper names of intellectual property become inevitably untranslatable in the very naming of innovation – patent, trade mark, design, copyright.

What's in a Name

> 'We name things and then we can talk about them: can refer to them in talk.' – As if what we did next were given the mere act of naming. As if there were only one thing called 'talking about a thing'.[28]

---

with the reading public, with the encouragement and spread of education, and the continued production of useful books.' See R Deazley, *On the Origin of the Right to Copy: Charting the Movement of Copyright Law in Eighteenth-Century Britain (1695–1775)*, Hart, Oxford, 2004: 226. On the economic justification of the patent system as an agreement for disclosure of the invention to the world, and as such, 'an agreement between society and the inventor', see Roughton et al., *Modern Law of Patents*, 1049. See further AS Oddi, 'Un-Unified Economic Theories of Patents: The Not-Quite-Holy Grail', 71 *Notre Dame Law Review*, 1996: 267.

27   Wittgenstein, *Philosophical Investigations*, §115.
28   Wittgenstein, *Philosophical Investigations*, §27.

It follows that a critical fascination of this inquiry is not to accept the value and meaning of the name in and of itself (intellectual property, innovation, social and so on), but the 'social operations of *naming* and the rites of institution through which they are accomplished'.[29] That said, for the purposes of initiating this adventure, it is nevertheless unavoidable that an impression or a notion of innovation is introduced here, at the beginning of the journey, simply to establish the mechanism for the discussion, even if to trouble it and do away with it at the end. This is certainly not to set out with an explanation of innovation with disclaimer, and by no means should this be the case. It is simply to identify the limits of the language in which it is necessary to direct attention, at the same time investigating the way in which processes of naming facilitate comparison and appropriation within an economy of innovation.[30] In this regard, innovation as an ideal is forced into a linguistic engagement with intellectual property. However, that said, 'the meaning of a word is its use in the language. And the *meaning* of a name is sometimes explained by pointing to its bearer.'[31]

Leaving aside the complexity of the different types of intellectual property (or indeed putting it right in the way), technological change, provocation and irritation mean that the same advances agitating for reform are the same advances confounding the grammatical devices of copyright, design, trade mark and patent that punctuate the intellectual property narrative. Fundamentally, whether understood as a catalyst or a tool, technology is nevertheless the vehicle for cultural and political transformation in the social and economic life of the subject matter itself. Technology is understood as implicated in changes in creative practice as well as changes in use, changes in protection as well as changes in appropriation, changes in the objects as well as the subject matter. This is the 'double-speak' of intellectual property and this is arguably what threatens to maintain an antagonistic and impossibly polarized subplot if it remains unaddressed. Indeed, the overarching term itself, 'intellectual property', is a portmanteau word, joining

---

29   P Bourdieu, *Language and Symbolic Power*, JB Thompson (ed), G Raymond & M Adamson (trans), Polity P, Cambridge, [1991]/1992: 105.

30   In this way naming is also part of the coding and privileging or ranking of information. On the classification of schools and groups within the fields of art and literature, Bourdieu explains: 'The names of the schools or groups which have proliferated in recent painting (pop art, minimal art, process art, land art, body art, conceptive art, *arte provera*, Fluxus, new realism, *nouvelle figuration*, support-surface, *art pauvre*, op art, kinetic art, etc.) are pseudo-concepts, *practical* classifying tools which create resemblances and differences by naming them; they are produced in the *struggle for recognition* by the artists themselves or their accredited critics and function as *emblems* which distinguish galleries, groups and artists and therefore the products they make or sell.' See P Bourdieu, *The Field of Cultural Production: Essays on Art and Literature*, R Johnson (ed), Polity P, Cambridge, 1993: 106.

31   Wittgenstein, *Philosophical Investigations*, §43.

the 'common' purpose of intellectual activity with the 'own' self (and 'owning self') of property.[32] What happens if we are faced with Jabberwocky?

Innovation is therefore not to be infused with notions of industry alone, although this has nevertheless become a troubling resonance of the term. Undoubtedly this 'industrial' notion of innovation that has persisted to an extent comes from a pragmatic and telescopic abbreviation within intellectual property, but it would not be an accurate one for this discussion. Intellectual property functions in a way that is merely indexical of innovation, of change, of creativity, of difference. Innovation is not synonymous with intellectual property, nor can intellectual property objects alone summarize innovative activity. Attention must be given to the action of the product, to the social life of commerce, the cultural life of use.

Therefore, recalling Schumpeter's notion of innovation's entrepreneurial role, as distinct from the creativity inhering in the product, innovation is perhaps more accurately understood in the context not of the material good or unit, but of use. Indeed, with the transformation of intellectual markets from the materiality of goods to the relationships with services, accounting for use would seem to be the only sensible way in which to render those markets within contemporary developments in intellectual property and its custom. The meaning of innovation is found in use.

Use will always bring with it a collateral risk: use of language, use of an invention, use of advice. In Wonderland, words are applied in multiple ways, but those with the sense to do so will be able to see the double pictures, the 'double-speak': *I seem to see some meaning in them after all.*[33] Language plays frequently and extensively throughout this tale, with the user taking responsibility for the meaning on each occasion: 'Sometimes you have to take an expression out of the language (withdraw <an expression from the language>), to send it for cleaning, – & then you can put it back <u>into circulation</u>.'[34]

In some ways, intellectual property frameworks provide the justification for extensive use by grounding the 'new' and 'inventive' and so on in a pedigree of prior art and published works, thereby initiating a 'market' of meaning (through use) for an otherwise unanticipated (and unrecognizable and incomparable) product. Use therefore, in some way, provides the circumstances for innovation, not through the banality of a market, but through a paradoxical remembering or recollection of the new, as it were: *What do* you *think it was?*[35] Innovation, invention, creativity cannot persist as a private affair; innovation must be rendered communicable and communicated: 'For an invention can never be *private* once its status as invention,

---

32   The bundled meaning of 'intellectual property' and the activity and process of portmanteau words more widely are explored in greater detail in '*Re* Use'.

33   *Wonderland*, 100.

34   L Wittgenstein, *Culture and Value: A Selection from the Posthumous Remains*, GH von Wright & H Nyman (eds), A Pichler (rev ed), P Winch (trans), rev ed, Blackwell, Malden MA, [1977]/1998: 44e.

35   *Through the Looking-Glass*, 198.

let us say its patent or warrant, its manifest, open, public identification, has to be certified and conferred.'[36] The very public nature of innovation, in all its disguises, is use.

Give it a Name

> Are you inclined still to call these words 'names of objects'?[37]

Intellectual property language thus transforms whatever is designated by the words, the names (patent, copyright, trade mark, design), into signs of innovation. Thus, the words of intellectual property (patent, trade mark, copyright, design) designate examples of innovation and creativity, and in so doing are themselves taken to reflect and to 'reveal' innovation: 'Words are not signs, but the moment a word appears, the designated object becomes sign. For an object to become sign means precisely that it conceals a "content" hidden within its manifest identity, that it withholds another side of itself for a different glance upon it, a glance that might never be taken.'[38] The words themselves thus become 'proper names' for the narrative of intellectual property.

For example, when it comes to discourse on technology and the patent industries, the invention is described, named (both in terms of the designation as patent and with respect to the 'name' of the invention) and becomes the subject of the innovation. However, innovation itself is suggested and commonly described as a something, an everything, a quantifier, assessable and assessed within economic models of innovation. It is presented as something to which we can refer as both the product of and catalyst for the patent system, both the justification and the reward, both the credit and the debt. In this way, not only the invention (in the wider Schumpeterian sense) but also the event of innovation can be subjects in this narrative. However, this tension between the qualifying, aspirational concept of innovation, and the quantified, measurable target of innovation, is critical to the landscape of the law as well as the policy surrounding its application, not only in patent law but also across the spectrum of regulation of creative enterprise by intellectual property rules and logic.

Nevertheless, this is not the same as suggesting that patent law, or indeed intellectual property more widely, proposes any qualitative evaluation of its subject matter. Rather, it is a logical system, based upon accepted or assumed premises from which valid identifications or conclusions of what is inventive, what is distinctive, what is new, what is original and so on will be made. Innovation has an existence and that existence must be measurable for the system to operate. An

---

36   Derrida, *Acts of Literature*, 315.

37   Wittgenstein, *Philosophical Investigations*, §27.

38   J-F Lyotard, *Discourse, Figure*, A Hudek & M Lydon (trans), J Mowitt (intro), U of Minnesota P, Minneapolis, [1971]/2011: 82.

invention possesses the qualities of an invention, but it does not actually exist other than through the patent, the quantitative measure of its innovation; that is, it does not exist other than 'in the name of' intellectual property. Somewhat controversially, therefore, inventions are arguably ideas manifest through the declaration of the patent. However, the product does not as such exist. Rather, it is the market that is heralded by the patent as distinct from the product itself. The product is an 'effect' of the patent, the patent as a catalyst – the interaction between the activity of naming invention and quantifying innovation.

Similarly, for a wide range of immaterial labour and products, intellectual property rights therefore denote a product in order to create the circumstances for market exchange. Innovation as such is unrepresentable without the devising and denoting of a product in order to create a market out of the infinite. That does not mean there is no innovation without intellectual property, but simply that it is perhaps less spectacular, less conspicuous, less perspicuous. In other words, innovation and technology drive the product, the good, through the creation of markets in use. Increasingly markets do not precede and demand new products and 'change does not result from adjustment of the product to the demand'.[39] Rather, innovation pushes new markets. In this way, it is therefore the innovation event that provides the conditions for innovative and creative activity described in the simple narrative of intellectual property and its markets. These markets are themselves choreographed by the way in which otherwise non-existent products might be declared, deduced, identified by use. In some ways, therefore, the use (and product) may not coincide as such with the proper names of intellectual property at all – the patent, the design, the trade mark, the copyright.

In this same way, new proper names and denominative devices have emerged to propel the 'products' of the intellectual property discourse and debate. This language of intellectual property is indeed crucial to the discussion as such terminology not only serves to abbreviate the industries (for example, 'creative industries' is itself a largely meaningless slogan in practice and in use, as will be considered in later discussion) but also acts to supply bywords that risk attenuating the creative and innovative process, and to subjugate it to a machinic or technical priority. At once the ringing of 'technophobia' is heard. But this is not a phobia of technology, it is a criticism of language and catchwords. To make language work in intellectual property discourse, it must not be left to waste away in linguistic shortcuts and empty slogans.

It is perhaps the very 'product-based' nature of intellectual property that has made it so vulnerable to the language of 'brands', as it is difficult to ascertain the same kind of product placement in other areas of law. Language such as 'digital', 'technology', the 'creative industries', the 'creative economy', the 'digital

---

39 P Bourdieu, *Sociology in Question*, R Nice (trans), Sage, London, [1984]/1993: 113. See further the discussion of the production of goods and production of tastes in Chapter 4 of the present work.

economy' and so on serves to abbreviate complexity to the point of deletion.[40] The policy buzz of intellectual property has indeed become a discourse of window-shopping, with no need to engage with the transaction, just the shiny motto. Intellectual property has succumbed to the mediocrity of its own celebrity.

Technology itself is both a grammar and a system in itself. While the shorthand 'technology' is ordinarily treated in respect of the industrial and technical aspects, the machinery of communication, it is at once also the discourse upon those industrial arts. As a collective within intellectual property discourse, it nevertheless to an extent has been stripped of that fuller meaning to the detriment not only of the innovation in technical areas, but also of the recognition and understanding of the wider cultural and social import of communication that utilizes technology, rather than lauding technology as in and of itself the value of communication. This perspective on technological innovation has altered to some degree the priorities and perspectives of the debate, and arguably has diminished the innovative and cultural life of the technical arts themselves. Technology has been elevated to a value in and of itself, as distinct from a means. *Technology is useless.*

The paradox of the policy discourse on creativity and innovation can be seen most clearly, perhaps, in the forced 'set' of 'creative industries'. Notably, this kind of generalization has not happened in the innovative or patent-based industries to the same extent. There does not seem to have been the same capture of value within the technology as distinct from the use in the innovative sector. In the context of discourse and debate on patent protection, it is remarkable in fact the way in which the various industries have sustained their identity (their use); for example, the pharmaceutical industry, the automobile industry, the software industry, the telecoms industry, the biotechnology industry, the pharmaceutical industry. Indeed, this is so explicit that the protection itself is often identified in this way ('software patents', 'telecoms patents', 'biotech patents', 'pharmaceutical patents' and so on). Within the very articulation of the law, the user is explicit in terms of expertise and adjudication of the object itself (the person skilled in the art);[41] and similarly in design (informed user) and trade marks (average consumer). However, there is no scope for the user in the limits of copyright, and it is copyright that has come to be taken as synonymous with creativity (and the creative industries). Although enterprises in the creative industries often necessarily engage the full range of intellectual property rights (including copyright, design, trade mark and patent), the policy rhetoric appears to continue to separate out along traditional intellectual property categories. As a result, creative industries (an enterprise) continue to be interchanged, frequently and erroneously, with copyright (a tool), even in terms

---

40   This combination and compression of meaning is considered later in '*Re* Use' in the context of 'intellectual property' and portmanteau words.

41   This role of the user is discussed later in this chapter, together with trade marks and designs.

of the funding of research itself.[42] And as later discussion will explore in greater detail, the user is absent from copyright. How can policy engage industry when industry is being supplanted by law, and law by technology?

The contrast is thus stark between the engagement with individual industry (and use) in the patent sector and the emphasis on policy product in the 'creative industries'. No such respect for individual industries has persisted in the creative sector. Instead, there is a preoccupation among policy-makers to define the 'creative industries', as distinct from the 'use' of individual industries. In an effort to abstract or fashion some meaning to the term, the UK government has proposed an ambitious plan to define an exhaustive list of creative occupations and creative industries.[43] The inclusions and omissions in the suggested lists range from the curious to the absurd. For example, commercial art galleries and antiques are relegated as 'retail activities'; however, sales director is nevertheless identified as a 'creative occupation'. Design and fashion are classified together; whereas 'IT software and computer services' is now a creative industry in its own right. Crafts are removed seemingly for reasons of insignificance and as a 'skilled trade'; however, information technology and computer services are newly included as 'readily identifiable' (and no mention of skilled trade). To an extent the list appears to be in furtherance of the proposition that the 'creative industries' comprise a hugely significant economic sector, and the evidence is perhaps contrived to support that proposition.[44] Further, as an umbrella term it is a case of the *in common of nothing in common*, providing a set of things where nothing is related, as it were. Attempts to list and define, and to devise policy in respect of an industry sector that is imagined but impossible, ignore the fundamental and insurmountable differences in policies and structures of wholly distinct and in many ways

---

42 For example, during 2011–12, Research Councils UK, led by the Arts and Humanities Research Council (AHRC), launched major funding into research into the creative industries, through a 'Centre for Copyright'.

43 UK Department for Culture, Media and Sport, *Classifying and Measuring the Creative Industries: Consultation on Proposed Changes*, April 2013. The Consultation had closed at the time of going to print, however results were not yet available. The Consultation arose as a consequence of the NESTA Report, *A Dynamic Mapping of the UK's Creative Industries*, prepared by H. Bakhshi et al., first published December 2012, which attempts to map the creative sector by 'creative intensities', which measures 'the proportion of workers in any given creative *industry* that are engaged in a creative *occupation*' (8). See further A Freeman, 'London's Creative Sector: 2004 Update', Greater London Authority, London, 2004.

44 For example, this is evident in the insistence in the consultation document on identifying 'digital' as creative in its own right, which is perhaps as much a contrivance as determining an industry creative if it uses pen and paper. Digital technology is a means of communication; it is not a creative value in and of itself. In this sense, the manipulation of language is not only serving a particular evidentiary purpose, but also resulting in some unanticipated absurdities and oddities with respect to the definitions. See further the discussion of evidence-based policy in Part III. In particular, see Chapters 11 and 12.

incompatible industries, such as broadcasting and the arts, for example. The term itself is simply a ludicrous brand, it is simply calculated nonsense.

Whither the digital? The 'digital' is in fact the proximate and experiential quality of innovation and communication, the fingers and the touch. It is both the tool (the finger) and the impression (the social, the meaning). What is the use value of the digital? In this respect, the data of the digital is the grasping, tensile dexterity of the technology, but digital in every sense is much wider than the mechanism alone. It is a manner in which the analog actually again becomes meaningful, where information and communication become resemblances, analogical, elsewhere, circumstantial, corresponsive, where the eye takes on a tactile, 'digital' function.[45] The 'digital' is thus not an abbreviation of high technology, but rather an emancipation of expressive thought with respect to the creation, use and benefit of innovation, regardless of the means or the media: 'This is an animality that can be seen only by touching it with one's mind, but without the mind becoming a finger, not even by way of the eye.'[46] It is 'to give the eye a digital function'.[47]

Where the analog takes us to different circumstances, the digital is part of our proliferative use of those opportunities. Reducing the digital to an abbreviation of technology eliminates this wider significance of the cultural and social transformation for which digital is not an antagonist, it is simply a corollary. It is to insist on the conceptualization of the problem within the language of 'products'.[48] But the 'digital' is not a product; it is an innovation upon the relationship to use and to production, with or without the technology. Digital technology is not transformative, and its appearance alone is not creative. Use is transforming digital technology. Therefore, at once the socio-cultural domain is analogous and digital, rendering mere resemblances proximate through experiential and experimental use.

So what 'use' is the digital?

---

45   In later discussion the concept of smooth and striated space is considered (see in particular Chapter 2, but discussed throughout), the former as the space of innovation and the 'digital' in the broader sense, the latter being organized and adjudicated space. Deleuze and Guattari explain: 'Smooth is both the object of a close vision par excellence and the element of a haptic space (which may be as much visual or auditory as tactile). The Striated, on the contrary, relates to a more distance vision, and a more optical space.' See G Deleuze & F Guattari, *A Thousand Plateaus: Capitalism and Schizophrenia*, B Massumi (trans), U of Minnesota P, Minneapolis, [1980]/1987: 493.

46   Deleuze & Guattari, *Thousand Plateaus*, 494.

47   Deleuze & Guattari, *Thousand Plateaus*, 494.

48   In other words, this is to overcode and 'striate' the potential of innovation. In this respect, Deleuze and Guattari note: 'Striated space, on the contrary, is defined by the requirements of long-distance vision: constancy of orientation, invariance of distance through an interchange of inertial points of reference, interlinkage by immersion in an ambient milieu, constitution of a central perspective. It is less easy to evaluate the creative potentialities of striated space.' See Deleuze & Guattari, *Thousand Plateaus*, 494.

Value in Use

Use is paramount

The concept of 'use' is generally aligned with notions of purpose and towards achieving that purpose for profit (in the broadest sense) or benefit (with all the attending meaning of beneficiary and use as creating debt, literal or otherwise). Usage results in wear, in depletion, in exhaustion. Indeed, capitalist models of production and economic strategy rely upon such depletion (along with scarcity) in order to generate a 'lack' in consumers.

Use in this sense emphasizes consumption in so far as use requires expenditure: that expenditure may be financial, in order to consume a commodity, or it may be an expenditure of effort, in order to exhaust a resource by employment. In either case, to use is to consume. And to consume is ultimately to be in debt; that is, to promise to discharge certain obligations with respect to the intellectual property. To consume is to waste so as to generate further demand.[49] In cooperation with this, the notion of ownership appears necessary to the preservation of property, the guarding against waste and the contribution to demand through 'scarcity'.[50] In fact in this respect waste is the dutiful counterpart to scarcity within conventional economic structures of capitalism. Debt is thus a personal accountability, subjectivity, responsibility.

Nietzsche explains this contractual rendering of the social, and the assumption of responsibility:

> To inspire trust in his promise to repay, to provide a guarantee of the seriousness and sanctity of his promise, to impress repayment as a duty, an obligation upon his own conscience, the debtor made a contract with the creditor and pledged that if he should fail to repay he would substitute something else that he 'possessed,' something he had control over; for example, his body, his wife, his freedom,

---

49   The contribution of 'waste' is considered further in the discussion of antiproduction, later in this chapter.

50   Arnold Plant explains: 'It is a peculiarity of property rights in patents (and copyrights) that they do not arise out of the scarcity of the objects which become appropriated. They are not a *consequence* of scarcity. They are the deliberate creation of statute law; and, whereas in general the institution of private property makes for the preservation of scarce goods, tending (as we might somewhat loosely say) to lead us "to make use of them," property rights in patents and copyright make possible the *creation* of a scarcity in the products appropriated which could not otherwise be maintained. Whereas we might expect that public action concerning private property would normally be directed at the prevention of the raising of prices, in these cases the object of the legislation is to confer the power of raising prices by enabling the creation of scarcity.' See A Plant, 'The Economic Theory Concerning Patents for Inventions', 1(1) *Economica* 1934: 30, at 33.

> or even his life ... Above all, however, the creditor could inflict every kind of
> indignity and torture upon the body of the debtor.[51]

In this way debt incorporates the moral language of judgment: 'It was in *this* sphere
then, the sphere of legal obligations, that the moral conceptual world of "guilt,"
"conscience," "duty," "sacredness of duty" had its origin.'[52] This relationship of
duty circulates in the narrative of production and use in intellectual property. The
very moral discourse of the debate is in concert with the construction of a 'debt'
of use to the 'credit' of the producer, the right-holder.[53] In the digital environment,
and through the cultural participation of social media, this debt becomes a debt
of life, where production is wholly and completely captured in its striation by
intellectual property and the terms of participation and contribution of content.[54]
Production, literally, becomes life.[55]

As well as the artificial scarcity in knowledge products (that would otherwise
have potential for unlimited reproduction), the creative and innovative industries
rely also on a certain inbuilt obsolescence, an exhaustion of the resource.
This obsolescence applies not only to the products themselves, but also to the
instruments for identification (and archiving) of intellectual property. For example,
the patent document therefore undertakes an important cultural and social role, by
preserving the 'argument' of innovation in market-based economies. The patent
'archive' actually necessitates its own obsolescence – the very core of innovation.
One presumes to see innovation, yet it is only the model of the invention, the
abstraction – innovation is what happens in its destruction. Not a single object
can persist through the archive, but its replacement in innovation becomes
instead certain, an inevitability of disconfirmation and improvement.[56] The patent
elucidates or clarifies a moment in innovation, but does not define innovation.
Its relationship is indexical rather than determinative. This has an interesting
resonance with the practicalities of the patent as well. Just as for the probability of
any viable theory, where more facts provide more chances to falsify that theory,
indeed the patent document, the scope of the patent (and the limitation of chance),
is an art in and of itself for drafters and litigators.

---

51   F Nietzsche, *On the Genealogy of Morals*, W Kaufmann & RJ Hollingdale (trans)
[1887], and *Ecce Homo*, W Kaufmann (trans and ed) [1908], Vintage-Random House, New
York, [1969]/1989: 64.

52   Nietzsche, *On the Genealogy of Morals*, 65.

53   Discussed in detail in Chapter 2.

54   This is considered in more detail later in this chapter.

55   A Negri, *Negri on Negri*, with A Dufourmantelle, MB De Bevoise (trans),
Routledge, New York, 2004: 62.

56   This question of obsolescence and disconfirmation is also relevant to the status
and quality of evidence in policy-making. Evidence itself works with respect to processes
of disconfirmation, discussed in more detail in Chapters 11 and 12.

Thus, whether in terms of use and using up (exhausting a resource, or person, or thing) or use and debt, use in its simplest connotation is configured according to a certain imbalance, defying the rhetoric of equilibrium,[57] efficiency and equality that ordinarily attends economic modelling of intellectual property transactions and policy 'balance'.[58] As conceived within conventional economic models, use (as consumption) is attached to the creation of necessity or demand (the market) whereby the scarcity (such as that created artificially by intellectual property rights) or lack (through making knowledge products limited, exhaustible and rivalrous) manipulates consumption so as to create desire or even a sense of exigency with respect to products (old and new).[59] *Use generates need.*

Use in debt

Immediately the language of 'use' and 'user' introduces the notion of a sustainable imbalance in the relationship between user (ordinarily indicating the more limited concept of 'consumer') and owner (or right-holder). Use introduces a debt to the owner, a responsibility and accountability.[60] However, this relationship of debt and credit is fundamentally asymmetrical: 'Far from being a pathological consequence, the disequilibrium is functional and fundamental.'[61] In other words, what continues to drive exchange in terms of the current business models for innovation industries built on intellectual property and the markets for intellectual property products is this disequilibrium. The debtor–creditor relationship in intellectual property models is thus not about equality or an equilibrium, as such; rather, it is concerned with the disequilibrium of debt and creditor as 'preconditions'.[62] Therefore, 'efficiency' of the current system comes precisely from a kind of 'dysfunction', a

---

57   See the discussion in '*Re* Use' on the 'Eggs' business model of consumption and use, and on the model of the 'Looking-glass cake'.

58   On the force of calculation and assessment in the intellectual property debate, see the further discussion in Chapter 1. See also the work of Maurizio Lazzarato, who suggests that contrary to admitting the asymmetry or power imbalance in the debt–credit relationship, 'Economists remove trade from the complexity of power relations and make it, along with utility, the origin of society and man ... Measure, evaluation, and appraisal all arise from the question of power, before there is any question of economics.' See M Lazzarato, *The Making of the Indebted Man*, JD Jordan (trans), Semiotext(e), Los Angeles, [2011]/2012: 80–81.

59   See also later in this chapter with respect to lack and the generation of want in consumers.

60   This raises particular issues of personal integrity when it comes to the complicated flow of rights in social media, as discussed further in Chapter 2. See also the discussion of the relationship between selfhood and property in J Gibson, *Creating Selves: Intellectual Property and the Narration of Culture*, Ashgate, Aldershot, 2006: 77–80.

61   G Deleuze & F Guattari, *Anti-Oedipus: Capitalism and Schizophrenia*, R Hurley et al. (trans), M Foucault (preface), U of Minnesota P, Minneapolis, [1972]/1983: 150.

62   Deleuze & Guattari, *Anti-Oedipus*, 187.

disequilibrium that drives change and exchange: 'it is *in order to function* that a social machine must *not function well*.'[63] The producer is always already in credit. The user is in infinite debt.

When considered in the context of tracing rights in social media, the 'social' becomes also an asymmetrical relationship of debt and credit. In other words the 'economic' relationship in social media (characterized through assignment of copyright in pictures and content, for example) is inextricable from the 'social' production of the subject (as debtor),[64] and the 'debt of existence'.[65] The producer/ consumer paradigm is therefore crucial to the architecture of the intellectual property system, but is complicated in the digital environment of social media. Importantly, even here use does not make creators (or producers) of users, as such, but rather such content is qualified and described as 'user-generated', maintaining the distinction between debtor (user) and creditor (producer).

Use also implies possession and indeed a certain 'trust or confidence', that is, a certain credit placed in the person that supplies the property from which the benefit or profit may be derived. The producer is always already in credit. The counterpoint to this, as discussed, is the notion of trust in the debtor to pay, the memory of the debt. However, increasingly the relationship of producer to user is mediated by technology not only in terms of the delivery of the product, but also in terms of ensuring the user's promise to pay (through technical protection measures and so on). Clearly such technical solutions for the legitimacy of the law simply reaffirm the conventional models of innovation and production, without necessarily addressing the relationship of trust (and thus legitimacy and enforcement). In this way, the debt–credit relationship in intellectual property is made to function without any action or input from the user. Use is literally ruptured and meaning in the system is abandoned.

Use and custom

Use is also associated with notions of habit, custom and practice, recalling the notion of duty and debt introduced above, but at the same time introducing the customary and familiar value of use that will be considered in more detail later in this chapter, and which is crucial to the concept of familiar production. Habit also begets habit, as it were, and pronounces a semblance of stability of manners but through the very censure of difference in repetition. In habit, use operates as a kind of observation (of custom) or preservation (of manners and stability). In other words, in habit, use operates as audience, as distinct from use as production. However, production in the social, in the digital, interrupts the commonplace

---

63   Deleuze & Guattari, *Anti-Oedipus*, 151.

64   Maurizio Lazzarato suggests: 'The modern notion of "economy" covers both economic production and the production of subjectivity.' See Lazzarato, *Making of the Indebted Man*, 11.

65   Deleuze & Guattari, *Anti-Oedipus*, 197.

conception of habit as a linear and passive process. In the digital, use repairs, makes familiar, becomes accustomed, habituates and occupies the digital. Participation in digital cultural life takes some getting used to.

In law, as well as its similar meaning in ordinary language, use has an etymology associated with land and the profit or benefit derived from lands or tenements, ultimately to its development as a mode of exercising ownership (through use and occupation).[66] Indeed, use can similarly contribute to the development and protection of rights in intellectual property.[67] Use not only generates or indicates certain rights or ownership, but also refers to the practice of those rights.

Thus, use is importantly, for the purposes of the present discussion, complicit with value not only in terms of meaning for users (as consumers) but also in terms of enforcement for users (as producers).[68] In this respect, the 'use' and the ability to use (whether in terms of access to justice[69] or manifestation of rights at the point of enforcement and so on) are implicated in both the use of the system and the use of the product. Therefore, use is not only in respect of a product, but also with respect to a custom, and in observation with a law. To use is, quite literally, to put the law into practice and to enforce its rule. In this context, the nexus between use, law and custom (as value) is played out not only in the observation of the law but also in the observation of its penalties, such as where use may also be in relation to infringing use, and value arising with respect to that use (through enforcement).[70]

---

66   As well as the historical foundations for the relationship between use and access to the land in feudal systems, the modern doctrine of adverse possession also recognizes a relationship between use and the creation of interests and rights, as distinct from ownership. See the discussion in M Bloch, *Feudal Society 1: The Growth of Ties of Dependence*, LA Manyon (trans), 2nd ed, Routledge, London, 1962: 68 and 115. See further the relationship between land, property and use in J Gibson, 'The Lay of the Land: The Geography of Traditional Cultural Expression', in CB Graber & M Burri-Nenova (eds), *Intellectual Property and Traditional Cultural Expressions in a Digital Environment*, Edward Elgar, Cheltenham, 2008: 182–201.

67   A trade mark, for example, either comes into being upon use (as in the United States, 15 USC §1051(a)(1) where a mark must be used in commerce to be registered) or requires use for its continued existence (as non-use is a ground of revocation: see for example in the EU, Directive 2008/95/EC to approximate the laws of the Member States relating to trade marks, art 10).

68   See, for example, the discussion of value in infringement and value in enforcement in Chapter 11.

69   For example, see the discussion of the UK Patents County Court and discursive justice as well as accelerated justice in Chapter 11.

70   Enforcement not only refines the scope of value but also 'accounts' for value (as in damages, account of profits and so on).

Use and production

However, as well as this relationship to things or to a particular purpose, use can also be taken to designate a purpose in itself, a service, or indeed a function. Thus, use imposes a certain 'character' upon the person to which it is applied. In this way, it is not only things which may be useful for a purpose (by means of their particular properties or qualities) but also people in terms of their office or class,[71] their function or indeed their service, immediately implicating the persistent relationship between property and self, and indeed property and the ability to be a producer. Nevertheless, consumers, as users, are indeed 'useful' to contemporary modes of production, including the possibly obvious example of social media. However, the role of users is changing notably throughout more revolutionary models of research and development, ownership and production that are emerging through the applications of new practices, new technologies and new environments (including not only the digital environment, but also unconventional sites of productions, such as cultural institutions, the domestic sphere and so on).[72]

At the end of the last century, Deleuze writes that developments in new technologies, rather than economic strategy in and of themselves, are instruments in the transformation of capitalism, from the factory as a 'site of confinement' to owning the means of production in a variety of other sites, such as homes, schools and so on.[73] At the start of this century, Hardt and Negri argue: 'The informationization of industry and the rising dominance of service production, however, have made such concentration of production no longer necessary. Size and efficiency are no longer linearly related; in fact, large scale has in many cases become a hindrance.'[74] Production has thus become distributed and diffuse, domesticated and ubiquitous:[75] 'While production is carried on through social networks and is closely connected with the processes of commodity-circulation, and while productive labour (which, though diffuse, is above all *socially integrative*) is to be found everywhere, *by means of the social, production and reproduction constitute a completely uniform, undifferentiated network.*'[76] Moving on from this, the emergence of 'clusters' of smaller entities has become the new 'organism' of production, where proximity

---

71   See the discussion of peasantry and property later in this chapter. See further the notion of class, property and selfhood, including the concept of the peasant and the propertied, in Gibson, *Creating Selves*, particularly chapter 1.

72   See further the discussion of network production in M Hardt & A Negri, *Empire*, Harvard UP, Cambridge MA, [2000]/2001: 294–300.

73   G Deleuze, *Negotiations: 1972–1990*, M Joughin (trans), Columbia UP, New York, [1990]/1995: 180–81.

74   Hardt & Negri, *Empire*, 294.

75   The introduction of 3-D printing technology, indeed, markedly expands the potential for 'industrial' production within the domestic and private sphere.

76   A Negri, *The Politics of Subversion: A Manifesto for the Twenty-First Century*, J Newell (trans), Polity P, Cambridge, [1989]/2005: 89.

is based not on size but on the personal and 'social power'[77] of innovation.[78] The digital becomes personal, as it were, in that in a very literal sense emerging models of innovation are based almost invariably upon principles of communication and the social. It is affinity and the social that is precisely the 'use' of social media for production. The capacity of production in the digital is precisely its familiarity.

Use and social

Thus, both production and use are distinctly social, but at the same time each becomes associated with a certain expectation as to social roles, obligations and indeed also status, even perhaps in spite of 'use' and production becoming more decentralized through the digital.[79] This meaning of use in relation to social status is found in ordinary language and, at the same time, this social and political value of 'use' is integral to dominant models of production. Indeed, in current economic models of intellectual property production and use, and in the way in which intellectual property regulates so much of our social, creative and cultural lives, as well as the way in which we engage with the resources of cultural life, use implies a type of social 'class' within the discourse of intellectual property.[80] The credit and accountability for creativity and innovation is situated with the producer, while the user is situated outside the value-making of the creative or innovative economy – the 'propertyless peasant'.[81]

Intriguingly this resonates with the notion of use in relation to social status and to fashion and taste, and the consolidation and institutionalization of intellectual property as a register of credit and public aesthetics.[82] In this sense, within the 'class' logic of intellectual property, the value of use is articulated in terms of

---

77   Discussed in more detail later in this chapter and throughout later chapters.

78   For example, in London, Digital Shoreditch and Tech City (or Silicon Roundabout, in honour of the large number of web-based businesses located near the Old Street Roundabout, considered the 'heart' of Tech City) are attempts to consolidate talent and innovative activity based upon the principle of proximity in community and the important relationship between place (and clusters) and innovation (in the digital). Further information is available at http://techcity.io.

79   See further the discussion in '*Re* Use'.

80   Gibson, *Creating Selves*, chapter 1.

81   'In the economy of intellectual property, therefore, the user is "propertyless", the "peasant" class.' See Gibson, *Creating Selves*, 75. See further the discussion of investment and creativity in *Creating Selves*, 13–15, and the discussion of the 'industrial-creativity complex of commercial models of innovation and progress' where a type of 'aristocracy' of producers is positioned against a distributed class of 'peasant folk', in *Creating Selves*, 22.

82   See Chapter 4.

exclusion (indeed, the 'peasant').[83] On the other hand, the 'social' production of the digital is predicated on use as inclusive.[84]

Of interest in this respect is the way in which regulation of activity (and associated business models) cooperates in re-inscribing the user within a logic of exclusion. Thus, in many social media platforms the user-consumer is effectively 'employed' (used) in business models to 'produce' surplus value (through their content, which can subsequently be 'made use of' and indeed turned to account for the purposes of advertising and other means of raising revenue through 'social power', including transactions in social data).[85] Therefore, despite the seemingly automated nature of contemporary capitalist production, with the apparent transition from human surplus value to that produced by machines,[86] arguably the human is indeed not tangential but in fact firmly at the centre of production. Capital has thus been enormously successful in capturing an increase in leisure time as another opportunity to organize human labour.

Use and meaning

It is thus important to recuperate the active, meaningful[87] role of use, that is, the very crucial role of the user not only in the sense of the terminology, but also in the legitimacy of the system itself: 'the meaning of a word is its use in language'.[88] And so 'Let the use of words teach you their meaning.'[89] As well as the purposeful

---

83   Hardt and Negri explain the notion of exclusion and class: 'Working class is fundamentally a restricted concept based on exclusions.' M Hardt & A Negri, *Multitude: War and Democracy in the Age of Empire*, Hamish Hamilton-Penguin, London, 2004: 106.

84   This shares aspects of Hardt and Negri's concept of the multitude as 'an open and expansive concept', while nevertheless remaining 'a class concept'. See Hardt & Negri, *Multitude*, 107 and 103 respectively.

85   This is discussed further in Chapter 2 and in '*Re* Use'.

86   Deleuze and Guattari note the problem of understanding 'how one can maintain human surplus value as the basis for capitalist production, while recognizing that machines too "work" or produce value, that they have always worked, and that they work more and more in proportion to man, who thus ceases to be a constituent part of the production process, in order to become adjacent to this process. Hence there is a machinic surplus value produced by constant capital, which develops along with automation and productivity, and which cannot be explained by factors that counteract the falling tendency – the increasing intensity of the exploitation of human labor, the diminution of the price of the elements of constant capital, etc. – since, on the contrary, these factors depend on it.' Deleuze & Guattari, *Anti-Oedipus*, 232.

87   In a sense, the meaning of use (to employ, to engage) also resonates with Wittgenstein's understanding of meaning through use: 'So one might say: the ostensive definition explains the use – the meaning – of the word when the overall role of the word in language is clear.' Wittgenstein, *Philosophical Investigations*, §30.

88   Wittgenstein, *Philosophical Investigations*, §43.

89   Wittgenstein, *Philosophical Investigations*, 220e.

engagement or action that is implied by the term, use also ushers in notions of the game, and the participation of the user in the game of intellectual property is crucial. The concept of the game is not only an element of *Alice's Adventures in Wonderland* and indeed frames the entirety of *Through the Looking-Glass*, but also it is crucial to the reimagining of innovation and intellectual property in the digital environment, the language-game of innovation: 'I shall also call the whole, consisting of language and the actions in which it is woven, the "language-game."'[90] *Use is language.*[91] *Start to play.*

Use is thus profoundly linguistic, as it were, giving utterance to words, to meaning, entirely complicit and necessary in the iterative and stuttering becoming and creativity of the digital: 'And this multiplicity is not something fixed, given once for all; but new types of language, new language-games, as we may say, come into existence, and others become obsolete and get forgotten.'[92] Use as talk, as speaking, as singing, guessing riddles and telling jokes: '"Speaking about something" can just mean so many things.'[93] Use is intrinsically and inextricably in dialogue, in play, in combat: 'to speak is to fight'.[94]

And this also returns the discussion once again to the figure of the consumer, and the literal use or consumption that figures throughout Alice's adventures. Alice is perhaps the ultimate consumer, the user driving the narrative journey through Wonderland by eating and drinking her way through change, catalysing transformations and events. Use is doing. This is not merely allegorical, but is entirely accustomed within the language of use and land, use and real property. In terms of 'use as doing', the jurisprudence of equitable and presumptive rights in land arising from use sets out the significance of the way in which use gives rise to certain responsibilities as well as entitlement in respect of property. Use is significant not only to the question of access, but also to the question of the value of the property and the performance of that value. *Use occupies meaning.*

In ordinary language, use is also applied to the notion of place, particularly in the sense of the habitual use of a place. Once again, the relationship between

---

90 Wittgenstein, *Philosophical Investigations*, §7.

91 Lyotard's interpretation of Wittgenstein's language-game, and his subsequent consideration of this in the context of law and justice, is illuminating with respect to language and use: 'What [Wittgenstein] means by this term is that each of the various categories of utterance can be defined in terms of rules specifying their properties and the uses to which they can be put – in exactly the same way as the game of chess is defined by a set of rules determining the properties of each of the pieces, in other words, the proper way to move them.' J-F Lyotard, *The Postmodern Condition: A Report on Knowledge*, G Bennington & B Massumi (trans), F Jameson (foreword), U of Minnesota P, Minneapolis, [1979]/1984: 10. The language-game is considered throughout the present discussion. The importance of the language-game and the use of language in the course of justice are explored in more detail in Part III.

92 Wittgenstein, *Philosophical Investigations*, §23.

93 Wittgenstein, *Philosophical Occasions*, 417.

94 Lyotard, *Postmodern Condition*, 10.

use and habit arises, but this time in the context of repetition and occupation. Repetition brings about changes in meaning. Use occupies meaning, gives a place to the name. Use is thus the locale of ethics in intellectual property reform,[95] the customary and familiar repair of production, as it were. Use is therefore a kind of repetition, a 'refrain',[96] a revisiting, an ethics of place: 'it is *ethos*, but the ethos is also the Abode.'[97] This resonates with the notion of use and creativity throughout one's home and indeed one's life, with the notion of 'living labour' and that of creativity as 'tied to the productive life of people'.[98]

Use is ethics.

User Names

It is difficult to exact in the term 'user' the precision that is necessary. All participants in the system are users of the system. However, in respect of use, the discourse and debate frequently understands 'user' to refer to the consumer and the characterization of that individual in relation to the producer in the traditional sense of the creator (and bearer of rights) or owner (as right-holder). In some respects, departing from the way in which user is used to qualify (or disclaim) creativity in the consumer (such as the use of terms like user-generated, user-innovation and so on, as though this type of creativity is somehow apart from other models of creativity and innovation) would be desirable. Particularly in the context of social media and various instances of distributed and collaborative production, a person who generates products through use is nevertheless still a creator. But retaining the term 'user-generated' still defers and displaces that status, that function. In many respects a term such as generator would be applicable, following the same principle at play in terminology such as actor or performer (both of which are synonymous with the notion of doing, doing to produce, using to produce). Indeed, rights such as the performer's right demonstrate the law's capacity to account for meaning (and production) through 'use', as it were.

Nevertheless, the term 'user' is retained, almost defiantly and combatively, because of its cooperation with notions of community, collaboration and diffusion. And also, fundamentally, user is retained because of its defiance of the unilateral model of use and production: 'Naming must therefore be a collective and common process … naming is perhaps the only process through which a form of decision can be imagined.'[99]

---

95   The relationship between use and ethics is considered throughout; however, in particular, please see the discussion in Chapters 1, 4 and Part III.

96   This notion of ethics and the refrain is considered further in Chapter 10.

97   Deleuze & Guattari, *Thousand Plateaus*, 312.

98   A Negri, *Art & Multitude, Nine Letters on Art, followed by Metamorphoses: Art, and Immaterial Labour*, E Emery (trans), Polity P, Cambridge, [2009]/2011: 93.

99   Negri, *Negri on Negri*, 120.

The Logic of Production

As an intellectual discipline, and as discipline of the intellect, logic is one of the most ancient systems. Languages such as mathematics and modern information technologies and processing are considered to be 'logical' systems. However, rather than an anathema to juridical language, is the logical language of modern technology wholly coincident with the principles of law and, more specifically, intellectual property law?

Arguably, the codification of law operates by a similarly logical process and indeed, intellectual property laws attempt to create a logical system for creativity and innovation not only through categorizing the products of innovative and creative endeavour, but also in documenting the process of change itself. In fact, intellectual property is built precisely on providing what will be acceptable justification for the monopoly – a good logical reason – and as such purports to simulate the map of reason and calculability for our creative and innovative societies. In order to chart the course of change, of innovation, what is indeed immediately implied is the quantification of innovation, of creativity, rendering it assessable, calculable. If one is to ask the question, what is innovation, the only logical response appears to be the indication towards a somewhat limited, and necessarily limiting, exemplary object. But that object is not necessarily the 'product', as we know and understand it, but rather the 'debt' to the creative.[100]

Thus, the 'social contract' of intellectual property appears to be an unsustainable rhetoric, as it currently stands. More accurately it is 'the contractual relationship between *creditor* and *debtor*',[101] which inevitably remains imbalanced in order to continue the flow of cultural 'debt'.[102] Balance and equilibrium are thus convenient myths to offset the credit of 'risk' assumed by creators, and paid by consumers,[103] and risk is always taken for granted as 'value'.[104] Indeed, within the World Trade

---

100   The notion of 'debt' in intellectual property is discussed in detail in Chapter 2.

101   Nietzsche, *On the Genealogy of Morals*, 63.

102   Lazzarato explains: 'The economy and society are organized according to power differentials, an imbalance of potentialities.' See Lazzarato, *Making of the Indebted Man*, 75.

103   In *Anti-Oedipus*, Deleuze and Guattari illustrate this function of disequilibrium (and so the inevitable interminable nature of the debt) through the dynamics of the Kula gift economy: 'debt is the actual direction of this movement, a kinetic energy that is determined by the respective paths of the gifts and countergifts' (149). In relation to balance and disequilibrium, they explain, 'If one postulates that somewhere there has to be a kind of equilibrium of prices, one is compelled to see in the manifest disequilibrium of the relations a pathological consequence … Far from being a pathological consequence, the disequilibrium is functional and fundamental' (149–50). See the more detailed discussion of debt and disequilibrium in Chapter 2. See further the discussion in Lazzarato, *Making of the Indebted Man*, 51–54.

104   See further the discussion in Chapter 5.

Organization (WTO)[105] Agreement on Trade Related Aspects of Intellectual Property Rights (TRIPs), the 'balance' to which that text refers in Article 8 is itself this relationship between risk (on the part of the producers or right-holders) and obligations on the part of the State to provide private rights in order to offset that risk, thus incurring on behalf of its citizens the 'debt' of intellectual property.[106] Not only are users therefore deferred, but also they are thus excluded from the language-game of international intellectual property rules and interpretation. In the proper names of intellectual property, as a starting point to meaning and interpretation, the user is nowhere to be seen.[107]

The Proper Names of Intellectual Property

And this is the logical response of intellectual property, to designate that which is inventive, new, distinctive, original, individual – that which is innovative. In order to enable society to recognize innovation, the intellectual property system provides examples which may be imitated, repeated and consumed – the patent, the design, the trade mark, the copyright. The very logical foundation of the system, the crucial mechanism for its sense, is imitation, or perhaps more accurately, repetition. However, returning to the question *what is innovation*, perhaps the serious limitation of current curiosity and debate into the operation of the intellectual property system and the incentive for innovation is an attempt at an always already attenuated version of innovation. The question is perhaps slightly to the side of this – namely, the interest in innovation itself, and the interest in the relationship between innovation and intellectual property. When this relationship performs well, the rest of the game takes care of itself. Thus, at the very outset it is necessary to understand that while it might be a case of simply accepting that the intellectual property system just tells us what is 'innovative' in certain exemplary cases, the intellectual property system does not and cannot explain innovation itself.

And so to revisit Wittgenstein's interlocutory:

> What is the meaning of a word?

---

105   The World Trade Organization (WTO) is an intergovernmental international trade organization, and therefore outside but in cooperation with the United Nations, 'whose primary purpose is to open trade for the benefit of all'. See www.wto.org. In the Results of the Uruguay Round of Multilateral Trade Negotiations (1986–94), the WTO was established in 1995 (succeeding the General Agreement on Tariffs and Trade (GATT)) by the Agreement Establishing the WTO to which is annexed the GATT (Annex 1A), the General Agreement on Trade in Services (GATS) (Annex 1B), and the Agreement on Trade-Related Aspects of Intellectual Property Rights (TRIPs) (Annex 1C).

106   See the discussion in Chapter 12.

107   For further discussion of treaty interpretation in this context, see Chapter 12.

Let us attack this question by asking, first, what is an explanation of the meaning of a word; what does the explanation of a word look like?

The way this question helps us is analogous to the way the question 'how do we measure a length' helps us to understand the problem 'what is length?'[108]

In other words, how do we measure innovation? This will help us to understand the problem – *what is innovation*? Crucially, calibration and assessment of innovation have been almost entirely subjected to machinic, empirical models of innovative progress. Regardless of the efficacy, accuracy, or reality of such models, it is important to reappropriate this language within the domain of 'use', as it is indeed the 'use' of innovation that qualifies and propels such modelling. Therefore value might be calibrated according to 'use' of the system (such as registries, applications and filings), 'use' of the innovation (such as through manufacturing of a patent) or indeed 'misuse' (of the invention, where value is understood in terms of the value of infringement, of a trade mark, where misuse is no use).[109] All of these 'uses' are articulated through the intellectual property system (through procedure, litigation and practice). However, the contention of this present discussion is that such modelling of 'use' is inadequate or incomplete. In other words, it is unclear whether, as users (whether consumers, producers or otherwise), we should accept the subjugation of the explanation of innovation to the 'machinery' of intellectual property rules or the calculation of economics alone.

Surplus to Requirement

Marx notes the distinction between the value produced by labour (variable capital)[110] and the cost of maintaining that labour, surplus value being produced where labour exceeds the costs of maintaining it: 'that part of capital which is turned into labour-power does undergo an alteration of value in the process of production. It both reproduces the equivalent of its own value and produces an excess, a surplus-value, which may itself vary, and be more or less according to the circumstances.'[111] Additional value is thus produced only through the

---

108   Wittgenstein, *Blue and Brown Books*, 1.

109   In US patent law, misuse is an affirmative defence (based on anti-trust law/ competition law). For a discussion of the rationale of this doctrine see TF Cotter, 'Four Questionable Rationales for the Patent Misuse Doctrine', 12(2) *Minnesota Journal of Law, Science and Technology*, 2011: 457–89. In relation to trade mark non-use see fn 68.

110   Variable capital is distinguished from constant capital: 'That part of capital … which is turned into means of production, i.e. the raw material, the auxiliary material and the instruments of labour, does not undergo any quantitative alteration of value in the process of production. For this reason, I call it the constant part of capital, or more briefly, constant capital.' See K Marx, *Capital: A Critique of Political Economy, Volume I*, E Mandel (intro), B Fowkes (trans), Penguin, London, 1976: 317.

111   Marx, *Capital*, 317.

commodification and exchange of time, and the manipulation of time, such as the working week, annual leave, production time and so on. In other words, additional value is produced through the commodification and exchange of labour-power: 'For this relation to continue, the proprietor of labour-power must always sell it for a limited period only, for if he were to sell it in a lump, once and for all, he would be selling himself, converting himself from a free man into a slave, from an owner of a commodity into a commodity.'[112] In other words, the effort or labour has value, but in respect of supply (and potential limits to that supply): '[The proprietor] must constantly treat his labour-power as his own commodity, and he can do this only by placing it at the disposal of the buyer, i.e. handing it over to the buyer for him to consume, for a definite period of time, temporarily. In this way he manages both to alienate his labour-power and to avoid renouncing his rights of ownership over it.'[113]

Therefore, there is a clear and important distinction, according to Marx, between capital and labour.[114] However, in a debt economy, the productive and innovative flow of labour is attributable to, commensurate with and logical only through capital; that is, the flow of debt and credit: 'Capital becomes filiative when money begets money, or value a surplus value.'[115] That is, value is acquired not in terms of use-value or even in terms of an indexical abstraction, that of money. In this manner, value (and meaning) of production is in terms of the infinite debt, of capital, where value is understood as value in process, and money is nothing other than money in process: 'In simple circulation, the value of commodities attained at the most a form independent of their use-values, i.e., the form of money. But now, in the circulation M-C-M,[116] value suddenly presents itself as a self-moving substance which passes through a process of its own, and for which commodities and money are both mere forms.'[117] Thus, the referencing of 'value' is not with respect to use as such but in terms of 'a private relationship with itself, as it were. It differentiates itself as original value from itself as surplus-value.'[118] Bank credit thus 'effects a demonetization or dematerialization of money' where 'it assumes, then loses, its value as an instrument of exchange, and where the conditions of flux imply conditions of reflux, giving to the infinite debt its capitalist form'.[119] *Money literally buys nothing.*[120]

---

112   Marx, *Capital*, 271.
113   Marx, *Capital*, 271.
114   Deleuze & Guattari, *Anti-Oedipus*, 11.
115   Deleuze & Guattari, *Anti-Oedipus*, 227.
116   The figure M-C-M is used to designate 'Money-Commodities-Money'. See Marx, *Capital*, 248.
117   Marx, *Capital*, 256.
118   Marx, *Capital*, 256.
119   Deleuze & Guattari, *Anti-Oedipus*, 229.
120   See further the discussion in Chapter 9.

In other words, capital calculates but does not acquire meaning. Rather, it manifests '*the transformation of the surplus value of code into a surplus value of flux*'.[121] Within the debt economy, thus, invention, production and innovation are rationalized purely by their quantification within capitalism. Meaning and therefore 'use' are not merely marginalized, they are altogether inconsequential. Thus, in the machinic subordination[122] of the user, the consumer, the debtor, surplus value is generated beyond the relationship of labour-power.

What then of value in the re-conceptualization and re-articulation of creativity and use through the digital? It would seem that from the mechanized production process of variable capital we are now within an era of 'a progressive increase in the proportion of constant capital'.[123] In other words, while this revolution in production might appear to liberate the user, in fact there is instead the risk of 'a new kind of enslavement: at the same time the work regime changes, surplus value becomes machinic, and the framework expands to all of society'.[124] This is the 'machinic enslavement'[125] of the debt economy and the displacement of meaning through use: 'the relationship between man and machine is based on internal, mutual communication, and no longer on usage or action.'[126] Thus, in addition to time and commodity, the very nature of consumer behaviour may be purchased and re-circulated within the debt economy.

In this way, consumption also becomes the very basis for social media business models, where the product is elsewhere, deferred, absent. Leisure time becomes not only an opportunity (for a specific product) but also a way in which to produce value over and above conventional production models. Social media and viral marketing therefore produce, literally, a surplus value of code, through the additional communication of data on use and traffic: 'There is no genetics without "genetic drift" … fragments of code may be transferred from the cells of one species to those of another … by viruses or through other procedures. This involves not translation between codes (viruses are not translators) but a singular phenomenon we call surplus value of code, or side communication.'[127] More

---

121   Deleuze & Guattari, *Anti-Oedipus*, 228.

122   This is understood in terms of the way in which the user's decision-making process might be removed from the process (automatic machines and so on) or where the user's work is in fact part of a machine or structure, or indeed where consumer data (purchasing history and similar information) is used to inform marketing and other strategies. See the discussion of machinic enslavement in Deleuze & Guattari, *Thousand Plateaus*, 457–58.

123   Deleuze & Guattari, *Thousand Plateaus*, 458.

124   Deleuze & Guattari, *Thousand Plateaus*, 458.

125   Deleuze & Guattari, *Thousand Plateaus*, 458.

126   Deleuze & Guattari, *Thousand Plateaus*, 458.

127   Deleuze & Guattari, *Thousand Plateaus*, 53.

importantly, however, is the production of socio-economic 'chimera'[128] and the transformation of a surplus value of code into a surplus value of flux.[129]

Thus, the entirety of social life, of existence, is transformed into labour power. The debt of the consumer is a *debt of existence*,[130] rendering the time of production both infinite and instantaneous. Therefore, any use outside the assessable and calculable data on traffic, preferences and so on is simply 'waste'. Social media becomes an instrument in a tremendous push towards economic rationality and efficiency. Indeed, in this way, perhaps a cultural revolution but certainly an invaluable invention of capitalism, social media is the greatest success in a drive to economize and economicize free time.[131] Therefore, expanding the coverage of social media and indoctrinating all citizens into the 'production' capacity of social media elaborates and proliferates more social production and social innovation, at all times, as it were, allowing for less and less 'free' time towards the seemingly infinite expansion of the economic sphere.[132]

Production is Life, the Social is Production

References to 'social production' are now commonplace in considerations of contemporary networked productivity and creativity in the digital space. Its first

---

128   In other words, there is an inter-species 'drift' of capital in terms of the use of the product and the products of that use (data on use).

129   Deleuze & Guattari, *Anti-Oedipus*, 227–29.

130   Deleuze & Guattari, *Anti-Oedipus*, 197.

131   On the 'economicization' of free time, and the possibility of a society of 'free time', see the discussion in A Gorz, *Critique of Economic Reason*, G Handyside & C Turner (trans), Verso, London, [1988]/1989. In this work, Gorz explains: 'Computerization and robotization have, then, an economic rationality, which is characterized precisely by the desire to *economicize*, that is, to use the factors of production as efficiently as possible ... For the moment, suffice it to say that a rationality whose aim is to *economicize* on these "factors" requires that it be possible to *measure, calculate and plan* their deployment and to express the factors themselves, whatever they may be, in terms of a single unit of measurement' (2–3). Gorz noted in later work: 'Where, in the absence of market and commodity relations, it has not yet established itself, "socialism" cannot set economic rationality in the service of a social project intended to transcend it. Where "socialism" is conceived as the planned development of as yet non-existent economic structures, it inevitably turns into the opposite: it subordinates society to the accumulation of capital, and posits economic rationality as the objective around which social life is to be reorganized. Such a society cannot assert its independence of economic rationality. It is "economicized" through and through.' See A Gorz, *Capitalism, Socialism, Ecology*, C Turner (trans), Verso, London, [1991]/1994: 69.

132   Compare the notion of 'holiday time' or 'leisure time' which, according to Gorz, 'has no utility, nor is it the means to any other end and the categories of instrumental rationality (efficiency, productivity, performance) are not applicable to it, except to pervert it.' See Gorz, *Critique of Economic Reason*, 4–5.

use in debates in intellectual property has been attributed to Yochai Benkler,[133] who describes it as production which 'first and foremost harnesses impulses, time, and resources that, in the industrial information economy, would have been wasted or used purely for consumption'.[134] However, it has a much earlier appearance than this, and much of what Benkler is considering is in debt to Marx's analysis of the social character in production.[135]

For Marx, production is always inherently social: 'Whenever we speak of production, then, what is meant is always production at a definite stage of social development – production by social individuals.'[136] In this sense, the 'social' maintains a relationship between property and person, that is, between property and sense of self: 'All production is appropriation of nature on the part of the individual within and through a specific form of society … That there can be no production and hence no society where some form of property does not exist is a tautology.' Even in its contemporary use, the term 'social production' emphasizes this relationship between self and production,[137] as distinct from the relations between selves in the course of production. What is particularly useful about the foundations of the term in Marxist theory is the nature of social production through ritual and use, that is, use and habit in the production of meaning and norms. This is especially relevant in the context of digital creativity and innovation (in the widest possible sense) and prefigures social production and the value of the language-game in Wittgenstein: 'to imagine a language means to imagine a form of life'.[138]

Despite this rich background, the term has become little more than shorthand for peer-based or communal production in the sense adopted by many intellectual property commentators today. In its current use it has little to say about the cultural circumstances of production itself, as such, only its capture (and indeed more specifically, the technological circumstances of that capture). In other words, it is assumed that the subject matter of social production would have happened whether or not captured within social media and related platforms which 'harness' and arguably 'organize' that production within the industrial information economy. In

---

133 The term itself became popular following its use by Yochai Benkler in Y Benkler, 'Coase's Penguin, or Linux and the Nature of the Firm', 112, *Yale Law Journal*, 2001: 369. Benkler provides some further exploration of social production in *The Wealth of Networks*, Yale UP, New Haven, 2006: 122–27.

134 Benkler, *Wealth of Networks*, 122.

135 See Marx, *Capital*, 173–77.

136 K Marx, *Grundrisse: Foundations of the Critique of Political Economy*, M Nicolaus (trans), Penguin, London, 1973: 85.

137 For instance, Benkler concentrates on market-based enterprises and although he acknowledges changes in consumer behaviour, it is nevertheless still in respect of a 'product' in terms of platforms and tools as distinct from finished goods, as well as the specific competitive challenges to present conventional models in the knowledge sector. Benkler, *Wealth of Networks*, 122–27.

138 Wittgenstein, *Philosophical Investigations*, §19.

the narrower contemporary sense of the term, the emphasis on dominant models of peer-based and open systems has been such as to abbreviate the potential of production to the digital 'systems' of production, rather than referring to the socio-cultural character and manifestations of production, as it were. Within the conventional rendering of the digital according to linear and analogous production models, social production produces difference, incompletion, 'uselessness', not least because of its seemingly inconceivable relationship with traditional forms of innovation, protection and dissemination (or exploitation). That is, 'use', in so far as it is attenuated within intellectual property, is seemingly irrelevant in social production as presently treated, and so this kind of production is apparently 'useless' at best, or illegitimate at worst, in respect of incumbent models of production and exploitation. Rather than adopting this more attenuated discussion of social production, there is a need to address the kind of social labour power underpinning social production, significant not only for suggestions of open and networked production, but also for the cultural and political transformation in production and ownership.

For this reason, earlier uses of 'social production' are insightful in order to address a wider 'remit' of the term and the limitations of the current use with respect to analysis of property relations and a restoration of use. Deleuze and Guattari build upon Marxist theories of production to examine social production: 'the forms of social production, like those of desiring-production, involve an unengendered non-productive attitude, an element of antiproduction coupled with the process, a full body that functions as a *socius*.'[139] Social production as a becoming is nevertheless coded extrinsically, but capital is all at once intrinsic to social production, capturing that flow (within the various 'terms of use', that is, rules of participation in the social).[140] This concept of antiproduction or waste,[141] considered in more detail later in this chapter, is fundamental to a reconsideration of the 'social' as a site of labour power and production in the conventional sense. Indeed, several decades before Benkler, Deleuze notes the shift from production of goods to the metaproduction of services: 'capitalism ... no longer sells finished products ... What it seeks to sell is services, and what it seeks to buy, activities.'[142]

---

139   Deleuze & Guattari, *Anti-Oedipus*, 10.

140   Paul Patton explains: 'capitalism has no need to mark bodies or to constitute a memory for its agents. Since it works by means of an axiomatic intrinsic to the social processes of production, circulation and consumption, it is a profoundly cynical machine.' See P Patton, *Deleuze and the Political*, Routledge, Abingdon, 2000: 95.

141   Antiproduction may be introduced here simply as consumption to no perceivable end; that is, as wasteful.

142   DeleuzeG, *Negotiations*, 181. Benkler identifies a very similar phenomenon in *Wealth of Networks*, noting also that 'The emerging businesses of the networked information economy are focusing on serving the demand of active users for platforms and tools' (126) as distinct from what he describes as passive consumption.

In the digital (that is, understood more broadly than the simply technical aspects, but rather to embrace a more unstructured and distributed environment for productivity beyond the 'social' of the individual), production as a becoming, as a social and transformative process, is characterized by significant cultural and political aspects. In particular, the term 'social' is immediately indicative of the communal and public nature of social media as a meeting place (and the way in which various ideologies of user-led approaches to innovation, production and consumption are transforming contemporary innovation strategy in firms). However, arguably that production is more complicated than simply communal production. While a simple communal model of conventional outputs is certainly the case in more traditional formats such as Facebook, which preserve the habitual identity of producers, there is a significantly different approach to innovations upon the 'social' itself in formats such as Twitter, where the traditional boundaries of identity, privacy and communication are abandoned in favour of an exponential and proliferative community and dialogue and incidental mention. Therefore, 'social production' in its adapted meaning demarcates a new 'source' of production and resources; however, it is crucial to take account of a new and paradoxical intimacy not only in the format of production in the digital, social sphere, but also in the socio-political momentum of changes to commercial strategies for research and development.

First, indeed production may be considered to be communal, but it is also mutual and reciprocal, iterative and unpredictable. There is a sense of the 'in common' as well as the 'inconstant'. Secondly, notwithstanding certain concessions to privacy, it is production in public; that is, production that may be communal, but it is perhaps more significantly also civic, national and supranational. Thirdly, production is communal in the sense that it is networked and organized, but at the same time that 'network' is largely associational and sociable, a kind of digital and filiative kinship. Finally, social production is indeed a kind of amiable production, as it were; that is, it is distinctly personal and gregarious, fundamentally welcoming and yet exclusive.

Social production, in this sense, describes a wider range of phenomena and relations in the course of social participation than that suggested by its usual meaning in technology and intellectual property discourse. The kind of production underway in the digital is a form of production without physical borders (displaced from the site of the 'factory' as such) and increasingly without dominion (a 'nomadic' sense of authorship, as it were, defined through kinship or use, as distinct from relations through things or products). This is the inconstant, iterative and proliferative 'smooth space'[143] of familiar production, an unmarked space without property and without enclosure. The producer is dislocated, dispossessed, without

---

143 The concept of 'smooth space' comes from Deleuze and Guattari and is introduced here as the aggregating, associational space of the social, as distinct from the more stable space of institutional norms and processes. The concept of smooth space is addressed in more detail in Chapter 2.

province, nomadic in what is an intensive space of innovation, of becoming clearly at odds with the conventional intellectual property recognition of authority and territory in production.[144] The very articulation of 'property' ushers in problematic notions of land, as distinct from territory.[145] Long before the inception of social media, the logic, personality and family of social production and distribution resonates in *Difference and Repetition*:

> [T]here is a completely other distribution which must be called nomadic, a nomad *nomos*, without property, enclosure or measure. Here, there is no longer a division of that which is distributed but rather a division among those who distribute *themselves* in an open space – a space which is unlimited, or at least without precise limits. Nothing pertains or belongs to any person, but all persons are arrayed here and there in such a manner as to cover the largest possible space. Even when it concerns the serious business of life, it is more like a space of play, or a rule of play, by contrast with the sedentary space and *nomos*. To fill a space, to be distributed within it, is very different from distributing the space.[146]

From this it can be seen that the logic behind any disregarding of the social in commercial legal frameworks, and indeed the discontinuous and nomadic play of social production from the domain of conventional innovation strategy and commercialization, has a considerable amount in common with the difficulties (technical and political) in reconciling traditional knowledge as innovation within intellectual property.[147] In both traditional innovation and in social production, the work is diminished by virtue of its activity 'outside' the conventional models

---

144   This shares much with the mechanisms of disavowal and dispossession of traditional knowledge-holders through conventional rules of intellectual property. For instance, see the discussion in J Gibson, *Community Resources: Intellectual Property, International Trade and Protection of Traditional Knowledge*, Ashgate, Aldershot, 2005: particularly chapter 2.

145   Deleuze and Guattari explain in relation to land, as distinct from territory: 'This is the very model of an apparatus of capture, inseparable from a process of relative deterritorialization. The land as the object of agriculture in fact implies a deterritorialization, because instead of people being distributed in an itinerant territory, pieces of land are distributed among people according to a common quantitative criterion (the fertility of plots of equal surface area). That is why the earth, unlike other elements, forms the basis of a striation, proceeding by geometry, symmetry, and comparison.' See Deleuze & Guattari, *Thousand Plateaus*, 441. The relationship between the smooth space of emergent becoming (territory) and the striated space of the State (land) is considered in greater detail in Chapter 2.

146   Deleuze, *Difference and Repetition*, 36.

147   See further the discussion in Gibson, *Community Resources*, chapter 2.

of production, and therefore outside the models of entitlement to property and ownership:[148]

> where there is no State and no surplus labor, there is no Work-model either. Instead there is the continuous variation of free action, passing from speech to action, from a given action to another, from action to song, from song to speech, from speech to enterprise, all in a strange chromaticism with intense but rare peak moments or moments of effort that the outside observer can only 'translate' in terms of work.[149]

In other words, conventional modes of production (and translated into the intellectual property framework itself) are directed towards the creation and distribution of discrete products, exemplified as momentous 'events' of patents, copyright and so on, as distinct from incremental or continuous innovation (as seen in traditional and social forms of production). This emphasis on the solution and finished product, as it were, betrays 'an enduring quality of traditional domestic production, that it is production of use values, definite in its aim, so discontinuous in its activity'.[150]

---

148   This perspective is influenced considerably by the so-called 'labour-desert' argument for a natural property right, that is, the notion of labour as deserving of property. For an introduction to the 'labour-desert' argument, see JW Harris, *Property and Justice*, Oxford UP, Oxford, [1996]/2001: 204–12. See further the dominant idea of property through labour as set out in LC Becker, *Property Rights: Philosophic Foundations*, Routledge & Kegan Paul, London, 1977; SR Munzer, *A Theory of Property*, Cambridge UP, Cambridge, 1990; and L Brace, *The Politics of Property: Labour, Freedom and Belonging*, Edinburgh UP, Edinburgh, 2004. See further the discussion by Donna Dickensen in respect of labour and property or self-ownership in the body: 'The liberal basis of a right to property is thus intimately linked to self-ownership; it derives from the connection between our value-creating labour and our agency, although not from our ownership of our physical bodies. That labour is an expression of our agency and not of our bodies as such; it derives its value from that agency, but it is done through the medium of our bodies', in D Dickensen, *Property in the Body: Feminist Perspectives*, Cambridge UP, Cambridge, 2007: 39. For the relationship between intellectual property and self, see Gibson, *Creating Selves*, chapter 1. See further the discussion of property and the notion of own self later in this chapter.

149   Deleuze & Guattari, *Thousand Plateaus*, 491. On indigenous knowledge, Deleuze and Guattari note further: 'It is commonplace of missionaries' narratives that there is nothing corresponding to the category of work, even in transhumant agriculture, with its laborious ground-clearing activities', *A Thousand Plateaus*, 573, n 25. On colonialism and land ownership, Laura Brace notes: 'Questions about whose labour and whose honour count as significant underpin not only self- and property-ownership, but also the sense of belonging to the nation and imagined community.' See Brace, *The Politics of Property*, 209. See further the discussion of the work of Deleuze in relation to indigenous title in Patton, *Deleuze and the Political*, 122–31.

150   M Sahlins, *Stone Age Economics*, Routledge, London, [1972]/2004: 86.

Familiar Production

Therefore, as a departure from (or development of) the term 'social production' and an introduction to the play of innovation, developed in the present inquiry is the concept 'familiar production'. Familiar production is introduced here and used throughout to encompass the social of production as well as the transformation in the social of 'contract', the sense of community and sense of self. The kinship and community of the digital and the individual's participation in the cultural and social life of the digital is not merely a new system of production; it is a distinctive cultural revolution of which production is perhaps merely collateral. We are always producing, as it were. Production is a form of life.

The use of the term 'familiar production' indicates in part the informal nature (or perhaps more accurately, the unstructured and unorganized nature) of this type of production, production which is intimate, unpredictable and yet at the same time usual in the 'form of life'. In a sense, the term 'customary production' might be similarly appropriate to aggregate the various component concepts of production that make up this wider concept, but in order to avoid semantic baggage of the traditional as well as the specific meaning of 'customary' in legal scholarship, the term 'familiar production' is favoured. Nevertheless, this term incorporates the similar characteristics of traditional and familiar in the sense of use, the production of meaning and value through use, and the role of ritual and habit in the 'social' aspects of production and in the architecture of the 'digital' (online or otherwise).

The concept of familiar production also ushers in particular ethical dimensions. The ethical dimension of this inquiry is explored in more detail in later discussions,[151] but for the purposes of introduction here, the ethics of familiar production will be especially relevant to characterizing reform and transformation in the legal and commercial framework for the digital environment as well as innovation more broadly conceived. The concept of 'familiar' draws in explicit notions of family and household, of neighbourhood,[152] and indeed a qualified 'openness' that is respected within such communities; that is, the 'abode' of ethics. In this way, the term 'familiar' is relevant to notions of use and selfhood such that the self is translated beyond conceptualizations in terms of 'own' property that otherwise dominate conventional renditions of creativity and self.[153]

Familiar production also incorporates the very fundamental changes to the question of self and integrity of self in the context of shifts in concepts of privacy and sharing in social media and digital environments. The 'familiar' concerns

---

151   See in particular Chapters 1, 4, 7 and Part III.

152   See further the discussion of neighbourhoods of practice and sharing of knowledge in Raqs Media Collective, 'To Culture: Curation as an Active Verb', in B von Bismarck et al. (eds), *Cultures of the Curatorial*, Sternberg P, Berlin, 2012: 99–115.

153   For a full exploration of the relationship between conceptualizations of creativity and the sense of self, through intellectual property as well as other frameworks, see Gibson, *Creating Selves*.

not only the individual but also that individual's relations (including 'relations' through a broader community of users in a digital environment). Rather than the somewhat tired (and also inaccurate) rhetoric of 'democratization' online, more interesting is the way in which the digital space puts everyone on a 'family footing', as it were, where communication (and use) engenders intimate associations and interactions. Thus, the nature of knowledge itself is intimate in that there is, in the case of familiar production, a literal investment of the self in the production of knowledge. The question for the way in which familiar production is translated within the market and commercial settings will thus have a direct impact on the notion of self as commodity.

The term familiar production indicates such production as domesticated production, that is, production which occurs outside the professional or industrial space, away from the site of the factory and the firm. It is production through association and through associational knowledge, that is, through the very process of familiarity, habit and ritual by which individuals acquire close acquaintance or intimate knowledge. The familiar is thus a sphere of affect: 'as a general rule the affection (*affectio*) is said directly of the body, while the affect (*affectus*) refers to the mind.'[154] In this way, familiar production is a sphere of becoming, 'Affects are becomings.'[155] Therefore, in familiar production the way in which the user interacts with others and forms alliances and assemblages is as important as any particular product of those assemblages: 'We know nothing about a body until we know what it can do, in other words, what its affects are, how they can or cannot enter into composition with other affects, with the affects of another body.'[156] Familiar production is therefore also production in experience: 'It is certain that the affect implies an image or idea, and follows from the latter as from its cause. But it is not confined to the image or idea; it is of another nature, being truly transitive, and not indicative or representative, since it is experienced in a lived duration that involves the difference between two states.'[157] Familiar production is thus continuous production: continuous in terms of the intimacy of constant exchange and interaction with relations and associates, and continuous in terms of the 'unfinished' or iterative character of the knowledge products of familiar production. Products in many respects are thus demarcated by use as distinct from 'finished' within conventional circumstances of production and development.

Familiar production is therefore also a sphere of manners and ritual, not only one's own manners but also those of 'relations' and 'things'. This indicates a kind of 'family' in the familiar, but one which is social and socialized in a field of innovation, as distinct from the inheritance of a tree of knowledge. In this way, familiar production acknowledges the connections and combinations of a creative

---

154   G Deleuze, *Spinoza: Practical Philosophy*, R Hurley (trans), City Lights Books, San Francisco, [1981]/1988: 49.

155   Deleuze & Guattari, *Thousand Plateaus*, 256.

156   Deleuze & Guattari, *Thousand Plateaus*, 257.

157   Deleuze, *Spinoza: Practical Philosophy*, 49.

assemblage of users (including both what might be conventionally identified as producers and consumers), 'a greater creativity, of the "virus" type',[158] a creativity of experiment and experience, an innovation of chance. In this way, familiar production is a creative force without the notion of a taxonomical and classificatory framework of innovation, property and 'self'.

The familiar thus ushers in notions of proximity, privity and territory in an otherwise landless, borderless environment and domain. The familiar is, quite literally, the ordinary and the usual, the creation of meaning and value through current and everyday use and occupation. It is the generation of understanding and meaning in the speaking of language, the value of the language-game: 'the *speaking* of language is part of an activity, or of a form of life'.[159] Familiar production is thus fundamental to and inextricable from the ordinary and everyday actions of social life.

Finally, the concept of the 'familiar' also resonates with a particular structural problem, as it were, in the relationship between familiar production and economic and legal models of the market. In familiarity and intimacy comes the notion of being free, that is, at liberty, open. One of the critical issues with the reconciliation of the nature of use through digital communities and cultural transformation is that it is simply at odds with the legal framework and thus 'taking liberties with' the rules, as it were. Familiarity breeds contempt. The counterpoint, however, is that familiar production is similarly all too easily disregarded as tangential to mainstream commercial activity, or worse, that it is free for the taking. Thus, the particular 'personal effects', as it were, of this kind of production need to be addressed not only in terms of the assumptions made with respect to the creative activity of the user but also in terms of the colonization of that production by other interests, including major technology and media firms.

Contemporary familiar production thus unrests conventional modes and modes of production, including the narrower remit of 'social production' as assimilated within capital. This 'ethics' of familiar production will be traced throughout this journey, including the challenge presented by familiar production to the reasoned and stable causality of conventional production models and ways of generating value, indeed, to the very rationale of capital;[160] the emergent space of familiar production and the destruction of stable systems in the realm of familiar production;[161] the embracing of chance and proliferative opportunity in the familiar production of the digital sphere;[162] the relationship between habit and intimacy in the digital sphere, and the relationship to institutionalized aesthetics and quality;[163] the revolution of 'self' and 'ownership' within the familiar, and the

---

158   Deleuze & Guattari, *Thousand Plateaus*, 564.
159   Wittgenstein, *Philosophical Investigations*, §23.
160   Chapter 1, Cause.
161   Chapter 2, Space.
162   Chapter 3, Chance.
163   Chapter 4, Taste.

'discharging' of the debt to accredited producers and 'risk' in conventional modes of production and within the oversight of intellectual property frameworks;[164] the intensive becoming in familiar production in the digital sphere;[165] the time of familiar production, and its continuous and instantaneous nature;[166] participation in the language-game of innovation and disruption of the rules of intellectual property;[167] innovation and production of infinite difference in the proliferation of reproduction and repetition in the digital sphere;[168] an ethics of chance as distinct from the moral certainty of conventional models of judgment and accreditation (including the moral language of intellectual property discourse);[169] an ethical accountability to the just as an iterative and discursive process;[170] and finally, the relationship between producer, user and observer in the 'evidence' of innovation and its translation within capital.[171] It is indeed ultimately *re use* that describes the possibility of an ethical and logistical reform from the perspective of the user, the pawn.[172]

This is not merely a demarcation of the digital environment as engaging in a specific form of production. Rather, it is to understand the way in which the 'digital' transforms production socially, culturally, linguistically and personally. Familiar production is not necessarily only the production of digital technology, it is not merely a phenomenon of the digital environment; rather, it is something which has become an inevitable agitation within the digital. Rather, the 'digital' is a cultural and social transformation of production. In this way, a literal (or analog) example of such familiar production might include the art of the readymade, where manufactured objects are taken from one context (of use, of waste, of obsolescence) and placed in another (mounted or inscribed). One of the most famous examples of readymade is perhaps Marcel Duchamp's *Fountain*, where the artist signed a men's urinal with a false name, Mr Mutt (contrary to the authenticity of the author), and exhibited it not in its usual orientation but instead lying flat on its back (contrary to use). Rather than interpreting the latter as simply removing its domestic use and ensuring it was elevated to the realm of art, this may also be understood in the context of familiar production, where the production becomes continuous, in the experience of the object, as distinct from value within the object itself: 'The endlessly ironic touches in Duchamp's narrative are also, even in their apparent irrelevance, the source of a strange exhilaration and brilliance.'[173] Therefore, the

---

164 Chapter 5, Risk.
165 Chapter 6, Change.
166 Chapter 7, Time.
167 Chapter 8, Rule.
168 Chapter 9, Blame.
169 Chapter 10, Reason.
170 Chapter 11, Account.
171 Chapter 12, Witness.
172 '*Re* Use'.
173 R Krauss, *The Optical Unconscious*, MIT P, Cambridge MA, 1993: 95.

'controversy' attaching to finding or identifying 'artistry' in the manufactured and readymade arises from the attachment to conventional modes of production and the logic of innovation as a progressive narrative. In other words, the familiar production of the readymade does not generate use value as such in the object (the representation), but use through meaning beyond use value (the appearances of the affect). It is the affective environment of the 'use' of the object that is the site of value.[174]

Similarly, appropriation art (of which readymades might be considered a part) also raises the questions of representation that resonate throughout this book and indeed are crucial to the understanding of familiar production. That is, the very act of appropriation resituates the representation and generates a proliferation of new use and thus new meaning, at the same time irritating notions of authorship and originality that underline traditional progression narratives in innovation: 'The role of the imagination, or the mind which contemplates in its multiple or fragmented states, is to draw something new from repetition, to draw difference from it.'[175] It is like the story of Pierre Menard, by Jorge Luis Borges. The '*visible* work' may be 'easily and briefly enumerated', but the repetition in use produces astonishing differences in meaning: 'Cervantes' text and Menard's are verbally identical, but the second is almost infinitely richer.'[176]

Another example might be the works of London artist Banksy[177] and the insight provided through the attempt to sell a particular work, *Slave Labour*, at a Miami auction in 2013. The work depicts a barefooted boy at a sewing machine making Union Jack bunting. The work had appeared on the wall of a shop in North London and was removed (apparently in a sale) and subsequently turned up in a Miami auction catalogue in February. Overwhelmed by public outcry, the work was replaced by six identical copies, 36 hours ahead of the Miami auction.[178] Following the complaints and public pressure on the auction house, the work was

---

174   Rosalind Krauss notes, 'Duchamp is not interested in redeeming the commodity for plastic values, for form, for "the artist's vision," for the logical moment. Because the commodity has always already been swept up into *form*, has already, by its very condition as an item of exchange, been rationalized.' See Krauss, *Optical Unconscious*, 142.

175   Deleuze, *Difference and Repetition*, 76.

176   JL Borges, 'Pierre Menard, Author of the *Quixote*', JE Irby (trans), in JL Borges, *Labyrinths: Selected Stories and Other Writings*, DA Yates & JE Irby (eds), New Directions, New York, [1962]/1964: 36, at 42.

177   Although Banksy's actual identity is somewhat a guarded secret, this London-based pseudonymous stencil graffiti artist is believed to have been born and raised in Bristol. His stencil works appear spontaneously throughout London (and internationally) and some London councils conscientiously preserve them with roving graffiti teams (for example, Islington Council). Banksy is cited as claiming 'All graffiti is low-level dissent but stencils have an extra history. They've been used to start revolutions and to stop wars. They look political just through the style.' Cited in T Manco, *Stencil Graffiti*, Thames & Hudson, London, 2002: 76.

178   'Banksy's Slave Labour mural auctioned in London', *BBC News*, 3 June 2013.

removed,[179] but appeared for sale again in London in May along with one of the copies which had been confiscated from a member of the public.[180]

The basis for the outcry, however, is not necessarily found in intellectual property law or in a strict claim of theft. Strictly speaking, in applying the work to a building, it essentially becomes a fixture and property of the owner of the building. However, Haringey Council (the relevant council for the shop wall) attempted to compel restoration of the work and ultimately launched a campaign, 'Bring back Banksy', to bring the work back to the UK.[181] The belief of the community and indeed of the council was that the 'legitimate' course of action was the persistence of the work in situ. The sense of justice and legitimacy subsisted in the affirmation of the work as a 'gift' to the community (notwithstanding the lack of actual legal grounds). Once the work was removed from that position, its frame of reference was literally displaced, and all meaning was lost.[182] Meaning and value is acquired literally through use, through experience, through interception between the public and the work. Banksy himself is claimed to have said, 'Painting the streets means becoming an actual part of the city. It's not a spectator sport.'[183] The community is not a passive audience, not a spectator, but a participant in the game of meaning, a contributor to the familiar production of the work.

In this sense, there is the question of whether or not the removal of the work (notwithstanding the rights of the owner of the building) is a derogatory treatment of the work, interfering with its integrity to the detriment of the moral rights of the artist (if the artist can be identified).[184] This raises the question of reputation and defamation, if the appearance of the work, out of context and in a Miami auction, is detrimental to Banksy's artistic identity: 'His position has been that if you take his work out of its context it's not his work any more, it's no longer a Banksy.'[185] Indeed, in response to a question posted on Banksy's website, the reply cites Henri

---

179   Z Fagenson, 'Banksy street murals pulled from Miami auction after controversy', *Reuters*, 23 February 2013.

180   'Banksy's Slave Labour mural auctioned in London', *BBC News*, 3 June 2013.

181   Haringey Council, www.haringey.gov.uk/index/news_and_events/banksy-campaign .htm. While restoration of the work was not achieved, the council agreed with the owners of the building a plan to restore a community art project to the wall.

182   Speaking to *The Guardian*, Marc Schiller of woostercollective.com notes that 'the work was worthless in an art auction because it was only ever intended as a piece of location-specific social commentary'. R Luscombe, 'Banksy mural: I'm being scapegoated, says Miami art dealer', *The Guardian*, 22 February 2013.

183   Manco, *Stencil Graffiti*, 79.

184   Notwithstanding any issues with respect to identity and Banksy himself, it should be noted that this argument is complicated further in this case by the fact that the work had already left the UK and was in a US jurisdiction, and so subject to a narrower perspective on moral rights.

185   Luscombe, 'Banksy mural'.

Matisse: 'I was very embarrassed when my canvases began to fetch high prices, I saw myself condemned to a future of painting nothing but masterpieces.'[186]

The auction house's claim was that remaining in situ was a 'waste', as it were, in that the work could not be adequately preserved. In other words, what arises is the fundamental difference between preservation and use. Use is treated as waste. But there is no value without use, so what is in fact being preserved?[187] Surely a more appropriate and effective practice of preservation would be to document the work and replicate the work? Indeed, perhaps the deterioration and interchange with its environment is part of the continuous artistic process of such a work of familiar production. The irony of the preservation argument is seemingly lost on the proprietor of the auction house: 'It's about conservation: here's a piece of art, [and] we are going to protect it ... It could have been destroyed. When you try to make an event with a speciality you want the best lots, and Banksy *is a part of the street art scene.*'[188]

The market, in this respect, becomes key. Indeed, it is not a question of the legality or otherwise of the sale, but rather the social and cultural character of the market: 'My argument is not that the sale shouldn't happen; it's that there shouldn't be a market for it.'[189]

A Time to Remember

Use takes time. However, departing slightly from the Marxist model of capital and labour-power (and the commodification of time), the crucial commodity for the digital knowledge economy is not so much time, as memory. Possessions become a collection of memory, rather than goods in space. Use becomes a form of 'recollection', as it were, but at the same time a new experience with each repetition: *remembering the new*. Thus, the product, the work, is never fixed, but is a new experience with each 'recollection': *the time is now*. Therefore, in terms of social, business and cultural models of the physical artefact, potential for the 'good' (the physical book, for example) is in terms not of its condition as a physical good, but rather of what it offers in terms of experience with use. This has nothing to do with the intellectual property comprised within the good, but in the culture of use of the good itself.

---

186   Cited in 'Give us our Banksy back: Haringey urges Miami auctioneers to halt sale of mural', *Evening Standard*, 23 February 2013.

187   This is similar to the arguments surrounding preservation of traditional knowledge and cultural expressions, as distinct from the ongoing use of artefacts within the community. See further Gibson, *Community Resources*, 106–109; MD Simpson, *Making Representations: Museums in the Post-Colonial Era*, London, Routledge 2011: 198.

188   Frederic Thut in Luscombe, 'Banksy mural' (emphasis added).

189   Marc Schiller in Luscombe, 'Banksy mural'.

The notion of the 'contractual' relationship of user-consumer (debtor) and producer (creditor) in intellectual property is indeed inseparable from this principle of memory. Use of intellectual property creates a debt, extracts a promise, from the user-consumer, a promise to observe the balance (between debt and credit) and discharge the trust placed in the consumer: 'It was here that *promises* were made; it was here that a memory had to be *made* for those who promised.'[190]

Indeed, when it comes to this notion of an economy of memory, contemporary business models for music transact not music as experience but music as memory. The language of 'go out and buy it now' has been replaced by 'download it now'. The first suggests a sequence of events, a relationship to consumption that is chronological and where those events (the actual movement itself) represent time. In the second, 'download it now', that moment of consumption is no longer deferred or moderated by an intervening act. The abode of the archive is no longer physical space but the size of the memory. Use occupies memory.

Antiproduction

However, in the contemporary knowledge economy of digital consumption and social media, the expenditure of the consumer upon innovation, the goal of that consumption, that which renders that consumption other than 'wasteful', is paradoxically not utilitarian. That is, it is not any utilitarian goal that ensures such consumption is not wasteful. Nevertheless, consumption sustains the capitalist system by ensuring collateral damage, as it were, the 'costs' of production that rationalize the proliferation of production; that is, a kind of 'antiproduction'.[191]

Antiproduction is understood as instrumental to generating further production, not through acquiring further capacity, but through creating further desire through waste, through lack: quite literally, consumption for consumption's sake.[192] However, in the digital environment, a proliferation of consumption and production related purely to the productive capacity of the digital is not necessarily escalating the 'costs' of production. Use without purpose is reappropriated as productive capacity. In other words, it is precisely the 'waste' that is reterritorialized within the intellectual property relationships set up to govern and regulate social media in so far as that which is expended is appropriated by the producer as further credit. Therefore, while consumption may appear to be associated with no goal, no productive output, this is 'production' at its most efficient.

The digital is, however, not a social or economic strategy in and of itself. Rather, digital technology is merely indexical of a departure from the industrial or mechanical context for capitalism: 'A technical machine is ... not a cause

---

190   Nietzsche, *On the Genealogy of Morals*, 64.
191   Deleuze & Guattari, *Anti-Oedipus*, see in particular the discussion in chapter 1.
192   Deleuze & Guattari, *Anti-Oedipus*, chapter 1.

but merely an index of a general form of social production.'[193] Therefore, understanding innovation in the digital environment is not necessarily assisted by, or even dependent upon, an anatomization of digital technology. Rather, digital technology is merely an instrument of social transformation in meaning and use. What is of particular relevance and importance is addressing these transformations at the social, political and cultural levels, rather than the technological alone.[194] It is through interrogating the aspects of production and antiproduction in the digital environment, rather than the technical means (or the technical solutions) that understanding of innovation and legitimacy of new business models might be achieved. Indeed, modern high technology markets do not create desire through lack (how can one desire a product not yet imagined?), but through the narrative of anticipation and want: *What is it you want to buy?*[195]

In explaining the way in which this 'antiproduction' operates both in social production and in the forces of desiring-production, Deleuze and Guattari in fact set out the transition that is being experienced today – that from goods (clear distinction between production and product) to services (product as part of production), the move from unit to experience, the movement from mechanical to digital: 'the regime of technical machines is characterized by a strict distinction between the means of production and the product; thanks to this distinction, the machine transmits value to the product, but only the value that the machine itself loses as it wears out. Desiring-machines, on the contrary, continually break down as they run, and in fact run only when they are not functioning properly: the product is always an offshoot of production, implanting itself upon it like a graft, and at the same time the parts of the machine are the fuel that makes it run.'[196] Thus, in the digital and through social media, pure consumption for consumption's sake generates a further opportunity for surplus value: that of 'social power'.

This concept of antiproduction thus creates a market among users precisely through use. The very proliferation of 'product', as it were, does not effect a market and desire through scarcity, but rather achieves this through ubiquity: this is the Looking-glass cake *writ large*.[197] Therefore, the modern digital knowledge economy defeats the conventional relationship between use value and exchange value, the traditional economic models of intellectual property based on scarcity, the contradictory digital business models based on exclusion. The digital knowledge economy is indexical of a transformation in creativity, innovation and productivity, to be rationalized not at the stage of the product, but to be understood

---

193   Deleuze & Guattari, *Anti-Oedipus*, 32.

194   For a critique of the way in which powerful technology lobbies are overwhelming policy-making in intellectual property, see A Orlowski, 'Cameron's "Google Review" sparked by killer quote that never was', *The Register*, 21 March 2012; J Warner, 'Intellectual Property reform cannot be dictated by Google', *The Telegraph*, 3 March 2011.

195   *Through the Looking-Glass*, 149.

196   Deleuze & Guattari, *Anti-Oedipus*, 31.

197   The Looking-glass cake model of distribution is examined in detail in '*Re* Use'.

as the very conditions for commercial innovation and exploitation that apply today. Thus, in the digital environment, antiproduction does not generate more production by ensuring adequate waste and therefore 'need' for further products. Antiproduction becomes itself the means of production. *Use creates value.*

Social Power

Social power thus brings about production that occurs through the very participation in social life itself. Therefore, social power derives from labour that is unwaged[198] and production that is decentralized: 'all forms of labor are today socially productive, they produce in common, and share too a common potential to resist the domination of capital.'[199] Critically the architecture and community of familiar production through explicit platforms (such as social media), as well as in less readily captured sites, itself presents a challenge to the goods-based material models of production underpinning conventional concepts within intellectual property.

In other words, familiar production produces not only conventional intellectual products but also social life itself. That is, familiar production leads both to products and to experiences, to 'forms of life'.[200] And this is one of the most startling confrontations to the logic of intellectual property and indeed the logic of innovation itself. Innovation thus is perhaps no longer adequately understood in terms of a clear distinction between the means of production (including capital, research and development) and the product itself; thus, innovation becomes inextricable from the march of technology for technology's sake, the relentless pursuit of communication.

Product Cycles

As seen thus far, invention, production and innovation are rationalized and quantified within the debt economy, the value of each being determined subject to measures (and to time) of use. As a result, use outside such systems of measures (whether through economic modelling, intellectual property logic, or similar) is either unrecognizable or simply 'waste'. The time of innovation, therefore, becomes shorter and shorter until it is instantaneous. The innovator is always already in credit.

---

198 Hardt and Negri note: 'our understanding of labor cannot be limited to waged labor but must refer to human creative capacities in all their generality.' See Hardt & Negri, *Multitude*, 105.

199 Hardt & Negri, *Multitude*, 106–107.

200 Hardt & Negri, *Multitude*, 108.

In defence of Apple's various and myriad intellectual property challenges in the last couple of years, Tim Cook, CEO of Apple, argued the following:

> I think the US Court system is currently structured in such a way that tech companies aren't getting the intellectual property protection they need. Our cycles are fast, the court system is very long and the foreign competitors in the US can quickly take IP and use it and ship products with it and they're to the next product as well. I would love to see conversations between countries and see protections between IP globally. For us, our intellectual property is so important, I would love the system to be strengthened in order to protect it.[201]

Product cycles in technology are faster than justice. If value is generated through the value of infringement, technology products are outside systems of value altogether. They are always already in credit. However, as has been argued earlier, innovation and new technologies create new markets and products throughout the field of intellectual property not only in terms of the conditions for production and reproduction, but also regarding the very fundamental conditions of social and cultural life in which we both create and consume knowledge and the resources of our intellectual, social and political identities. So what is left of the user?

The User

As discussed thus far, the intellectual property system delimits the scope of that property through the very important function of use and indeed the very important figure of the user, but it does so with a significant exception – that of copyright. In patents, this logical ideal is the nominal person skilled in the art. In designs it is the informed user. In trade marks it is the average consumer (or relevant consumer). While this might be seen as a 'sliding scale' of expertise (and of engagement, from operator through to audience) translated within the overall intellectual property framework, the rationale and indeed the justice of each seems much clearer than that of the copyright system.

Thus, the relationship between new products and the patent (and the invention), the design right (and the design), the trade mark (and the mark), is perhaps a much more straightforward relationship than that between new products, copyright developments and new markets. This is arguably for the key reason identified: in copyright, the user is absent. The trace of the user persists in certain limitations and exceptions, but in terms of the subject matter itself and the scope of that subject matter – that is, the very heart of copyright (and the very parameters of value) – the user is at best deferred. The system, literally, has no meaning.

---

201   Cited in Y Heisler, 'Tim Cook on the state of IP protection: our product cycles move much quicker than the court system', *The Unofficial Apple Weblog (TUAW)*, 21 May 2013.

Therefore, does it follow from the preceding discussions that traditional copyright products, in a digital market, simply do not exist because copyright is incapable of revealing those products through use and through the user? It would seem that, faced with the challenges of the digital trading environment, the key to re-inscribing copyright law with meaning and legitimacy would be to address this conceptual and logical absence of use, and the particular implications for the digital knowledge economy.

This differentiation or taxonomy of use within the intellectual property system resonates with the differentiation of artistic forms according to genre. Aristotle provided one of the first systems of generic categories, which included the classification of artistic production (identifying poetics as a 'productive' discipline) according to lyric, epic and drama,[202] according to the narrative voice in each. Interestingly, these categories indeed might provide some insight into the distinctions drawn between categories of intellectual property, according to the 'narrator' of innovation within each category: 'Character in a play is that which reveals the choice of the agents.'[203] In this sense, it might be understood that the epic of the patent is narrated by the authority of the patent specification (the first person as it were), but the characters of the invention are able to speak through experimentation, research and the like. Therefore, the community of the patent resonates with the use of the patent. The drama of designs and trade marks allows for the characters to speak for themselves through the informed user or average consumer. However, the lyric of copyright is told only through the first person, the producer. The characters, the users, are absent. Of interest to the present discussion is the way in which the lyric is nevertheless expectant of accompaniment, the musicality of use. How might this refrain be composed?

While a relative hierarchy of the categories might have applied as above, it is possible to recognize an enticing coincidence with certain assumptions that may be said to operate with respect to the categories of intellectual property, in terms of the limitation of scope (and quality) through the principle of use. For example, patents might receive an emphasis on inevitable social and economic benefit (the noble epic), while copyright material may encompass lower or even amoral creative endeavour (as use is not a factor), relegating it to the sidelines of economic and social development. This is an issue not only for the industries seeking to protect established business models and rights, but also for those attempting to emphasize the importance to society, development and expression of the material comprising copyright. This is not to advocate the introduction of a role of moral censure into copyright, indeed, far from it. But it serves to illustrate the way in which use relates not only to quality but also to the 'social' in patents, trade marks and design.

---

202   Aristotle, *Poetics*, in *The Complete Works of Aristotle: Volume II*, J Barnes (ed), rev Oxford trans, Princeton UP, Princeton, 1984: 2316–40.

203   Aristotle, *Poetics*, 2321.

A peculiar symptom (or effect) of the absence of the user in copyright is therefore the difficulty of refining copyright industries to new business models and, in particular, the paradox of devising a 'new' business model for a practice already underway. How can this model be 'predicted', as it were? Is its destruction already underway? Surely the way in which technology is translated into new markets and new products is largely through referencing what has gone before?

Genre and a Taxonomy of Use

Genre does not merely categorize and organize innovation in ways perhaps wholly compatible with the intellectual property system; it also assists in recognizing and appreciating the new. It is not static, but rather modernizes and changes so as to be nevertheless recognizable and acceptable. Genre and the conventions of genre therefore resonate with the same processes of authoritative, regulatory and institutionalized taxonomy that take place within the iconography of intellectual property. Just as the categories and mechanisms within the intellectual property system differentiate between the new, the inventive, the distinctive, the different, the rules of genre similarly allow the spectator, the audience, the consumer to differentiate and organize expectation according to seemingly consistent, coherent, predictable and reliable rules. Indeed, perhaps much more importantly in terms of harnessing a particular consumer or use value in innovation, genre cooperates with user expectations and understanding, making the unavoidable value of, and need for, the product inevitable in any new market thus created.

Similarly, new business models are not necessarily predictive and revolutionary; rather, the emphasis on modelling relies rather disingenuously on re-inscribing sensible and meaningful narratives based on what has gone before. The prediction is possible precisely because of generic conventions, in that users can recognize the narrative and hypothesize the outcome. Genre therefore becomes the very 'aesthetic' vehicle for introducing new technology for technology's sake, consumption for consumption's sake. Capturing and putting into circulation the value of innovation therefore becomes much more about marketing strategy than about resolving actual problems. This might seem clear in copyright, trade marks and designs, but how might this be resolved within the patent narrative (for example, the problem and solution approach[204])? In literature, genre is itself a problem and solution approach to meaning – that is, it operates according to the creation and resolution of problems within narratives. The pleasure derived

---

204   This is the approach used by the European Patent Office (EPO) to determine whether a claimed invention involves an inventive step. The test originated in T 1/80 *Carbonless copying paper/BAYER* [1981] OJ 206. It involves identifying the objective technical problem, identifying the closest prior art and then determining whether the skilled person would come to the solution to that problem from the prior art. For a more detailed explanation, see Roughton et al., *Modern Law of Patents*, 63–74.

from genre comes from both the familiar repetition of meaning as well as the sufficient difference from what has gone before – indeed, this is the fundamental nature of innovation and new technology markets themselves. New technology is therefore not necessarily about a specific solution to a problem, but rather presents an innovative or disruptive event in a narrative that builds sense on the value of change.

Therefore, both the architecture of the intellectual property system itself (for example, the way that the patent system undertakes to situate innovation within the state of the art) and innovation more broadly can be understood as relying upon conventions of genre in order to rationalize the new, to 'remember' the new, to archive the new. In other words, new technological innovations must realize new products by making unanticipated needs or desires appear familiar, commonsensical, and always already in search of a solution: *ah, so that's what I wanted*. In other words, it is to convince the consumer that the 'lack' is real. The act of innovation itself (in the broadest sense) makes it necessary to devise an otherwise non-existent market and product. In order to make sensible the need for that product, an intertextuality with that which has gone before must be designed. Thus markets in technology must become sensible to consumers through referencing certain expectations that have been created in the sector.

For example, buying a new piece of technology (a new phone, a new tablet, a new widget) is almost like watching a new film and all the merchandising that goes with it – genre in film encapsulates and informs not only the production, but also the marketing and reception or consumption of the film. That is, genre is implicated immediately in use. We understand what we are going to see through practices that have been built up over time, through habits, through custom, through familiarity. This does not render a film unoriginal (at least not all), in the same way that new technology is not simply repetitive, because genres also adapt and transform. So all at once we can recognize what we need to in order to desire the new phone, the new indefinable, but we can still take an inordinately long time to find a use for half the things it can do.

In other words, just as new films respond to expectations of both the industry and the audience, so does the innovation narrative respond to expectations of both the industry (in terms of new technology for technology's sake, as it were) and the consumer (in terms of referencing previous markets, repeating certain proven models). Familiarity may breed contempt, but without it relevance becomes a bigger question. Innovation is thus not only innovation in products, but also innovation in production and consumption processes themselves: 'Constant revolutionalising of production, uninterrupted disturbance of all social conditions, everlasting uncertainty and agitation distinguish the bourgeois epoch from all earlier ones. All fixed, fast-frozen relations, with their train of ancient and venerable prejudices and opinions, are swept away, all new-formed ones become antiquated before they can ossify. All that is solid melts into air.'[205]

So, returning to the organizing principle of use, new economic models in copyright and the digital environment cannot be proposed, delivered and won on technology alone. That is, in order to achieve the legitimacy of the system (and the relevance of that system to users) the codification of exceptions within the law becomes an instrument of reform that is beneficial to industry as well as consumers. Normative and technical developments do not address any crisis of legitimacy. Indeed, a substantive language of the limitations and exceptions debates in copyright is premised upon an inaccurate assumption (that of incentives, as discussed earlier), which undermines the actual validity of the debate with respect to greater efficiency and effectiveness of the relationships and transactions themselves.

Which Way?

Importantly, fundamentally, logically the intellectual property system merely provides a language with which to mimic innovation. At this it is arguably reasonably effective. Where relationships become problematic is where the system fails to be meaningful in contemporary social, cultural and commercial use. This is not a banal antagonism between proponents and opponents of intellectual property, a wholly unhelpful characterization of the landscape and arguably one which is instrumental to maintaining the status quo as understood by 'both' poles. Rather, it is what happens outside the system that interferes with the rules of the (language) game; on the face of things, it is towards meaningful consensus on the value and choices in the game.

> Meaning is not a process which accompanies a word. For no *process* could have the consequences of meaning.
>
> (Similarly, I think, it could be said: a calculation is not an experiment, for no experiment could have the peculiar consequences of a multiplication.)
>
> … The familiar physiognomy of a word, the feeling that it has taken up its meaning into itself, that it is an actual likeness of its meaning – there could be human beings to whom all this was alien. (They would not have an attachment to their words.) – And how are these feelings manifested among us? – By the way we choose and value words.[206]

In their earliest imaginings, incarnation and development, the various rules of intellectual property arguably did not cover goods as such. Goods became merely

---

205   K Marx & F Engels, *The Communist Manifesto: A Modern Edition*, E Hobsbawm (intro), Verso, London, [1848]/2012: 38.

206   Wittgenstein, *Philosophical Investigations*, 218e.

the vehicle in which to limit the meaning of intellectual property and provide expression for our social life. Rather, intellectual property rules were instituted to ensure use through the resilience of the mechanisms of production and circulation of those tools, the circumstances and conditions of our social and cultural life. And so it is use that must be repaired, restored and remembered.

In a digital environment, we have perhaps not retained all those same tools of social life. It is perhaps not merely the ease of reproduction driving change but rather the increasingly individual nature of participation, the solitary and perhaps even isolated act of consumption, despite the paradoxical nature of connectivity and collaboration that is facilitated and proclaimed by the digital. Instead we are receiving only a proliferation of decontextualized messages. The dialectical integrity and subtlety of knowledge as a communicative event is paradoxically both threatened and liberated. The overwhelming question is therefore how do we use all this information, all this dialogue? How do we enshrine use and achieve meaning in intellectual property development and reform? There is no such thing as digital intellectual property. All intellectual property is digital. The question is what can we do with that?

Our guide for this journey is Alice: from her initial negotiation of depths, bodies, the physical, the world of 'goods', through to the giddy heights of the virtual, the experience, to the resolution and renewal at the surface of things,[207] through to what is hoped is a revolution through the Looking-glass.[208] It is an exploration, an adventure, where Carroll 'is an explorer, an experimenter'.[209] And this book maps and is framed by Alice's journey, at the same time changing and propelling her journey. To wonder: Part I deals with the depths of the debate, the 'interminable fall' of the combat and battle of intellectual property debate; Part II journeys through to the apparent promise of the digital, the battle between goods and experience,

---

207   In *The Logic of Sense*, Deleuze notes the three parts of Alice and explains their identification in the text by their changes in location: 'The first part (chapters 1–3), starting with Alice's interminable fall, is completely immersed in the schizoid element of depth … But the second part (chapters 4–7) seems to display a change of orientation … now it is drinking which brings about growth and eating which causes one to shrink … a question of choosing between depth and height. In the third part (chapters 8–12), there is again a change of element … The surface is burst, "… the whole pack rose up into the air, and came flying down upon her."' Deleuze, *Logic of Sense*, 234–35.

208   Deleuze notes that 'In *Through the Looking Glass* there is instead a surprising conquest of surfaces (no doubt prepared by the role of the magic cards at the end of *Alice's Adventures*): one no longer sinks, one slides; it is the flat surface of the mirror or of the game of chess; even the monsters become lateral. For the first time literature thus declares itself an art of surfaces, a measurement of planes.' G Deleuze, *Two Regimes of Madness: Texts and Interview, 1975–1995*, D Lapoujade (ed), A Hodges & M Taormina (trans), Semiotext(e), New York, 2006: 64.

209   Deleuze, *Two Regimes of Madness*, 63.

'a question of choosing between depth and height';[210] and in Part III, the problems and solutions are 'raised' to the surface. Finally, in '*Re* Use', imagined 'through the Looking-glass',[211] the user enters the game, with everything to play for at the surface of things.

> And ever, as the story drained
>   The wells of fancy dry,
> And faintly strove that weary one
>   To put the subject by,
> "The rest next time – " "It *is* next time!"
>   The happy voices cry.
>
> Thus grew the tale of Wonderland:
>   Thus slowly, one by one,
> Its quaint events were hammered out –
>   And now the tale is done,
> And home we steer, a merry crew,
>   Beneath the setting sun.[212]

Next time!

210　Deleuze, *Logic of Sense*, 235.
211　With respect to *Through the Looking-Glass*, which frames '*Re* Use', this play at the surface is paramount. See further the discussion in Deleuze, *Logic of Sense*, 9–11.
212　*Wonderland*, 2.

WONDER

PART I
Of Properties

Chapter 1
Cause

Down the Rabbit-Hole

> Alice was beginning to get very tired of sitting by her sister on the bank, and of
> having nothing to do: once or twice she had peeped into the book her sister was
> reading, but it had no pictures or conversations in it, "and what is the use of a
> book," thought Alice, "without pictures or conversations?"[1]

A book without pictures or conversations is a book without use, a useless book.
These 'conversations' are integral to the methodology of inward speech and
interlocutories[2] that informs not only philosophical method, but also a more
conventional narrative 'cure' of the current tensions (political, social, cultural,
ideological, digital) within the intellectual property system. In the present
inquiry, the interlocutory with Carroll structures the literary reimagining of the
intellectual property debate. The paradox of inward speech is immediate when
Alice hears the rabbit's own voice, speaking to itself: 'nor did Alice think it so
very much out of the way to hear the Rabbit say to itself, "Oh dear! Oh dear! I
shall be too late!"'[3]

Cause and Effect

> "Oh dear! Oh dear! I shall be too late!"[4]

From the very beginning, setting out on the present journey there are the
elements of chance and time that will inform the discussion throughout. Bataille
has noted that the origin of the word *chance* in French is the same for deadline
(*échéance*):[5] *I shall be too late*. There is thus immediately an entanglement of the

---

1   *Wonderland*, 3.

2   See further the discussion of the interlocutory as philosophical method later in
this chapter.

3   *Wonderland*, 3. This is remarkable considering that throughout the rest of the tale
Alice speaks to herself and manages to do so privately.

4   *Wonderland*, 3.

5   G Bataille, *On Nietzsche*, B Boone (trans), S Lotringer (intro), Paragon House,
New York, [1945]/1992: 70.

unpredictability of chance, on the one hand, and notions of causality, measurability and accountability on the other.[6]

The conventional narrative of intellectual property industry and policy[7] (and indeed, the narrative ordinarily applied to innovation) is linear, causal, chronological – it is curiosity in pursuit, solutions in search of a problem. Causality presupposes an end, a finishing point, a plan: 'Now, in order to act, we begin by proposing an end; we make a plan, then we go on to the detail of the mechanism which will bring it to pass.'[8] Intellectual property institutionalizes this narrative of progress, in concert with the overarching economic organization of innovation: 'Chronological time is thus altogether structured by the logic of the functional relations between economic structures.'[9] The logic of the relations between the conventional modes of production, commercialization and exploitation is thus chronological: 'What is produced is then sold and the income from this is shared out. But the circulation, exchange and distribution of this income presuppose this production of what is to circulate and to be exchanged and shared. Logical relations are therefore at the same time chronological ones, in so far as the logical moments correspond to the different moments of time in the economic process.'[10]

As introduced earlier,[11] the obligation of 'use' incorporates these very same notions of purposeful depletion and wear when applying resources. Such depletion not only contributes to regulation of access (and artificial scarcity constructed by the rules of intellectual property) but also addresses economic or market concerns of a 'lack' (that is, a necessity, a desire in the user-consumer). Use and consumption is expenditure. In conventional economic models, the lavish expenditure of antiproduction concentrates resources in the sovereign: 'his sovereignty in the living world identifies him with this movement; it destines him, in a privileged way, to that glorious operation, to useless consumption.'[12]

However, what of the virulent sharing and consumption that might fuel familiar production? The question, and one to which the discussion will return in later chapters, is whether it is possible through use to reconcile the antiproduction

---

6   Building upon this relationship between directionless and immeasurable chance in familiar production, as distinct from the quantified and measurable risk narrative of conventional modes of production, see further the more detailed discussion of the smooth space of innovation and the striated space of the State in Chapter 2.

7   See further the discussion of this principle in the context of evidence-based policy-making in Chapters 11 and 12.

8   H Bergson, *Creative Evolution*, A Mitchell (trans), The Modern Library, New York, [1911]/1944: 50.

9   M Godelier, *Rationality and Irrationality in Economics*, B Pearce (trans), Verso, London, [1966]/2012: 146.

10   Godelier, *Rationality and Irrationality in Economics*, 146.

11   See the discussion in 'Use'.

12   G Bataille, *The Accursed Share: An Essay on General Economy, Volume I, Consumption*, Zone, New York, [1967]/1988: 23.

of capital with the expenditure of sharing: 'It is necessary at this point to note a dual origin of moral judgments. In former times value was given to unproductive glory, whereas in our day it is measured in terms of production: Precedence is given to energy acquisition over energy expenditure. Glory itself is justified by the consequences of a glorious deed in the sphere of utility.'[13] Consumption is to be recuperated for production.[14] This is indeed social power, as distinct from prestige and rank: 'Prestige, glory and rank should not be confused with *power*.'[15] The consumption of use, of familiar production, is therefore also glorious not in the sense of possession, but in the sense of energy, creativity and becoming: 'a movement of senseless frenzy, of measureless expenditure of energy, which the fervor of combat presupposes'.[16]

In addition to the relationship of use to waste, and property to preservation, the populist notion of intellectual property and its relation to innovation as causal is consequently 'pharmacopeiac' – that is, intellectual property operates as a kind of naming register of recognition and comparison,[17] an archive, a 'book', a depository, and a recipe for cultural progress. This may be formal, in terms of requirements in order to access rights (patent application and grant; trade mark registration; design registration in the case of registered design rights) or informal, in terms of the cultural 'accreditation' attributable to protection (such as through copyright or unregistered design right). This intellectual 'catalogue' spans the walls of the 'rabbit-hole', but the shelves are always already empty, their contents apparently obsolete as the intellectual journey progresses:

> Either the well was very deep, or she fell very slowly, for she had plenty of time as she went down to look about her, and to wonder what was going to happen next. First, she tried to look down and make out what she was coming to, but it was too dark to see anything; then she looked at the sides of the well, and noticed that they were filled with cupboards and book-shelves: here and there she saw maps and pictures hung upon pegs. She took down a jar from one of the shelves as she passed; it was labelled "ORANGE MARMALADE," but to her great disappointment it was empty: she did not like to drop the jar for fear of killing somebody, so managed to put it into one of the cupboards as she fell past it.[18]

A conventional and literal narrative of innovation relies upon such an obsolescence of answers, so much so that the shelves of past innovation are 'empty', 'used

---

13   Bataille, *Accursed Share: Volume I*, 29.

14   This model finds its ultimate incarnation in the form of the Looking-glass cake. See '*Re* Use'.

15   Bataille, *Accursed Share: Volume I*, 71.

16   Bataille, *Accursed Share: Volume I*, 71.

17   This may occur through either formal requirements or social norms.

18   *Wonderland*, 4.

up', as it were, by the market of innovation.[19] This is the 'debt' of innovation.[20] Nevertheless, the relics remain, despite being empty of relevance to technological progress; traditional progressions of innovation are not simply directed towards a product, but are necessarily subjected to a narrative of obsolescence and replacement, of waste and antiproduction.

> ... when suddenly, thump! thump! down she came upon a heap of dry leaves, and the fall was over.[21]

The notion of getting to the root of a problem, to the bottom of things, is not merely colloquial: 'For there seemed to pertain to logic a peculiar depth – a universal significance. Logic lay, it seemed, at the bottom of all the sciences. – For logical investigation explores the nature of all things. It seeks to see to the bottom of things and is not meant to concern itself whether what actually happens is this or that.'[22] In this linear conceptualization of the time of innovation, getting to the 'bottom' of a problem, of a question, of the well, is not about acquiring new knowledge; rather, it is about understanding what is already there: 'it is, rather, of the essence of our investigation that we do not seek to learn anything *new* by it. We want to *understand* something that is already in plain view. For *this* is what we seem in some sense not to understand.'[23] In other words, there is a presumption that the tools of assessment will inevitably provide the answer of 'what is already there', *a heap of dry leaves*.

The Probability of Estimates, the Deliberation of Guessing

> Undoubtedly quality is worth more than quantity,
> But one can discuss quality forever;
> Quantity, however, is unquestionable.[24]

---

19   This is perhaps a lesser interest of the 'scarcity' achieved by intellectual property, but in concert with a market economy it is crucial.

20   The concept of the debt–credit relationship in intellectual property was introduced in 'Use' and is discussed in further detail in Chapter 2.

21   *Wonderland*, 5.

22   Wittgenstein, *Philosophical Investigations*, §89.

23   Wittgenstein, *Philosophical Investigations*, §89.

24   M Tournier, *The Mirror of Ideas*, JF Krell (trans), U of Nebraska P, Lincoln, [1994]/1998: 97. Tournier also adds a little joke here, attributing this short verse to Edward Reinrot (in fact, Tournier spelt backwards, minus the 'u'). The author thus 'invents' the authority for this, and at the same time highlights the impossibility of self-awareness in the representation, the calculable (minus the 'u'), that is, the problem of the observer.

This purposeful search for answers reveals only the meaninglessness of the questions. This is the fallacy of depth, of looking inwardly. And arguably, this is the fallacy of constructing innovation as a calculable problem. Wittgenstein illustrates this with the example of the right hand giving the left hand money: 'My right hand can write a deed of gift and my left hand a receipt. – But the further practical consequences would not be those of a gift. When the left hand has taken the money from the right, etc., we shall ask: "Well, and what of it?"'[25] In other words, understanding comes not from some pretence of depth, but rather from looking at the surface: 'And the same could be asked if a person had given himself a private definition of a word; I mean, if he has said the word to himself and at the same time has directed his attention to a sensation.'[26]

This preoccupation with depth betrays a peculiarly Utopian presumption concerning meaning, innovation and, indeed, concerning the law itself and its explanation and modelling function, as well as its submission itself to models and calculation (mathematical, logical, economic). This conceptual and rhetorical strategy infuses not only the discourse of legal reform, but also the discourse of technological 'advancement' (always advancement, never simply change). The prerequisite of depth as the source of knowledge and freedom is thus the presumption of a linear narrative of meaning, progress and innovation; in other words, one must advance in order to appreciate depth. This is the fallacy of 'inner language'[27] and the bias towards 'depth': 'How should we counter someone who told us that with *him* understanding was an inner process? – How should we counter him if he said that with him knowing how to play chess was an inner process?'[28] In other words, this is the paradox of representation. Representation is always already constrained within language: 'a person who had never heard of chess' need not be troubled, as it were, with the 'secondary meanings attaching to *queen* and *rook*. There would be many crimes and errors which it would be beyond his power to commit, simply because they were nameless and therefore unimaginable.'[29]

Therefore, representation, and indeed any modelling, is relative, but this relativism 'constitutes not a relativity of truth but, on the contrary, a truth of the relative'.[30] This is the paradox and deception of calculation and explanation: 'We

---

25   Wittgenstein, *Philosophical Investigations*, §268.

26   Wittgenstein, *Philosophical Investigations*, §268.

27   As Wittgenstein explains, '– We should say that when we want to know if he can play chess we aren't interested in anything that goes on inside him. – And if he replies that this is in fact just what we are interested in, that is, we are interested in whether he can play chess – then we shall have to draw his attention to the criteria which would demonstrate his capacity, and on the other hand to the criteria for the "inner states."' See Wittgenstein, *Philosophical Investigations*, 181e.

28   Wittgenstein, *Philosophical Investigations*, 181e.

29   G Orwell, *Nineteen Eighty-Four*, Secker & Warburg, London, 1949: 312.

30   Deleuze & Guattari, *What is Philosophy?* H Tomlinson & G Burchill (trans), Verso, London, [1991]/1994: 130.

must do away with all *explanation*, and description alone must take its place. And this description gets it light, that is to say its purpose, from the philosophical problems. These are, of course, not empirical problems; they are solved, rather, by looking into the workings of our language, and that in such a way as to make us recognize those workings: *in despite* of an urge to misunderstand them.'[31]

Economic modelling is thus perhaps not so much about the seductive power of empirical models, but rather the reassurance provided by the circumstances created in such modelling. First, such models necessarily make the presumption that the phenomena (in this case innovation), or perhaps more accurately the propositions to which economic science is applied, are stable and precise. Secondly, there is a remarkable temptation and captivation about the apparent possibility of a phenomenal explanation of those propositions. And finally, from a policy perspective, there is the invitation and attraction of seemingly irrefutable bases on which to mobilize a subsequent force of public opinion, resting on the basis of those explanations but without any need to engage with the subject matter. On this point, repeated calls for education of the public are less about awareness and capacity and more about indoctrination of the 'public' (and thus public opinion, or worse, moral panic) as the arbiter of the conventions or norms of intellectual property discourse.[32]

The numbers on intellectual property apparently do not lie, but perhaps that is because it would be illogical to suggest such a thing, as they lack the engagement to do very much at all, other than promote abbreviation: 'And what can possibly be opposed to the "facts"? The law, which is given in numbers and in data (that is, in terms fabricated by technicians) but presented as the manifestation of the ultimate authority, the "real," constitutes the new orthodoxy, an immense discourse of the order of things.'[33] Appearing as transparent and accountable, economics and statistics become the perfect vehicles for public opinion in the intellectual property debate, without any necessary engagement with the 'use' of intellectual property rules and language. Reducing the language-game to a repeatable, assessable and predictable interaction does not describe value in its measurement, but rather, remakes it, evaluates it and opines upon it: 'The problems are solved, not by giving new information, or calculating the incalculable, but by arranging what we have always known.'[34] *What do* you *think it was*?

---

31   Wittgenstein, *Philosophical Investigations*, §109.

32   For example, see the UK Government Intellectual Property Office (IPO) materials on 'IP for education', at www.ipo.gov.uk/whyuse/education.htm and further the emphasis of the independent Intellectual Property Awareness Network (IPAN) on intellectual property education in primary and secondary school children, as well as across disciplines in universities, IPAN Brief No 14, www.ipaware.net/node/71.

33   M de Certeau, *Heterologies: Discourse on the Other*, B Massumi (trans), W Godzich (foreword), U of Minnesota P, Minneapolis, [1986]/1993: 207.

34   Wittgenstein, *Philosophical Investigations*, §109. See further the discussion in Bataille, *Accursed Share: Volume I*, 20 and the need 'to study the system of human

Don't Think, But Look!

> An immense industrial network cannot be managed in the same way that one
> changes a tire … It expresses a circuit of cosmic energy on which it depends,
> which it cannot limit, and whose laws it cannot ignore without consequences.
> Woe to those who, to the very end, insist on regulating the movement that
> exceeds them with the narrow mind of the mechanic who changes a tire.[35]

Like Alice, we must look all about us. We must look, not think: 'don't think,
but look!'[36] That is, search through the interlocutory, as it is what happens at the
surface of language that is crucial: 'Thinking and inward speech (I do not say
"talking to oneself") are different concepts.'[37] This is the fundamental problem
of representation and explanation, the infinite regress of accountability to the
explanation within the explanation, the representation within the representation,
that is, to the assumptions and conditions of explanation: '"The meaning of a
word is what is explained by the explanation of the meaning." I.e.: if you want to
understand the use of the word "meaning", look for what are called "explanations
of meaning."'[38]

Bataille argues for the decoupling of production and consumption, rather than
the notion of the economy (and indeed the creative or innovative economy) as 'an
isolatable system of operation'.[39] Bataille's thesis on the general economy thus
provides significant insight into the overturning of the traditional conceptualization
of 'use' within a rethinking of intellectual property development and reform and
towards a concept of familiar production: 'the extension of economic growth itself
requires the overturning of economic principles – the overturning of the ethics that

---

production and consumption within a much larger framework'. Similarly, in *Just Gaming*,
Jean-François Lyotard notes that 'The question of justice for a society cannot be resolved
in terms of models. This is very important because I think that we are always tempted,
whenever the question of justice arises, to go back to a model for a possible constitution,
to be drawn up by a possible constitutional convention.' See Lyotard & Thébaud, *Just
Gaming*, 25.

35   Bataille, *Accursed Share: Volume I*, 26.

36   Wittgenstein, *Philosophical Investigations*, §66.

37   L Wittgenstein, *Last Writings on the Philosophy of Psychology: The Inner and the
Outer 1949-1951, Volume 2*, GH von Wright & H Nyman (eds), CG Luckhardt & MAE Aue
(trans), Blackwell, Malden MA, 1993: 18e.

38   Wittgenstein, *Philosophical Investigations*, §560. In *What is Philosophy?*, Deleuze
and Guattari explain that 'the role of a partial observer is to perceive and to experience'.
However, 'ideal partial observers are the perceptions or sensory affections of functions
themselves. Even geometrical figures have affections and perceptions … without which the
simplest problems would remain unintelligible' (130 and 131 respectively).

39   Bataille, *Accursed Share: Volume I*, 19.

grounds them.'[40] This ethical dimension to innovation is indeed what underpins the 'logic' throughout this journey.

The relationship between the 'depth' of intellectual property and the 'sublime' character of logic is therefore perhaps tenuous at best. The intellectual property system cannot propose to understand, it merely indicates creativity and innovation, but this is indeed the logic of innovation within the intellectual property system. In rendering expressions of innovation (the patent, the design, the trade mark, the copyright work) the intellectual property system appears to be burdened with the responsibility of identifying, classifying and archiving the nature of all things. But instead, and more remarkably, it is confirming what is already there. While the intellectual property system might in one respect archive and report on innovation (which is considered in the context of its operation as an objective record of that endeavour), that archive is in itself a mechanism for 'forgetting' in order to remember the 'new'. Therefore, as the contemporary 'image' and rhetoric (the pictures and conversation) for innovation and innovative behaviour, and the implications of that for the 'imaging' of innovative behaviour, the intellectual property system espouses a kind of normative aesthetics for creative and innovative endeavour.[41]

Much of the discourse informing intellectual property law assumes a determinable sense of direction. This is found even in the language of the law itself, where patent law explicitly relies upon a sense of direction in the concept of inventive step and the notion of a 'threshold' for invention. Inventive step promotes a logical and perceptible progression which disguises innovation as finite progression, stops and starts, which are applied, paradoxically, at the end of the story (the patent) in order to restore the semblance of natural regress (prior art). This pragmatic presumption of a single trajectory of 'progress' in innovation then is unable to take account of radical contradictions to that causality. Most notably in this respect, patent law cannot account for much of the incremental innovation in traditional knowledge (through patent law) and yet at the same time, business and innovation strategy is able to manipulate the 'steps' of innovation recognized in patent law in order to perform the more controversial practices of evergreening[42] through improvement patents. While the protection of incremental innovation on already existing patented inventions may be within the scope of the law, provided the necessary 'step' is demonstrable,[43] there is considerable

---

40   Bataille, *Accursed Share: Volume I*, 25.

41   See further the discussion of the relationship between aesthetics, ethics and the law in Chapter 4.

42   Evergreening can encompass a range of activities and strategies, not limited to those facilitated by intellectual property rules, but for the purposes of this discussion, only the specific evergreening practices associated with intellectual property are noted here.

43   The Supreme Court of India recently held, in relation to Novartis's Gleevac patent, that a special standard for inventive step for pharmaceuticals (set out in Patents Act 1970 s. 3(d)) was appropriate in international law. The impact of the decision is to

criticism that such activity provides a de facto extension of monopoly protection of an existing product, particularly controversial in the area of pharmaceuticals where evergreening may frustrate access to medicines through delayed market entry of generic equivalents.

Digitations

Similarly contentious is the potential of intellectual property in the digital environment. While the rules and conventions of intellectual property laws might be able to govern behaviour, in practice the laws do not appear to have the legitimacy for social practice that they might. Without the book, the CD and indeed any object to grasp and retain, the physical constraints to a particular knowledge artefact, and indeed to knowledge as a resource, are no longer applicable. The digital environment characterizes almost completely the distinct difference between mere laws and rules (and the imposition of 'morality') and a desire for justice and ethics. The latter is inextricably bound up with questions of legitimacy and belief, while the former is simply an order, a prescription, a dictate. The challenge for intellectual property reform is to overwhelm the persistent rhetoric of moral certainty and attend to the questions of legitimacy and ethics that question not only the everyday application of the laws, but also the normative and legitimate justice in taking responsibility for utterances and for their repetition. Justice is incalculable, beyond the models of exchange and exchange-value that have come to overwhelm and radically simplify the environment of intellectual property. Reform of the intellectual property system (laws, practice, culture) must begin to look beyond the law. And for that, one cannot count on numbers.

As considered earlier,[44] digital as a mechanism is a significant reinvention of the way in which society organizes and redistributes knowledge. The digital environment (both technical and cultural) is directionless, instantaneous, gluttonous, and both confounds the causality proposed and conquered by the languages of intellectual property and traditional economics, as well as demonstrates the possible fragility of the norms of the game. Despite the presumed trajectory, in many senses intellectual property is not necessarily a narrative of 'improvement', but it is almost certainly always one of change. Contrary to the self-consciousness and supposed accountability of the expansive rhetoric of progress (and the reference to what has gone before), the value central to intellectual property is

---

prevent new patents being granted without there being enhancement of known efficacy. The Court further felt that the amendment was a good way of encouraging the development of the Indian pharmaceutical industry: see *Novartis v Union of India* (Appeal No. 2706-2716 of 2013) (Indian Supreme Court, 1 April 2013).

44   See the discussion in 'Use'.

privatistic, archipelagian change.[45] In the instantaneity of the digital environment, change is the only certainty. The notion of depth, of digging beneath, of looking to the bottom of the well, is a distraction. It is therefore a curious fascination with consolidating knowledge, but ultimately, like Alice, the system can only tumble along with innovation.

Before the Intellectual Property Law

> She was close behind it when she turned the corner, but the Rabbit was no longer to be seen: she found herself in a long, low hall, which was lit up by a row of lamps hanging from the roof.
>
>   There were doors all round the hall, but they were all locked; and when Alice had been all the way down one side and up the other, trying every door, she walked sadly down the middle, wondering how she was ever to get out again.[46]

While the desire to get to the bottom of things has caused Alice to fall, her desire is rescued by the rhizomatic potential of the passageway in which she finds herself.[47] Intellectual property reform is confronting a philosophical problem: 'A philosophical problem has the form: "I don't know my way about."'[48] Alice is literally faced with the passageway of the living rhizome within the rabbit-hole itself, with its many shelves and passageways: 'Perhaps one of the most important characteristics of the rhizome is that it always has multiple entryways; in this sense, the burrow is an animal rhizome, and sometimes maintains a clear distinction between the line of flight as passageway and storage or living strata.'[49]

The rhizomatic potential of the passageway shares much with the complexity of language and linguistic connections: 'Language is a labyrinth of paths. You approach from *one* side and know your way about; you approach the same place from another side and no longer know your way about.'[50] Innovation similarly cannot know its way about, but is necessarily a pure becoming. Logically it cannot become anything else; that would be the end of innovation. Innovation is a process,

---

45   This is very clear in high technology industries where arguably the rapidity of innovation and the speed of obsolescence suggest most clearly that there is 'innovation for innovation's sake' well in advance of any determined or identified markets. The market is the 'brand' but not the innovation as such (brand's aura).

46   *Wonderland*, 6.

47   Deleuze and Guattari explain: 'Whenever desire climbs a tree, internal repercussions trip it up and it falls to its death; the rhizome, on the other hand, acts on desire by external, productive outgrowths.' Deleuze & Guattari, *Thousand Plateaus*, 14.

48   Wittgenstein, *Philosophical Investigations*, §123.

49   Deleuze & Guattari, *Thousand Plateaus*, 12.

50   Wittgenstein, *Philosophical Investigations*, §203.

not an object, but one which is made sensible and familiar through intellectual property. Innovation is thus always already familiar.

Not only does Alice no longer know her way about, but also in this passageway the many doorways are locked and the rhizomatic potential of the passage is seemingly lost or at best deferred. Similarly, an unlimited potential of innovation is diced by intellectual property in order to render it fixed, marked, directional, just so. But this limitation is temporary. In the passageway the burden of a logical, linear causality inhibits the creative potential of an otherwise infinite interconnectedness, and it will take all of Alice's creative powers to unlock it.

Telescopic Depth

> "Oh, how I wish I could shut up like a telescope! I think I could, if I only knew how to begin." For, you see, so many out-of-the-way things had happened lately, that Alice had begun to think that very few things indeed were really impossible.
>
> There seemed to be no use in waiting by the little door, so she went back to the table, half hoping she might find another key on it, or at any rate a book of rules for shutting people up like telescopes.[51]

Shutting up like a telescope, Alice's interaction with depth becomes ambiguous. The distant is now close, and yet deferred, through an arrangement of mirrors, reflection and representation. The journey is compressing, combining and collating a number of contradictory and conflicting concepts and meanings. Indeed, throughout the journey, the telescopy of language will emerge, and the portmanteau nature of meaning manifests through use.[52] This is the paradox of remembering and recognition, of false comparisons. Even a gigantic telescope cannot have an eyepiece bigger than an eye.[53] This is the nature of innovation and its representation, and the attenuation of the innovation in familiar production precisely through such processes of representation: 'As if they had a telescope with which they can't possibly reach the moon, but can see what is ahead of the mathematician who is flying there.'[54]

---

51 *Wonderland*, 6–7.

52 In '*Re* Use' portmanteau words are explored in more detail, themselves sometimes referred to also as 'telescope words' or 'telescopic words'.

53 'It's a matter of the grammar of the words "mental image", as opposed to the grammar of "things". [(A strange analogy could arise from the fact that the eyepiece of even the most gigantic telescope mustn't be any bigger than our eye.)]' L Wittgenstein, *The Big Typescript TS 213*, CG Luckhardt & MAE Aue (eds & trans), Blackwell, Malden MA, 2005: 335e.

54 L Wittgenstein, *Wittgenstein's Lectures on the Foundations of Mathematics, Cambridge, 1939*, C Diamond (ed), U of Chicago P, Chicago, 1976: 13.

The Rules of Life

> It was all very well to say "Drink me," but the wise little Alice was not going
> to do *that* in a hurry. "No, I'll look first," she said, "and see whether it's marked
> '*poison*' or not"; for she had read several nice little histories about children who
> had got burnt, and eaten up by wild beasts, and many other unpleasant things,
> all because they *would* not remember the simple rules their friends had taught
> them: such as, that a red-hot poker will burn you if you hold it too long; and
> that, if you cut your finger *very* deeply with a knife, it usually bleeds; and she
> had never forgotten that, if you drink much from a bottle marked "poison," it is
> almost certain to disagree with you, sooner or later.[55]

Alice thus attempts to anchor the unfamiliar through referencing familiar rules and
conventions. At the beginning of her journey, Alice is conscious of the rules of life,
of foresight, of accountability. At the beginning of her journey, Alice is composed,
responsible and indebted to her surroundings. However, she is at once faced with
unlimited potential and at the same time subject to a measured 'forgetfulness' such
that she may preserve her 'psychic order, repose, and etiquette'.[56] In other words,
Alice renders herself '*calculable, regular, necessary*' as guarantor for her own
future in this journey.[57]

So at the commencement of this journey, one might mitigate the risk and
'close the doors and windows of consciousness for a time'[58] so as to 'forget' the
desire of the journey. Or one might navigate the enormous rhizomatic and non-
chronological potential of chance. Which way?

> "What a curious feeling!" said Alice. "I must be shutting up like a telescope."[59]

And You May Ask Yourself

> "Come, there's no use in crying like that!" said Alice to herself, rather sharply.
> "I advise you to leave off this minute!" She generally gave herself very good
> advice (though she very seldom followed it), and sometimes she scolded herself
> so severely as to bring tears into her eyes; and once she remembered trying
> to box her own ears for having cheated herself in a game of croquet she was
> playing against herself, for this curious child was very fond of pretending to be
> two people.[60]

---

55   *Wonderland*, 7.
56   Nietzsche, *On the Genealogy of Morals*, 58.
57   Nietzsche, *On the Genealogy of Morals*, 58.
58   Nietzsche, *On the Genealogy of Morals*, 57.
59   *Wonderland*, 7.
60   *Wonderland*, 8.

What is emerging is the nature of the relationship of the user (and consumer) to the producer, and the importance not of the mere economic transactional value of this relationship, but rather of the production of subjectivity. Rather than misunderstanding this as searching deep within oneself for meaning, it is instead to demonstrate the production of meaning through use, through self-questioning. Alice's introspection literally demonstrates Wittgenstein's method of explanation, of arriving at meaning, of interlocutories; that is, a 'philosophical investigation': 'I am only trying to recommend a certain sort of investigation. If there is an opinion involved, my only opinion is that this sort of investigation is immensely important.'[61] Philosophical investigation, that is, is just like *pretending to be two people*.

Nevertheless, at first, in undertaking these interlocutories Alice attempts to make an objective measurement and assessment of her surroundings. She attempts to render visible what is around her, by evaluating her conditions according to objective measures, rather than, as will come to pass, her more reliable subjective assessments. But as will be seen, both in the progress of Alice's journey and in the exploration of the logic of innovation, any attempts to assess and rationalize would appear to be fraught.

Safety in Numbers

A recurring concern in the present discussion is the current momentum of policy and practice towards an economics universe for intellectual property, and the conviction that all innovative and creativity activity, as well as the law itself, might be rendered wholly assessable and accountable to objective measures alone. In this worldview, relationships within the system are 'balanced' and equitable perhaps only if understood from a quantitative calculation.[62] Further, the illusory nature of an objective assessment or measure of intellectual property, within an economics framework, is denying the overwhelming social nature of users, producers, consumers.[63] The flows of finance control directions not only of markets but also

---

61 Wittgenstein, *Lectures on the Foundations of Mathematics*, 103. This structure of philosophical investigation and the language-game is indeed applied literally in *Just Gaming* where Jean François Lyotard and Jean-Loup Thébaud examine the problem of justice through Platonic dialogue.

62 Brian Massumi notes, for example, that the relationship between a product and its value is 'equivalent' only if judged by numerical criteria: 'It equates elements that are obviously heterogeneous – a desired body (which is perhaps even desired for its unique intrinsic qualities) and a piece of paper bearing a recognized denomination.' See B Massumi, *A User's Guide to Capitalism and Schizophrenia: Deviations from Deleuze and Guattari*, MIT P, Cambridge MA, 1992: 129.

63 This is discussed further in Chapter 2 in the context of a more detailed examination of the debt–credit relationship in the digital environment.

of the means of subjectivity. Risk mitigation, insurance, modelling of innovation and so on, all attempt to control the flows of innovation (and limit the decisions and behaviour associated with innovation) at the same time compromising the potential of accident and chance.[64] To be deprived of risk puts innovation 'at risk', as it were, through failing to take a chance.[65]

The Laws of Death[66]

The relationship between intellectual property law and competition (or antitrust) laws is often described as a kind of 'balance' to intellectual property rights and the potential negative external effects on the market caused by such rights. Indeed, international discussion on trade and development has sought to introduce a focus on the potential for international harmonization of competition law.[67] While this might appear to be a balancing of the 'public interest'[68] against the private rights of corporations, nevertheless competition law is constituted in order to facilitate business interests and is wholly concerned with competition between commercial undertakings. The benefit to consumers is deferred or indirect, as it were, in that it is a benefit suggested to flow from fair competitive practices and the presumed benefit to the economy. Presenting it as concerned with the public interest or public benefit reconfigures the 'public' and thus 'use' as a function of those interests. In other words, if intellectual property frameworks are to be regulated or 'balanced' by competition, this sustains the interests of intellectual property within financial and indeed business contexts. Use becomes a function of business, as it were. Thus, the 'social' aspects of intellectual property are dominated by business and competition aspects.

---

64   See further the discussion in Chapter 3 as well as Lazzarato, *Making of the Indebted Man*, 138–42.

65   The value and role of chance in innovation is discussed further in Chapter 3.

66   This comes from John Ruskin, who states: 'Government and cooperation are in all things the Laws of Life; Anarchy and competition the Laws of Death.' See J Ruskin, *'Unto this Last': Four Essays on the First Principles of Political Economy*, John Wiley & Son, New York, 1879: 87.

67   For example, see J Malinauskaite, 'Harmonisation of Competition Law in the Context of Globalisation', 21(3) *European Business Law Review* 2010: 369–97; M Eechoud, PB Hugenholtz, S van Gompel, L Guibault & N Helberger, *Harmonizing European Copyright Law: The Challenges of Better Law Making*, Information Law Series 19, Kluwer Law International, Alphen aan den Rijn, 2009.

68   Joseph Stiglitz argues that 'Globalization of monopolies requires a global competition law and a global competition authority to enforce it, allowing both criminal prosecution and civil action in any case in which anti-competitive behavior affects more than one jurisdiction.' See JE Stiglitz, *Making Globalization Work*, Penguin, London, 2006: 203.

In many respects, this problem of representing balance or oversight of the system is indeed the problem of representation. That is, the system of representation (of communication, of information) cannot at once account for itself and stand to one side to represent that accountability. Further, in calculating 'benefit' in terms of economic benefit to commerce and international trade, and in necessitating 'balance' (or its supposition) and precepts through international responsibility, familiarity of use is suspended in favour of a type of internationalism. Under the auspices of international harmonization, the relevance of the nation-state in trade appears to diminish, and an 'anarchic' society emerges in significance; that is, a society where harmony is achieved not through government but through international trade agreements.

It might be suggested that international trade agreements and the WTO operate 'beyond the law', that is, beyond the sovereignty of nation-states: 'antitrust laws, adopted by the most dominant countries, aimed at defending competition in the national economy are weakened and subverted in order to allow monopoly practices and destroy competition on the international level.'[69] Nevertheless, the promise of contractual agreement might present an aspiration towards an ethics beyond judgment and the possibility of a resurrection of use. That is, rather than subjection of all desire to the sovereign (and an infinite debt), there is potential for justice in use, a cosmopolitanism in intellectual property development and reform, an ethics *to come*.

Which Way?

> Soon her eye fell on a little glass box that was lying under the table: she opened it, and found in it a very small cake, on which the words "EAT ME" were beautifully marked in currants. "Well, I'll eat it," said Alice, "and if it makes me larger, I can reach the key; and if it makes me smaller, I can creep under the door; so either way I'll get into the garden, and I don't care which happens!"
>
> She ate a little bit, and said anxiously to herself, "Which way? Which way?" holding her hand on the top of her head to feel which way it was growing, and she was quite surprised to find that she remained the same size: to be sure, this generally happens when one eats cake, but Alice had got so much into the way of expecting nothing but out-of-the-way things to happen, that it seemed quite dull and stupid for life to go on in the common way.
>
> So she set to work, and very soon finished off the cake.[70]

Alice does not care which happens, as long as there is change. What is emerging is a confidence which is perhaps not so much confidence from a deference to a higher judgment, a moral certainty, to habit, as it is confidence in the ethics of

---

69   Hardt & Negri, *Multitude*, 172.

70   *Wonderland*, 8–9.

innovation itself. Alice's decision to *set to work* and *finish off the cake* shows the very confidence that flows from and through to action. What Alice is interested in is what is *to come*. Rather than proceeding from a determined moral certainty, she is commencing on an ethical journey, undetermined, incomplete, inconstant and in common. Her very act of consumption is also production: *use is production*. All at once, familiar production *sets to work* and *finishes off the cake*. What is significant in the reform of intellectual property discourse from that of the depths of moral certainty to the surface of ethical language is whether the capacity for familiar production is simply redirected through again as debt to the producer, or whether there is the promise of genuine recuperation of an ethics of innovation.

Which way? Which way?

Chapter 2
Space

Curiouser and Curiouser![1]

> "Curiouser and curiouser!" cried Alice (she was so much surprised, that for the moment she quite forgot how to speak good English); "now I'm opening out like the largest telescope that ever was! Good-bye, feet!"[2]

Alice is literally losing her footing, her social and cultural bearings, her reference through manners, habits and custom, through people: 'We cannot find our feet with them.'[3] In a strange land with strange traditions, Alice is disconnected from the usual means by which to construct meaning and sense: 'We also say of some people that they are transparent to us. It is, however, important as regards this observation that one human being can be a complete enigma to another. We learn this when we come into a strange country with entirely strange traditions; and, what is more, even given a mastery of the country's language. We do not *understand* the people. (And not because of not knowing what they are saying to themselves.) We cannot find our feet with them.'[4] The conventions and logic of intellectual property, so firmly grounded in notions of the propertied self, the 'own' of creativity, the mechanical of the product, are literally at odds with the strange traditions of the digital landscape. The producers do not understand the consumers, as it were: *We do not understand the people*. Similarly, for Alice, the trauma of the disruption and confusion of the fall down the rabbit-hole is finally too much for her and she is reduced to (or liberated by) tears:

> "You ought to be ashamed of yourself," said Alice, "a great girl like you," (she might well say this), "to go on crying in this way! Stop this moment, I tell you!" But she went on all the same, shedding gallons of tears, until there was a large pool all around her, about four inches deep and reaching half down the hall.[5]

In making sense, one confers with the conventions and references available in order to chart, regularize and normalize the information at hand. Alice tries to determine who she is by recalling, remembering and reciting, but in so doing her

---

1   *Wonderland*, 9.
2   *Wonderland*, 9.
3   Wittgenstein, *Philosophical Investigations*, 223e.
4   Wittgenstein, *Philosophical Investigations*, 223e.
5   *Wonderland*, 10.

repetition changes her reality, so much so that she becomes convinced that it has changed her nature altogether. The very process of observation has undermined her subject. Alice attempts to apply the conventions of her ordinary language to orientate herself in this uncharted world. But in many ways, the result is arbitrary, nonsense, sensible purely by agreement.

> "I'm sure I'm not Ada," she said, "for her hair goes in such long ringlets, and mine doesn't go in ringlets at all; and I'm sure I can't be Mabel, for I know all sorts of things, and she, oh! she knows such a very little! Besides, *she's* she, and *I'm* I, and – oh dear, how puzzling it all is! I'll try if I know all the things I used to know. Let me see: four times five is twelve, and four times six is thirteen, and four times seven is – oh dear! I shall never get to twenty at that rate! However, the Multiplication Table doesn't signify: let's try Geography. London is the capital of Paris, and Paris is the capital of Rome, and Rome – no, *that's* all wrong, I'm certain! I must have changed for Mabel! I'll try and say 'How doth the little – '" and she crossed her hands on her lap as if she were saying lessons, and began to repeat it, but her voice sounded hoarse and strange, and the words did not come the same as they used to do.[6]

Memorials and Inscriptions

This is precisely the concern of the inscription of the intellectual property system upon innovation. While it is entirely necessary and proper to the system that a sea of innovation is navigated and mapped in some way, at the same time it is important to remind oneself of the effect upon our perception of innovation that this might have. How might the process of regulation affect the object itself?

It is noteworthy that in Alice's attempt to make sense, to plot her surrounds, she interprets her change as diminishment, a limitation, the result being a possible deterioration not only in her intellectual status ('she knows such a very little!') but also perhaps her social status, a particular condemnation given the period of Victorian England in which *Wonderland* was written:

> "I must be Mabel after all, and I shall have to go and live in that poky little house, and have next to no toys to play with, and oh! ever so many lessons to learn! No, I've made up my mind about it; if I'm Mabel, I'll stay down here! It'll be no use their putting their heads down and saying 'Come up again, dear!' I shall only look up and say 'Who am I then? Tell me that first, and then, if I like being that person, I'll come up: if not, I'll stay down here till I'm somebody else' – but, oh dear!" cried Alice, with a sudden burst of tears, "I do wish they *would* put their heads down! I am so *very* tired of being all alone here!"[7]

---

6   *Wonderland*, 11.
7   *Wonderland*, 12.

So in retelling, something is lost (more accurately, forgotten). Alice is literally positioned outside the centre to which she is socialized, with threat to property, intellectual and social value.[8] But the reality she has described is simulacral, objective and a replacement for her presumed identity. The potential is in that very difference. And intriguingly, Alice declares she will remain down in Wonderland until she consents to her identity: "'Who am I then? Tell me that first, and then, if I like being that person, I'll come up: if not, I'll stay down here till I'm somebody else.'"[9]

Thus, Alice's process of description, of relating to the world, is not a fantastical figment of Wonderland but the accepted process of conventional representation. However, what is forgotten in the process is the very potential of this difference. In a digital environment, the conceptual process of intellectual property is paramount, as the subject matter for value can no longer rely upon a tangible good as a frame. This difference produced by perspective, difference, repetition, is a preoccupation throughout this book: how might the conceptual process of the intellectual property system be understood in order to remedy the apparent practical, social and enforcement problems in a digital environment? In earlier discussion[10] the difficulty of displacing entrenched connections between social value and function and property relations highlighted not only the structural or practical difficulties in the digital sphere, but also the cultural and social challenges in moving from a relationship with goods to a field of relations through services and the affect. How might this be translated in the context of value constructed through use and indeed through infringement, as distinct from value in assets, as such? Why is it presumed that the recognition of value in the intellectual property system requires a property relation at all?

The Debt to Society

A man is no longer a man confined but a man in debt.[11]

What is intrinsic to the contention, as it were, with value within the intellectual property system is the way in which users of intellectual property are 'indebted' through the encounter with intellectual property (over and above any model of remuneration). This notion of the individual as 'indebted'[12] (morally, socially and

---

8   The discussion in *Creating Selves* notes the way in which the dominant economic model of creativity and the 'creative economy' risks positioning the user as a propertyless peasant and that asymmetry is fundamental to exchange in this model. See Gibson, *Creating Selves*, 15.

9   *Wonderland*, 12.

10   See the earlier discussion in 'Use'.

11   Deleuze, *Negotiations*, 181.

12   See further the notion of indebtedness in Lazzarato, *Making of the Indebted Man*.

politically as well as literally) in the modern debt economy means that credit and debt become the very basis of social relations in contemporary society: '*Credit* is the *economic* judgment on man's morality. In credit, *man* himself instead of metal and paper has become the *medium* of exchange, but not as man, but rather as the *existence of capital* and interest.'[13]

Debt is at once a promise and a lack in the user, an imbalance of contribution. This relationship of debt and credit demonstrates the disequilibrium at the heart of not only economic but also social exchange: 'the community, too, stands to its members in that same vital relation, that of the creditor to his debtors.'[14] The disequilibrium of the creditor–debtor relationship ('repaying is a duty but lending is an option'[15]) informs the 'morality' of the intellectual property system, and indeed the basis for moralizing rhetoric that frequently attaches to discussions of misappropriation, illegitimate access and similar.[16]

Indeed, the frequent rhetoric of 'balance' and 'social contract' that attaches to discourse on intellectual property is itself inaccurate with respect to the disequilibrium of exchange in the creditor–debtor relationship: 'Ideologies and philosophies that give privileged status to concepts of equivalence ("social equality") and exchange ("social contract") mask a caste or class-stratified society.'[17] A relationship of debt both obligates the user and renders the user at fault, guilty, suspect, going to the heart of the moralizing language of property and benefit: 'the major moral concept *Schuld* [guilt] has its origin in the very material concept *Schulden* [debts].'[18]

The Debt of the Author[19]

The ethics of the intellectual property system has thus come to be embodied in the economics of the system through the very 'subjectivity' of debt: 'a debt of

---

13   K Marx, *Writings of the Young Marx on Philosophy and Society*, Hackett, Indianapolis, [1967]/1997: 270.

14   Nietzsche, *On the Genealogy of Morals*, 71. Deleuze writes, 'Nietzsche's greatness lies in having shown, without any hesitation, that *the creditor-debtor relation was primary in relation to all exchange*', in G Deleuze, *Essays Critical and Clinical*, DW Smith & MA Greco (trans), Verso, London, [1993]/1998: 127.

15   Deleuze & Guattari, *Anti-Oedipus*, 198.

16   Such 'morality' discourse includes the labour-desert theory of property noted in the earlier discussion in 'Use'.

17   Massumi, *User's Guide*, 189. See also the earlier discussion in 'Use'.

18   Nietzsche, *On the Genealogy of Morals*, 62–63.

19   This is a play on the title of Roland Barthes's famous essay 'Death of the Author', where Barthes challenges the notion of the author as the final arbiter of meaning in the text: R Barthes, 'The Death of the Author', in *Image, Music, Text*, S Heath (trans), London, Fontana, 1977: 142–48. Barthes's work on the death of the author resonates clearly with the

existence of the subjects themselves'.[20] In this way, the relationship governed by intellectual property is one of debt, not only in terms of an actual debt in order to gain legitimate access, but also with respect to the social or ideological relationship that is arguably purely that of debt and credit. In other words, it is a relationship through a distribution of intellectual assets and one which, in the digital environment, amounts to a virtual economy based upon relations,, as distinct from a real economy, as it were, based on goods. Consumption, in that sense, is therefore more clearly a social affair: 'Through consumption, we maintain an unwitting relationship with the debt economy.'[21] Intellectual property, in this sense, becomes thus an instrument for producing credit.[22] And, as introduced in earlier discussion, our personal lives have become commodities.

This is perhaps nowhere more clear than in the digital environment with respect to the dominant business models of large-scale social media platforms. The architecture of social media immediately indebts the participant in a relationship to the platform (whether it is Facebook, Instagram, Twitter and so on) as the producer of credit. It does this explicitly through terms and conditions which almost invariably require the granting of a permanent non-exclusive licence to the platform,[23] although rarely if at all requiring a complete assignment of rights

---

presumed 'authority' of the producer (in credit) in the debt economy, and the iconography of rights and the 'creator' in intellectual property discourse: 'The Author is thought to *nourish* the book, which is to say that he exists before it, thinks, suffers, lives for it, is in the same relation of antecedence to the work as a father to his child.' The 'Debt of the Author' is thus a confrontation to the authority of the producer, and signals the importance of the concept of use as meaning that is central to the present discussion, re-instating the user in the ethics and justice of intellectual property. Rather than the 'past' and credit of the producer in the value and meaning of intellectual property, familiar production is instantaneous and continuous: 'the modern scriptor is born simultaneously with the text, is in no way equipped with a being preceding or exceeding the writing, is not the subject with the book as predicate.' See Barthes, 'Death of the Author', 145.

20   Deleuze & Guattari, *Anti*-Oedipus, 197. Maurizio Lazzarato notes 'The modern notion of "economy" covers both economic production and the production of subjectivity.' Lazzarato, *Making of the Indebted Man*, 11.

21   See Lazzarato, *Making of the Indebted Man*, 20.

22   On the use of credit as an instrument of exploitation, see Lazzarato, *Making of the Indebted Man*, 20.

23   For example, Facebook requires compliance with the following terms and conditions: 'For content that is covered by intellectual property rights, like photos and videos (IP content), you specifically give us the following permission, subject to your privacy and application settings: you grant us a non-exclusive, transferable, sub-licensable, royalty-free, worldwide license to use any IP content that you post on or in connection with Facebook (IP License). This IP License ends when you delete your IP content or your account unless your content has been shared with others, and they have not deleted it.' Facebook Statement of Rights and Responsibilities available at www.facebook.com/legal/ terms . Most platforms provide something similar, for example Twitter requires: '5. Your

in any content.[24] In uploading photographs and other content, the user, despite ostensibly 'creating', is nevertheless immediately instated in the position of debtor to the platform, which requires unfettered use and re-use in order for the user to participate.

Thus, this almost obligatory and socially ubiquitous form of cultural participation (and one which is almost geographically ubiquitous, subject to access to technology) at the same time maintains the participant in debt in order to take part in social and cultural life. The logic and culture of social media is thus in many respects exploitative and, in this way, individual subjectivity inevitably is reduced to a commodity.[25] Rather than an uncomplicated panacea, social media is in and of itself not necessarily a disruption to the debt economy, but rather has been thus far indoctrinated by it, whether through personal sharing platforms or through the more ostensibly 'productive' relationships of crowd-sourcing and crowd-funding. While it is true that production has been liberated from the place, the anchor of the factory, the commercial site and so on, this 'displacement' of production in fact seemingly ensures the incorporation of all creativity within the debt economy: 'There has been a transformation of valorization processes that witnesses the extraction of value no longer circumscribed to the place dedicated to the production of goods and services, but that extends beyond factory gates so to speak, in the sense that it enters directly into the sphere of the *circulation* of capital, that is, in the sphere of the exchange of goods and services. It is a question of extending the processes of value extraction to the sphere of reproduction and

---

Rights: You retain your rights to any Content you submit, post or display on or through the Services. By submitting, posting or displaying Content on or through the Services, you grant us a worldwide, non-exclusive, royalty-free license (with the right to sublicense) to use, copy, reproduce, process, adapt, modify, publish, transmit, display and distribute such Content in any and all media or distribution methods (now known or later developed). You agree that this license includes the right for Twitter to provide, promote, and improve the Services and to make Content submitted to or through the Services available to other companies, organizations or individuals who partner with Twitter for the syndication, broadcast, distribution or publication of such Content on other media and services, subject to our terms and conditions for such Content use. Such additional uses by Twitter, or other companies, organizations or individuals who partner with Twitter, may be made with no compensation paid to you with respect to the Content that you submit, post, transmit or otherwise make available through the Services.' Twitter Terms of Service available at https://twitter.com/tos.

24   Although there were some cases of conditions of assignment in the early social media business models, examples of assignment are now almost impossible to find, with most platforms using licensing models to facilitate use.

25   See further the discussion of the commodification of subjectivity in Lazzarato, *Making of the Indebted Man*, 34.

distribution.'[26] Increasingly the debt economy is the entire sphere of life, the 'debt of existence'[27] the 'bio-economy'[28] of innovation.

Therefore, even the seemingly 'democratized' world (for want of a less tiresome cliché) of social media is harnessed to the logic of consumption and thus debt. It is thus imposed as ostensibly illogical to stand outside notions of property and debt for one's sense of self, personality and subjectivity, where the regulation and governance of such space continues to be articulated through the architecture of credit and debt. In other words, whether in defiance or in compliance, intellectual property governs relations in the digital, and these are relations of power. Social media platforms therefore continue to manufacture the 'debt' of their users. Where the creditor is the privileged,[29] the user is perhaps now being indoctrinated as the 'peasant'.[30]

I Owe Therefore I Am[31]

'This debt will simply swallow all,
And make my life a life of woe!'[32]

The 'social' of social media would thus appear both to estrange and to unsettle in ways remarkably similar to Marx's analysis of credit in 1844: 'Within the credit relationship ... *Human individuality* and human *morality* have become an article of trade and the *material* in which money exists. Instead of money and paper, my very personal existence, my flesh and blood, my social virtue and reputation is the matter and the substance of the *monetary spirit*. Credit no longer reduces monetary value to money, but to human flesh and the human heart.'[33] The legal and customary architecture of social media would seem to be instrumental in capitalizing upon the skills of participants such that despite the language of user-generated content and user-led innovation, the user is marginalized, indebted, and the user's social power is incorporated efficiently within the machinery itself.[34]

---

26 C Marazzi, *The Violence of Financial Capitalism*, K Lebedeva & JF McGimsey (trans), new ed, Semiotext(e), Los Angeles, 2011: 48.

27 Deleuze & Guattari, *Anti-Oedipus*, 197.

28 Marazzi, *Violence of Financial Capitalism*, 49.

29 Lazzarato, *Making of the Indebted Man*, 29.

30 See the earlier discussion in 'Use'.

31 International Necronautical Society (INS), 'Interim Report on Recessional Aesthetics', in T McCarthy, S Critchley et al., *The Mattering of Matter Documents from the Archive of the International Necronautical Society*, Sternberg P, Berlin, 2012: 238, at 240.

32 *Sylvie and Bruno*, 136.

33 Marx, *Writings of the Young Marx on Philosophy and Society*, 270.

34 With respect to the more general force of technology, Negri argues: 'What is magical is technology's power (which is more mysterious the more sophisticated it is) to

A way of life, as it were, becomes inextricably bound within an economic landscape and any discursive logic of rights is replaced by the 'debt' to the economic infrastructure governing and regulating social media relationships. This debt is even clearer when considering the construction of subjectivities with respect to intellectual property, including through the inclusion of intellectual property on school curricula. Intellectual property becomes a life skill, a way of life, a part of personhood.

Indeed, this explains in some way the discourse on rights in intellectual property, where any notion of the 'rights' of users will be ordinarily understood as exceptions, limitations, defences. While an exception might be understood as a 'balance' of power with respect to the dynamic between 'owner' (creditor) and user (debtor), nevertheless users are not described as 'right-holders' in the ordinary sense within the logic and language of the intellectual property framework.

Everyone an Owner![35]

Free as in freedom, not free as in beer:[36] 'The debtor is "free," but his actions, his behaviour, are confined to the limits defined by the debt he has entered into. The same is true as much for the individual as for a population or social group. You are free insofar as you assume the *way of life* (consumption, work, public spending, taxes, etc.) compatible with reimbursement.'[37] In other words, this is free as in freedom, not free as in beer: 'The freedom in liberalism is always and primarily the freedom of private ownership and owners.'[38] In this way, the property in the intellectual becomes the ultimate incarnation of individual, subjective property.

---

furnish accumulated value which has been sucked out of society. Technology holds onto this value secretly, discreetly and richly. Therefore, we are operating within a framework which is entirely, and in all respects, Marxian and we are interpreting the productive power of machinery (even more so if it is social), bringing it back to the concrete, determinate and demiurgical determination of human labour.' See Negri, *Politics of Subversion*, 90–91.

35   From the 2007 French leadership election campaign of Nicholas Sarkozy, cited in Lazzarato, *Making of the Indebted Man*, 113. See further the observation of Michel Rocard (French Prime Minister, 1988–91) in interview with Caroline Meledo: 'the resilience of the system to economic hiccups is linked precisely to the purchasing power of employees, therefore attempts to get rid of employees and make "everyone an owner" completely fail.' M Rocard & C Meledo, 'Capitalism, Crisis, and Ethics: An Interview with Michel Rocard', 12 *BC Journal*, Spring 2009: 5–15, at 12.

36   This comes from the definition of free software provided within the GNU Operating System (Richard Stallman), cited also in L Lessig, 'Free, as in Beer', *Wired*, September 2006, 94.

37   Lazzarato, *Making of the Indebted Man*, 31.

38   Lazzarato, *Making of the Indebted Man*, 108.

This proprietary logic of personhood is re-inscribed by the language liabilities of 'beneficiary' and 'consumer' that sustain the user as indebted, obliged, a borrower.[39] Notably, such language also informs the discourse of human rights, and indeed the very 'right to benefit'[40] upon which much justification for intellectual property is rested. Indeed, this language of beneficiaries and benefit is the language of contract, the language of promise, the dynamic of debtor and creditor. In the context of the Welfare State, this has been described as 'a progressive transformation of "social rights" into "social debts"'.[41] The creditor–debtor relationship thus 'brings the issue of property to the fore',[42] transforming everyone into owners.[43]

As introduced in the previous chapter, competition law might be seen, therefore, as 'social' in nature in that it regulates advantage, acting as a 'social' law. Intellectual property law, on the other hand, promotes competition in a sense in that it delivers advantage for speed to market. In limiting and even excluding social and familiar production, intellectual property is a kind of law of the 'anti-social'. The interesting ruptures in this 'balance' occur where intellectual property might actually be appropriated in order to protect the social. For instance, intellectual property rights are necessary to guarantee obligations to share in open source models. A more specific, although perhaps less unambiguous, instance of the use of intellectual property in the public interest is in the removal of counterfeit medicinal and health products from the market.[44] Although product liability might be relevant, in many cases it is more difficult to make out as the originator of the product is unknown and the remedy provides only for a sanction after the product has already caused damage. In contrast to this, enforcement of intellectual property rights can prevent the damage occurring in the first place, by removing the products from the market.

Remembering the New

Returning to the lessons from Alice, her recollection becomes the object of her identity. Similarly, the recollection of the intellectual property (patent, design,

---

39   See further the discussion in Lazzarato, *Making of the Indebted Man*, 38 where the author examines the subjectivity of the 'indebted man' also in language such as 'worker', 'entrepreneur' and 'unemployed'.

40   ICESCR, 15(1). See discussion in Gibson, *Creating Selves*.

41   Lazzarato, *Making of the Indebted Man*, 103.

42   Lazzarato, *Making of the Indebted Man*, 107.

43   Lazzarato, *Making of the Indebted Man*, 111.

44   This may occur either at the customs border, as required by TRIPs, art 51 (in the EU, see Regulation (EU) No 608/2013 of 12 June 2013 concerning customs enforcement of intellectual property rights and repealing Council Regulation (EC) 1383/2003 (which comes into force 1 January 2014, replacing the existing regime)) or by taking action in the criminal courts (criminal sanction being required by TRIPs, art 61).

trade mark, copyright) becomes the object of the creative or innovative event. In other words, intellectual property has taken on a phenomenological quality for creativity and innovation and as such an intentionality all its own. We speak of entire businesses based upon intellectual property, putting undue pressure on a system of recognition to provide for business and industry the model for commercial transactions. Nevertheless, thus far, in all corners of the contested meanings over the 'value' of intellectual property, there is nevertheless consent to this intentionality. All users (producers, consumers and so on) have consented to the expressions that constitute intellectual property, but the criteria for protection (patentability, originality, distinctiveness, individual character and so on) do not determine the meaning of intellectual property (and so the legitimacy), that is, its 'use' in the wider social sense.

The crisis of meaning for the intellectual property system, if it may be called that, is that consent to the meaning has been deferred and displaced through the structure itself (and the various silences of the user). The relationship between users together forms the grammar and language of the intellectual property system. However, one cannot consent to meaning if not present in order to engage fully in the language-game of intellectual property. Users are being asked to consent to a convention of meaning that is adjudicated elsewhere. For each creative and innovative utterance, those taking responsibility are not the speakers (the users) but the adjudicators for the system, speaking on behalf of the system and on behalf of users. In copyright this deferral is at its most extreme, in that the user is absent in defining the scope of the work (and so the infringement). There appears to be the trace of the user in certain applications of limitations and exceptions (including fair use and fair dealing); however, each application is still defined in terms of the work (substantial part) and not in terms of the user. What might be reasonable for an ordinary person skilled in the art of expression? The informed reader? The average blogger? This is in no way suggesting that the scope of protection against infringement should necessarily be attenuated or abrogated, but it is to note the way in which the dynamics of the system always already exclude the user. Even if the outcome in practice might be the same (the work is infringed), the outcome in terms of the legitimacy of the law might be very different indeed.

Returning to Alice, she attempts to disentangle this new world and weave Wonderland into her conventional and familiar concepts, to render it sensible through recollection and recognition: seeing is remembering, and so that is how it has always been. It is the very character of 'recognition' that is integral to processes of harmonization. Harmonization of the system is in effect (and in reality) mutual recognition and assimilation. Similarly, the process of recognition in intellectual property examination (patents, trade marks, designs) is remarkably similar. First, the system must assume the command of innovation, the ability to recognize and validate innovation. It claims as much because in doing so it recollects, remembers what has gone before (prior art). In other words, paradoxically, no new knowledge is required, as such, in order to recognize the inventive, the distinctive, the individual. The intellectual property system simply 'comes to know' that which is already in its memory; this is indeed how one 'gets to know'.[45] On the other hand, copyright is excluded from this process of recognition, it is simply self-proclaiming. As well as the exclusion of the user from the adjudication of validity if that arises, copyright has no memory. Copyright is both meaningless (useless) and at the same time an unmapped and infinite space.

This 'memory' of intellectual property has important resonances with the notion of the 'archive'. The archive is considered in later chapters, however for the benefit of the present discussion, it is useful to note the archival operation of intellectual property recognition and the emphasis on rendering the narrative of innovation to which it occurs logical and accountable through patents, trade marks, designs and, to an extent, copyright. For example, if the patent is to depict the 'reality' of innovation, then certain additional mechanisms will be deployed, including depositories (of samples). In the international discussions and debate concerning the protection of the traditional knowledge of indigenous and traditional peoples,[46] this mechanism of intellectual property has been appropriated by traditional communities in order to reflect the mechanism back, as it were, in the form of 'mandatory disclosure of origin'.[47] It is perhaps no surprise, then, that some more powerful nations in the World Intellectual Property Organization (WIPO) Intergovernmental Committee[48] have objected to the imposition of disclosure, arguably recognizing the power of such a narrative.

---

45   Wittgenstein, *Blue and Brown Books*, 23.

46   The UN World Intellectual Property Organization (WIPO) defines traditional knowledge (TK) as the 'knowledge, know-how, skills and practices that are developed, sustained and passed on from generation to generation within a community, often forming part of its cultural or spiritual identity'. Available at www.wipo.int/tk/en/tk/. Within WIPO an intergovernmental committee (IGC) was established in 2001 in the 26th (12th Extraordinary Session) of the WIPO General Assembly, Geneva, 25 September to 3 October to address the issues for traditional knowledge exploitation, misappropriation and protection, and it is currently progressing towards an international text-based treaty solution. See further The Protection of Traditional Cultural Expressions: Draft Articles, WIPO/GRTKF/IC/22/REF/FACILITATORS TEXT and The Protection of Traditional Knowledge: Draft Articles (Rev. 2), WIPO/GRTKF/IC/24/FACILITATORS DOCUMENT REV. 2 both available at www.wipo.int/tk/en/igc. See also the further discussion of the WIPO IGC in the context of international process in Chapter 12.

47   In the context of the WIPO IGC negotiations, see the treatment of disclosure in the Consolidated Document Relating to Intellectual Property and Genetic Resources Rev. 2, WIPO/GRTKF/IC/23/WWW/230222 available at www.wipo.int/tk/en/igc/. See further the WIPO Technical Study on Patent Disclosure Requirements Related to Genetic Resources and Traditional Knowledge, Study No. 3, 2004, prepared for the seventh meeting of the Conference of the Parties (COP) to the Convention on Biological Diversity (CBD), Kuala Lumpur, Malaysia, 9–20 February 2004 (in response to a request made at its sixth meeting).

48   For instance, note, in the Report of the Twenty-Third Session of the IGC, held 4–8 February 2013, in particular the interventions on disclosure made by the Delegations Japan (38, 190), the United States of America (57), the Russian Federation (196), all of which have

What the intellectual property system provides, therefore, is a perception of 'quality' or 'properties', as it were, in terms of the expression (not quality in terms of the content). Any representation of innovation (a patent, a design document, a copyright work and so on) cannot actually 'show' innovation or capture the ideal. As such the patent will always be a model of ideal of 'invention', the trade mark of the mark and so on. Despite the language of originality, other than in the exceptional case of unique works of art, the intellectual property has nothing to do with originals and copies; it is all about repetition and difference.[49]

Family Differences

If the intellectual property system relies upon our ability to accommodate completely unanticipated innovation and nevertheless render it familiar, sensible and relevant within the system, then how might this be achieved? By way of example, consider the process of the patent. There is no single example of what is patentable, no common property to which one can refer to explain patentability. Despite the apparent clarity and certainty for which the principles are revered, it is in practice beyond the means available to analyse the concept of 'patent' in such a way as to provide an infallible definition (and therefore prediction) of a valid patent.[50] We need 'room for innovation' and 'room for judging' in the system itself.

This necessary adaptability, or so it seems, is explained to an extent by Wittgenstein's concept of family resemblances or overlapping similarities that allow us to recognize the similarities without necessarily requiring a definition or limit to eligibility for protection, that is, to patentability in the present example:

> Consider for example the proceedings that we call 'games'. I mean board-games, card-games, ball-games, Olympic games, and so on. What is common to them all? – Don't say: 'There *must* be something common, or they would not be called "games"' – but *look and see* whether there is anything common to all. – For if you look at them you will not see something that is common to *all*, but similarities, relationships, and a whole series of them at that. To repeat: don't think, but look!'...

---

also co-sponsored a proposal calling for further study into disclosure requirements. See Intergovernmental Committee on Intellectual Property and Genetic Resources, Traditional Knowledge and Folklore, Report of the Twenty-Fourth Session, 22–26 April 2013, WIPO/GRTKF/IC/24/8 (15 July 2013).

49   See in detail the discussion of the simulacrum and intellectual property in Chapter 9.

50   An example of this is the ongoing work of TELES, a German telecommunications company based in Berlin, which is attempting to introduce a process of determining and regularizing the scope of the inventive concept for patents. See the press release of 9 July 2013, available at www.fstp-expert-system.com.

> And the result of this examination is: we see a complicated network of similarities overlapping and criss-crossing: sometimes overall similarities, sometimes similarities of detail.[51]

Indeed, the items in the catalogue of patentability appear to have nothing in common except the very unifying character of having nothing in common. This remembering as seeing is critical to the intellectual property system as a whole, but is clearly demonstrated in the prosecution of a patent:

> I can think of no better expression to characterize these similarities than 'family resemblances'; for the various resemblances between members of a family: build, features, colour of eyes, gait, temperament, etc. etc. overlap and cross-cross in the same way. – And I shall say: 'games' form a family.[52]

This landscape of overlapping similarities that characterizes the intellectual property system is important. Rather than the traditional, linear, causal taxonomical model that tends to be associated with (or indeed imposed upon) the intellectual property system, what is available rather is the capacity for a network of innovative events that is paradoxically heterogeneous in ways complicit with 'digital' models of innovation, and indeed with familiar production. Rather than a characteristic of innovation, causality is merely the form given to the claims and expressions of the intellectual property system: 'The law of causality is not a law but the form of a law.'[53] It is a reassurance, just as for Alice, where the catalogue and referencing of her adventures provides some reassurance of progression towards the garden.[54] But as Alice discovers, she cannot rely on simple causality to understand her surroundings: 'There is no causal nexus to justify such an inference. We *cannot* infer the events of the future from those of the present. Superstition is nothing but belief in the causal nexus.'[55]

In other words, what is presented is a seemingly ordered, tree-like catalogue of shelves and narratives of prior art, but this is simply a contrivance. What is actually available within the intellectual property system is the capacity for rhizomatic, de-centred, 'digital' maps of innovation: 'any point of a rhizome can be connected to anything other, and must be. This is very different from the tree or root, which plots a point, fixes an order.'[56] This is not a tree-like and variegated model of the familiar, as it were, but rather an aggregating, budding kinship of productivity and

---

51   Wittgenstein, *Philosophical Investigations*, §66.

52   Wittgenstein, *Philosophical Investigations*, §67.

53   L Wittgenstein, *Tractatus-Logico-Philosophicus*, DF Pears & BF McGuinness (trans), B Russell (intro), Routledge, London, [1921]/1974: 6.32.

54   Compare the rapid journey into the garden on the 'other side' of the dominant paradigm, that is, on the other side of the Looking-glass in '*Re* Use'.

55   Wittgenstein, *Tractatus-Logico-Philosophicus*, 5.136–5.1361.

56   Deleuze & Guattari, *Thousand Plateaus*, 7.

connection. In ideal encounters with the language of intellectual property, as soon as a bud is extracted through the intellectual property tools (and transacted) it itself continues to bud if the system is able to operate to its full potential. The problem, however, is that such encounters have become perhaps less than ideal.

Therefore, it is merely an error or an artifice that the intellectual property system has come to be taken for innovation, and has had imposed upon it the linear, hierarchical character of the tree, as distinct from the rhizome.[57] Nevertheless, this is the rhetoric that attaches to the system and is part of its undoing, providing a classic battle narrative of epic proportions. Fundamentally and logically, however, the intellectual property system does *not* evaluate knowledge, nor does the difference and change identified in the system progress in a simply linear way. In fact, intellectual property rules are already strictly non-hierarchical and this system of 'similarities' ruptures the conventional picture of order that has been imposed and replaces it with the rhizomatic structure of resemblance: 'Perhaps one of the most important characteristics of the rhizome is that it always has multiple entryways.'[58] Rather than inherent to the expression of intellectual property, it is indeed in the conditions in which we are able to apply the rules imposing artificial barriers to participation, to use, that meaning is restored. In order to remedy the practical problems facing the system (including crises in legitimacy, adherence and enforcement) it is important not to follow slavishly the model of stratification and hierarchization (apparent in conventional, taxonomical accounts of intellectual property), but rather to understand how it might be possible to give Alice, the user, keys to every door. It is in use (and access) that the meaning of the system breaks down, and its legitimacy founders. And it is here where we need to look for meaning.

Smooth Sailing

> As she said these words her foot slipped, and in another moment, splash! she was up to her chin in salt water. Her first idea was that she had somehow fallen into the sea, "and in that case I can go back by railway," she said to herself. (Alice had been to the seaside once in her life, and had come to the general conclusion, that wherever you go to on the English coast you find a number of bathing machines in the sea, some children digging in the sand with wooden spades, then a row of lodging houses, and behind them a railway station).[59]

So as well as the common indications of familiarity (Alice believes there is nothing new to see at the sea), there is the presumption that one can be found 'at sea' through its striation by the railway. But perhaps it is not so simple for Alice to

---

57   Deleuze & Guattari, *Thousand Plateaus*, 12.
58   Deleuze & Guattari, *Thousand Plateaus*, 12.
59   *Wonderland*, 13.

navigate the enormous potential of her pool of tears in the way she has expected: 'she soon made out that she was in the pool of tears which she had wept when she was nine feet high.'[60] That said, the potential of the sea space makes itself known to Alice:

> Just then she heard something splashing about in the pool a little way off, and she swam nearer to make out what it was: at first she thought it must be a walrus or hippopotamus, but then she remembered how small she was now, and she soon made out that it was only a mouse that had slipped in like herself.
>
> "Would it be of any use now," thought Alice, "to speak to this mouse? Everything is so out-of-the-way down here, that I should think very likely it can talk: at any rate, there's no harm in trying." So she began: "O Mouse, do you know the way out of this pool? I am very tired of swimming about here, O Mouse!"[61]

Indeed, *everything is so out-of-the-way down here*, way out. But this is precisely the creative power of Wonderland, and a potential into which Alice is becoming immersed. So what at first seems like a loss of meaning instead begins to present itself as opportunity to Alice, and eventually she finds herself swimming in a pool 'getting quite crowded with the birds and animals that had fallen into it'.[62] When earlier, Alice attempts to chart her territory through familiar processes of recollection and referencing, she becomes confused and despondent. She, literally, loses her way. But upon encountering the smooth space of the sea, unencumbered by railway tracks and other signposts, Alice acknowledges its *out-of-the-way* potential and ventures to speak to a mouse.

The pool of tears, the sea of tears, is thus an unlimited, innovative, *out-of-the-way* space of becoming. In the social organization of space, the pool has much in common with what Deleuze and Guattari would term 'smooth space', where social roles (Alice/Mabel; producer/consumer) and indeed species hierarchies (a mouse that can not only talk but also make itself understood) are unanchored. Building upon the work of composer, Pierre Boulez, Deleuze and Guattari adapt the concepts of smooth and striated space to present the smooth space of emergent becoming in concert with the regulated and measured striated space of the State apparatus: 'Boulez says that in a smooth space-time one occupies without counting, whereas in a striated space-time one counts in order to occupy.'[63] The smooth space of continuous, directionless, rhizomatic, qualitative familiar production is thus overlaid by the taxonomical, directional, arboreal, measured production of striated space.

---

60   *Wonderland*, 13.

61   *Wonderland*, 13–15.

62   *Wonderland*, 17.

63   Deleuze & Guattari, *Thousand Plateaus*, 477. See further the discussion of the smooth space-time of innovation in Chapter 7.

It is in the fluid, smooth space of the familiar that social, cultural, knowledge and even species borders can be transgressed, and new assemblages of creativity and innovation are possible: the digital 'war machine'.[64] The digital war machine appears to be 'irreducible to the State apparatus, to be outside its sovereignty and prior to its law ... In every respect, the war machine is of another species, another nature, another origin than the State apparatus.'[65] It follows, to reiterate earlier discussion, that the digital must be understood over and above any mere technical elements, or technologies of reproduction and repetition. The digital is over and over a social and cultural transformation, a connection and a kinship of the flows of familiar production: 'It is the machine that is primary in relation to the technical element: not the technical machine, itself a collection of elements, but the social or collective machine, the machinic assemblage that determines what is a technical element at a given moment, what is its usage, extension, comprehension, etc.'[66]

The digital is thus contrary to the striated space of the map, the railway, the organization of property, and indeed propriety and etiquette (that preoccupies and hinders Alice throughout her journey):

> For the sea is a smooth space par excellence, and yet was the first to encounter the demands of increasingly strict striation. The problem did not arise in proximity to land. On the contrary, the striation of the sea was a result of navigation on the open water ... It is as if the sea were not only the archetype of all smooth spaces but the first to undergo a gradual striation gridding it in one place, then another, on this side and that ... This is undoubtedly why the sea, the archetype of smooth space, was also the archetype of all striations of smooth space: the striation of the desert, the air, the stratosphere (prompting Virilio to speak of a 'vertical coastline,' as a change in direction). It was at sea that smooth space was first subjugated and a model found for the laying-out and imposition of striated space, a model later put to use elsewhere.[67]

---

64   Deleuze and Guattari identify the 'war machine' as the abstract machine that disrupts the equilibrium of social formations so as to continue the intensive becoming of smooth space. In this sense the 'digital' is a war machine, an assemblage capable of challenging the dominant system and so whenever and wherever conventional models attempt to enforce the usual modes of production upon familiar production, then the digital war machine is proven irreducible: 'Could it be that it is at the moment the war machine ceases to exist, conquered by the State, that it displays to the utmost its irreducibility, that it scatters into thinking, loving, dying, or creating machines that have at their disposal vital or revolutionary power capable of challenging the conquering State?' Deleuze & Guattari, *Thousand Plateaus*, 356.

65   Deleuze & Guattari, *Thousand Plateaus*, 352.

66   Deleuze & Guattari, *Thousand Plateaus*, 398.

67   Deleuze & Guattari, *Thousand Plateaus*, 479–80.

This distinction between smooth and striated space articulates the relationship between intellectual property and innovation. The smooth space of innovation is arguably striated and regulated by the divisions and classifications of the intellectual property system. But importantly, as seen in the rhizomatic nature of the system of similarities in intellectual property, the relationship between smooth and striated space is not a simple and polarizing opposition. Rather, the intellectual property system is indeed a mixture of the two:

> No sooner do we note a simple opposition between the two kinds of space than we must indicate a much more complex difference by virtue of which the successive terms of the oppositions fail to coincide entirely. And no sooner have we done that than we must remind ourselves that the two spaces in fact exist only in mixture: smooth space is constantly being translated, transverse into striated space; striated space is constantly being reversed, returned to smooth space.[68]

In mixture, therefore, there is potential for transgression, deterritorialization, in what Deleuze and Guattari would call lines of flight: 'Territorialities, then, are shot through with lines of flight testifying to the presence within them of movements of deterritorialization and reterritorialization'[69] – the *out-of-the-way, way out.* Thus, addressing the ecosystem of intellectual property (all users) and finding fundamental ways to achieve the symbiotic potential of the system will rely upon understanding the smooth space of innovation.

It is the dynamic quality of innovation that is celebrated in the digital and articulated through moments of familiar production, with arguments that over-regulation, while successfully striating the digital smooth space, will necessarily diminish its social, political and creative potential. However, arguably this picture is too simplistic and impossibly naive to account for 'today's digital societies where, after the sea and the air, capitalism has appropriated information to create *false smooth spaces* in order to control the circulation of people and of goods'.[70] An intriguing illustration of this regulation and mainstreaming of potential is perhaps provided by the case of Fatboy Slim, who became the first DJ to perform at the UK Houses of Parliament:[71] 'This was just what the organisers wanted: greater rapport between MPs and the creative industries, not to mention intellectual property

---

68   Deleuze & Guattari, *Thousand Plateaus*, 474.

69   Deleuze & Guattari, *Thousand Plateaus*, 55.

70   VA Conley, 'Of Rhizomes, Smooth Space, War Machines and New Media', in M Poster & D Savat (eds), *Deleuze and New Technology*, Edinburgh UP, Edinburgh, 2009: 32, at 35.

71   'Fatboy Slim DJs at House of Commons', *BBC News*, 7 March 2013. See further the comment of Michael White, 'Fatboy Slim – review', *The Guardian*, 7 March 2013: 'Everyone was on their best behaviour and it was all cool and fantastic, as everyone kept saying, even Colonel Bob Stewart, one of the MPs as square as an Oxo cube.'

rights.'[72] Similarly, the dislocation of a Banksy discussed in 'Use' is a further example of the striation of the smooth space of familiar production, diminishing the creative and political potential of street art and reconfiguring it as a 'spectator sport'[73] for the rich US auction market.

> Any society whatsoever has all of its rules at once – juridical, religious, political, economic; laws governing love and labor, kinship and marriage, servitude and freedom, life and death. But the conquest of nature, without which it would no longer be a society, is achieved progressively, from one source of energy to another, from one object to another. This is why *law* weighs with all its might, even before an object is known, and without ever its object becoming exactly known. It is this disequilibrium that makes revolutions possible. It is not at all the case that revolutions are determined by technical progress. Rather, they are made possible by this gap between the two series, which solicits realignments of the economic and political totality in relation to the parts of the technical progress.
>
> …
>
> The technocrat is the natural friend of the dictator – computers and dictatorship; but the revolutionary lives in the gap which separates technical progress from social totality, and inscribes there his dream of permanent revolution.[74]

Thus, it is not the ability to copy or indeed the advancements in technologies of reproduction that advance the digital revolution. It is not the mechanics of the digital but the spectacular changes in the social, the face as brand, the private as commodity, the public as private. It is a social revolution nonetheless, not a technological one. It is a revolution of the familiar as distinct from the technological.

Therefore, the smooth space of the digital is not necessarily to be read as the utopic garden, and it would be banal to suggest this: 'the smooth itself can be drawn and occupied by diabolical powers of *organization*.'[75] And indeed 'the sea is a smooth space par excellence, and yet was the first to encounter the demands of increasingly strict striation.'[76] The digital landscape is both liberating and intoxicating, pretending at once both the tools for socialization and communication, and preparing a smooth space readily striated and obliterated for the purposes of surveillance, overcoding and manipulation.[77] It is a mixture. And the answer to the

---

72   White, 'Fatboy Slim – review'.

73   Recalling Banksy's own comments on the way in which graffiti is not a spectator sport, stencil graffiti is an important example of the collaborative, communal process of familiar production. See further the earlier discussion of Banksy and familiar production in 'Use'.

74   Deleuze, *Logic of Sense*, 49.

75   Deleuze & Guattari, *Thousand Plateaus*, 480.

76   Deleuze & Guattari, *Thousand Plateaus*, 479.

77   Paul Virilio suggests that technology has somehow evaded a discourse of questioning or criticism: 'all criticism of technology has just about disappeared and we have

relationship between intellectual property and the digital also lies in the mixture, not in banalities: 'Never believe that a smooth space will suffice to save us.'[78]

> [I]t may be that the sound molecules of pop music are at this very moment implanting here and there a people of a new type, singularly indifferent to the orders of the radio, to computer safeguards, to the threat of the atomic bomb. In this respect, the relation of artists to the people has changed significantly: the artist has ceased to be the One-Alone withdrawn into him- or herself, but has also ceased to address the people, to invoke the people as a constituted force. Never has the artist been more in need of a people, while stating most firmly that the people is lacking – the people is what is most lacking.[79]

In the sea of innovation, there is a flow, a movement akin to life and creative potential – like innovation, it is a becoming, a form of life, a familiar production. This shares much with the momentum of Alice's journey. She is preoccupied with reaching the garden, and yet it is the journey itself that is the story, not any destination. The instantaneity of smooth space is in stark contrast to the causal, destined, linear time imposed upon conventional renditions of innovation.

The smooth space of the sea of familiar production is a most spectacular symbol of the crowd,[80] a swell of emergent creativity, a continuous wave of becoming, the tide of familiar production: 'There are also the individual drops of water ... They only begin to count again when they can no longer be counted, when they have again become part of a whole.'[81] Indeed, they 'count' only when they are part of the immeasurable smooth space of familiar production: 'The sea has a *voice*, which is very changeable and almost always audible ... But what is most impressive about it is its persistence. The sea never sleeps; by day and by night it makes itself heard, throughout years and decades and centuries. In its impetus and its rage it brings to mind the one entity which shares these attributes in the

---

slid unconsciously from pure technology to techno-culture and, lastly, to the dogmatism of a *totalitarian techno-cult* in which everyone is caught in the trap not of a society and its moral, social or cultural laws and prohibitions, but of what these centuries of progress have made of us and *of our own bodies*.' P Virilio, *The Information Bomb*, C Turner (trans), London, Verso, [1998]/2000: 39.

78   Deleuze & Guattari, *Thousand Plateaus*, 500.

79   Deleuze & Guattari, *Thousand Plateaus*, 346.

80   Elias Canetti explains: 'Crowd symbols is the name I give to collective units which do not consist of men but which are felt to be crowds. Corn and forest, rain, wind, sand, fire and the sea are such units. Every one of these phenomena comprehends some of the essential attributes of the crowd. Although they do not consist of men, each of them recalls the crowd and stands as symbol for it in myth, dream, speech and song.' E Canetti, *Crowds and Power*, C Stewart (trans), Phoenix P, London, [1960]/2000: 75. See further the discussion of crowds and crowd symbols in Chapter 5, Chapter 11 and again in '*Re* Use'.

81   Canetti, *Crowds and Power*, 80.

same degree; that is, the crowd.'[82] The wisdom of the crowd, the swell of familiar production, the smooth space of the sea is always in progress, always becoming: 'The sea is all-embracing; nor can it ever be filled.'[83] Further, the sea is familiar, the kinship of familiar production: 'The sea has no interior frontiers and is not divided into peoples and territories ... all life flows into it and it contains all life.'[84]

And in a swell of movement and social power, led by Alice, the whole assemblage, 'the whole party swam to the shore'.[85]

---

82   Canetti, *Crowds and Power*, 80–81.
83   Canetti, *Crowds and Power*, 81.
84   Canetti, *Crowds and Power*, 81.
85   *Wonderland*, 17.

Chapter 3
Chance

Intellectual Property: The Driest Story of All

> They were indeed a queer-looking party that assembled on the bank – the birds
> with draggled feather, the animals with their fur clinging close to them, and all
> dripping wet, cross, and uncomfortable.
>
> The first question of course was, how to get dry again: they had a consultation
> about this, and after a few minutes it seemed quite natural to Alice to find herself
> talking familiarly with them, as if she had known them all her life.[1]

They were indeed a strange but familiar assemblage on the bank. How to dry out?
Tell the driest story of all.

> At last the Mouse, who seemed to be a person of authority among them, called
> out, "Sit down, all of you, and listen to me! *I'll* soon make you dry enough!"
> They all sat down at once, in a large ring, with the Mouse in the middle. Alice
> kept her eyes anxiously fixed on it, for she felt sure she would catch a bad cold
> if she did not get dry very soon.
>
> "Ahem!" said the Mouse with an important air. "Are you all ready? This is
> the driest thing I know. Silence all round, if you please!"[2]

How is one to dry out, to dry up and be consumed, to dry-cure after the smooth
space of the sea of innovation? Tell the driest story of all: intellectual property.
That is, the eventful and impassioned sea of innovation might be traversed and
preserved by rendering the assemblage weary, literally satiated, dried up. At the
same time, the striation of the smooth space by intellectual property ensures that
the constant exchange and intensive becomings of innovation might be rendered
flat, even tasteless.

However, as well as organizing the sea of innovation with the stability of the
driest of stories, there is also the opportunity of embracing the space and engaging
in the speed of the race, the infinite potential of chance. In other words, it is
'change' that is significant, whether slowing down in the story circle or speeding
up in the Caucus-race.

---

1   *Wonderland*, 17–18.
2   *Wonderland*, 18.

"How are you getting on now, my dear?" it continued, turning to Alice as it spoke.

"As wet as ever," said Alice in a melancholy tone: "it doesn't seem to dry me at all."

"In that case," said the Dodo solemnly, rising to its feet, "I move that the meeting adjourn, for the immediate adoption of more energetic remedies –"[3]

Rather than the classificatory tedium of the driest story of all, more energetic remedies are the race of pure chance, the familiar production of continuous innovation. All at once the 'mixture' of smooth and striated space[4] is apparent.

"What I was going to say," said the Dodo in an offended tone, "was, that the best thing to get us dry would be a Caucus-race."[5]

But what *is* a Caucus-race?

The Caucus-race

"What *is* a Caucus-race?" said Alice; not that she much wanted to know, but the Dodo had paused as if it thought that *somebody* ought to speak, and no one else seemed inclined to say anything.

"Why," said the Dodo, "the best way to explain it is to do it." (And, as you might like to try the thing yourself some winter day, I will tell you how the Dodo managed it.)

First it marked out a race-course, in a sort of circle ("the exact shape doesn't matter," it said), and then all the party were placed along the course, here and there. There was no "One, two, three, and away," but they began running when they liked, and left off when they liked, so that it was not easy to know when the race was over. However, when they had been running half an hour or so, and were quite dry again, the Dodo suddenly called out "The race is over!" and they all crowded round it, panting, and asking, "But who has won?"

This question the Dodo could not answer without a great deal of thought, and it sat for a long time with one finger pressed upon its forehead (the position in which you usually see Shakespeare, in the pictures of him), while the rest waited in silence. At last the Dodo said, "*Everybody* has won, and all must have prizes."[6]

---

3   *Wonderland*, 19.

4   The concept of a mixture of smooth and striated space in the apparatus of intellectual property was introduced in Chapter 2.

5   *Wonderland*, 19.

6   *Wonderland*, 19–20.

Alice is introduced to the Caucus-race, a race apparently without object, rules, limits, providing for the affirmation of chance where 'nothing is exempt from the game'.[7] This race is a crucial radicalization of the competitive game to the game where winning (and innovation) is paradoxically maximized through the elimination of competition with other players and through the affirmation of chance: 'The most difficult thing is to make chance an object of *affirmation*, but it is the sense of the imperative and the questions that it launches ... When chance is sufficiently affirmed the player can no longer lose ... The whole of chance is then indeed in each throw.'[8]

The Caucus-race, however, is not arbitrary. To provide for a game based on chance as distinct from competition and risk similarly does not render the game of innovation necessarily arbitrary: 'Once chance is affirmed, all arbitrariness is abolished every time. Once chance is affirmed, divergence itself is the object of affirmation within a problem.'[9]

Time to Market

Contrary to the affirmation of chance, the traditional objective assisted by intellectual property tools has been to mitigate risk, to facilitate the creation of markets through the limiting of chance, providing for conditioned and disciplined access by others (both to the intellectual property object and to the market). In this way, the value of the product is in terms of its 'market', that is, its 'time', meaning that the duration or term of protection is the source of value in the product, as distinct from extrinsic measures of benefit. Time to (and of) market is thus a critical aspect of value and distinction between traditional modes of production (and exploitation of intellectual property) and familiar production. While time in the former actually contributes towards the construction of the intellectual property object (its position in the market, and indeed its actual form in the context of digital environments), the time of familiar production does not persist in any one object or instance of innovation, as such. Rather, it is an indefinite and inconstant event: 'the indefinite time of the event, the floating line that knows only speeds and continually divides that which transpires into an already-there that is at the same time not-yet-here, a simultaneous too-late and too-early'.[10] It is not only difficult (or even impossible) but also perhaps not an advantage (commercially or socially) to try and locate the object in familiar production.

For quite some time debate has raged over the efficacy and legitimacy of this system in a post-industrial age, and in a time of incredibly accelerated innovation. The speed of the Caucus-race is a race without competition and without winners (or

---

7   Deleuze, *Difference and Repetition*, 283.

8   Deleuze, *Difference and Repetition*, 198.

9   Deleuze, *Difference and Repetition*, 198.

10   Deleuze & Guattari, *Thousand Plateaus*, 262.

with all winners). This resonates with critiques of the patent system, for example, as dominating innovation as a 'race', leading to overinvestment in R&D in the race to the finish. However, the interest in winning is only that of the individual (or corporate) and not of society, where only speed may be of interest.[11]

The question that is perhaps common to the concerns throughout these different forms of intellectual property industry is whether it is possible to create a market where chance is affirmed, where 'everything is possible'.

A Relationship of Chance

The striation of smooth space, its codification, organization and containment through legislative over-writing, may nevertheless be overcome by the circuit of infinite proliferation of chance. That is, the machinery of intellectual property is made contingent by the dispersement of authority and authorship, the unlimited potential of the Caucus-race, the race with no winners and no losers, the rhizomatic track of the 'digital', the 'race' of innovation. This is a race with no beginning or end, simply a seemingly arbitrary declaration when the time comes, the counter-intuitive space of the digital., In an ordinary race there is a determinable beginning and end, a credit and debt, a recognizable winner (creator) and a verifiable prize (object). That is, in an ordinary articulation of the intellectual property race, the game divides chance with a logic of wagers and winners. But in the Caucus-race there is just speed.

Nevertheless, this is not a mere relationship of opposition,[12] as such, between the potential of chance and the stability of the law, of models, of practice. Rather, there is an important disequilibrium that ensures an interchange and mutually constitutive relationship, as it were. Indeed, the law stands to gain quite a lot from chance: 'the two spaces in fact exist only in mixture: smooth space is constantly being translated, transversed into a striated space; striated space is constantly being reversed, returned to a smooth space.'[13]

In the game of intellectual property, the rules traverse and apportion chance, but the expressive chaos of the digital interferes with the ordinary apportionment of chance and indeed with the conventional hypotheses of intellectual property in its rendition of change; namely, that change is progressive, referenced and remarkable. Change in the digital ecosystem of innovation is not merely manifest in the banal terms of the technology itself, but more importantly, and more intensively, in terms of social and cultural becoming.[14] The very architecture

---

11   P Dasgupta, 'The Economic Theory of Technology Policy: An Introduction', in P Dasgupta & P Stoneman (eds) *Economic Policy and Technological Performance*, Cambridge UP, Cambridge, 1987: 7–23.

12   Note Chapter 2 and the discussion of the mixture of smooth and striated space.

13   Deleuze & Guattari, *Thousand Plateaus*, 474.

14   See in particular the discussion in Chapter 6 of 'change' and 'becoming'.

of the digital is one of proliferative, multiplicative, symbiosis and rhizomatic connectivity, at times potentially at odds with traditional concepts of authorship and attribution (precluded either by the limits of the technology itself, or by the changes in 'use' associated with the digital), and frequently in constant antagonism with the more extensive rules and framework of intellectual property. Intellectual property rules mitigate risk and temper the gamble by assuring a value attributed to that risk (through the specific rules of protection, term and so on attaching to the recognition of intellectual property rights). The digital, on the other hand, puts everything to chance, overturning risk, as it were, putting everything into play.

But this does not mean that the digital environment is a pure game, since it is radically and strategically overcoded with respect to information (as value). Skill and chance have become inextricably linked (confused) and the digital has no effect (no physical goods) and yet the effects are felt everywhere. There is no consumption, and yet everyone is engorged; there is no entertainment, and yet everyone is watching. This is the game of information for information's sake, consumption for consumption's sake,[15] the game of familiar production. The digital affirms chance, but many are also taking unspeakable risks.

Therefore, both the driest story of all and the freedom of the Caucus-race have a place in making everyone 'quite dry again'. And indeed, this is key to understanding the dimensions of intellectual property discourse and politics. Immediately that it is reduced to a polarized discussion, much of the important information (and meaning) is lost.

Returning to the mix, what *is* a Caucus-race in intellectual property?

A Game without Frontiers

The Caucus-race in *Alice in Wonderland* appears to be a curious game of confusion and arbitrariness. Nevertheless, Alice recognizes and accepts it as a game, despite the fact that it is not possible to identify the characteristics by which it qualifies as a game. How meaning is generated such that participation is possible provides an important insight for any possible strategies or programmes to reform practice and thus meaning in the intellectual property system:

> Consider for example the proceedings that we call 'games'. I mean board-games, card-games, ball-games, Olympic games, and so on. What is common to them all? – Don't say: 'There *must* be something common, or they would not be called "games"' – but *look and see* whether there is anything common to all. – For if you look at them you will not see something that is common to *all*, but similarities, relationships, and a whole series of them at that. To repeat: don't think, but look!'[16]

---

15   Recall the earlier discussion of antiproduction in 'Use'.

16   Wittgenstein, *Philosophical Investigations*, 66.

A striking feature of the Caucus-race is the Dodo's declaration that everybody wins. It is a game without frontiers, as it were, drawing an enticing comparison with the spirit of innovation as a socio-economic concept. It is a game without value and yet everything has value; it is a game without waste and yet everything is consumed. Innovation, by its very nature, cannot have a logical limit or endpoint. It is a beginning without a tail. Intellectual property therefore is not the expression of innovation. it is but perhaps some evidence of it. It is a set of criteria by which we consent to the expression of innovation in a social, economic, legal and cultural setting. Innovation through familiar production is a concept without frontiers. Familiar production is a game without frontiers, a war machine without tears.[17]

> For I *can* give the concept 'number' rigid limits in this way, that is, use the word 'number' for a rigidly limited concept, but I can also use it so that the extension of the concept is *not* closed by a frontier. And this is how we do use the word 'game.' For how is the concept of a game bounded? What still counts as a game and what no longer does? Can you give the boundary? No. You can *draw* one; for none has so far been drawn. (But that never troubled you before when you used the word 'game'.)
>
> 'But then the use of the word is unregulated, the "game" we play with it is unregulated.' – It is not everywhere circumscribed by rules; but no more are there any rules for how high one throws the ball in tennis, or how hard; yet tennis is a game for all that and has rules too.[18]

Intellectual property rules therefore circumscribe aspects of innovation in order to render them perspicuous and memorable for the structural purposes of social and economic convention, but this does not render inauthentic those expressions of innovation outside those criteria:[19] 'we can draw a boundary – for a special purpose. Does it take that to make the concept usable? Not at all! (Except for that

---

17   *Games Without Frontiers*, Lyrics P Gabriel, Music P Gabriel, Charisma-Universal, UK, 1980.

18   Wittgenstein, *Philosophical Investigations*, §68.

19   For instance, conditions of incremental innovation in traditional knowledge and traditional cultural expressions have previously been disregarded by legal frameworks as unoriginal and completely outside the intellectual property (and authentic innovation) framework. This anomaly of the intellectual property system is the subject of the IGC of the UN World Intellectual Property Organization (WIPO) which is currently working towards a possible international treaty on aspects of the relationship between traditional forms of innovation and expression and the intellectual property system (see the discussion in Chapter 2). However, there is also some tension between the intellectual property system (as an abbreviated or abridged representation of innovation) and attempts to assimilate traditional knowledge within the socio-economic confines of intellectual property concepts. For further discussion see Gibson, *Community Resources*.

special purpose.)'[20] In other words, while the physiognomy of innovation may appear to be represented by the intellectual property system and its artefacts, this cannot map the entirety of innovation and innovative process: 'That would almost be as if someone were to believe that because only the actors appear in the play, no other people could usefully be employed upon the stage of the theatre.'[21] In this way, the grammar of intellectual property merely abridges innovation, appearing to stand for that which it purports to document, namely authentic creativity and innovation, its 'super-expression': 'You have no model of this superlative fact, but you are seduced into using a super-expression. (It might be called a philosophical superlative).'[22] This provides some pictures for the book, but the conversation in meaning is perhaps lacking in this representation of innovation: 'A *picture* held us captive. And we could not get outside it, for it lay in our language and seemed to repeat it to us inexorably.'[23]

This then is the problem of representation – what is innovation? – representing innovation is representing the unrepresentable: 'Representation must encompass an expression which it does not represent, but without which it itself would not be "comprehensive," and would have truth only by change or from outside.'[24] There is no language available to represent that world of which it is both inside and outside at once; that is, the problem of representation is at once and at the same time the accountability of the perspective itself in that representation: 'In giving explanations I already have to use language full-blown (not some sort of preparatory, provisional one); this by itself shews that I can adduce only exterior facts about language.'[25] This is the infinite regress of representation: 'Yes, but then how can these explanations satisfy us? – Well, your very questions were framed in this language; they had to be expressed in this language, if there was anything to ask!'[26]

The question here is that of innovation, so it is necessary to speak of innovation ('Your questions refer to words; so I have to talk about words'[27]), however, 'the point isn't the word, but its meaning, and you think of the meaning as a thing of the same kind as the word, though also different from the word. Here the word, there the meaning. The money, and the cow that you can buy with it. (But contrast: money, and its use).'[28] Money literally buys nothing:[29] 'In a word, money – the

---

20   Wittgenstein, *Philosophical Investigations*, §69.

21   L Wittgenstein, *Remarks on the Foundations of Mathematics*, GH von Wright et al. (eds), GEM Anscombe (trans), 3rd ed, Macmillan, New York, 1956: 173e.

22   Wittgenstein, *Philosophical Investigations*, §192.

23   Wittgenstein, *Philosophical Investigations*, §115.

24   Deleuze, *Logic of Sense*, 145.

25   Wittgenstein, *Philosophical Investigations*, §120.

26   Wittgenstein, *Philosophical Investigations*, §120.

27   Wittgenstein, *Philosophical Investigations*, §120.

28   Wittgenstein, *Philosophical Investigations*, §120.

29   See further the discussion in Chapter 9.

circulation of money – *is the means for rendering the debt infinite*[30] and 'the abolition of debts or their accountable transformation initiates the duty of an interminable service to the State that subordinates all the primitive alliances to itself (the problem of debts).'[31] *Money literally buys nothing.*

Alice consents to the recognition of the Caucus-race as a game, participating readily in its performance (*the best way to explain it is to do it*), relying upon remembered concepts in order to recognize the new. Similarly, the intellectual property system articulates innovation relying upon what is already known (for example, prior art and existing copyright). Paradoxically, the intellectual property system does not identify new knowledge as such. Everything is simply recollection. Indeed, it is amusing and enticing in this regard that it is the extinct Dodo that both calls and ends the race, declaring everyone winners. The challenge is therefore the wave of innovation beyond the extinct recollection of intellectual property. How does one get to know? Which way?

> On the other hand it obviously makes use of the word 'to know' in a new way. If you wish to examine how this expression is used it is helpful to ask yourself 'what in this case is the process of getting to know like?' 'What do we call "getting to know" or, "finding out"?'[32]

So the intellectual property system is a tool for bringing meaning to the innovative process within everyday life, allowing for and providing the circumstances necessary for certain expressions. In other words the legal right links the non-linguistic phenomenon of innovation with the criteria (or evidence) for intellectual property protection (and so the criteria or evidence for innovation and creativity thus defined). The crucial issue remains one of the wider legitimacy of the system. If users (whether producers, consumers and so on) feel disenfranchised through lack of 'consent' on meaning (lack of participation in the game), then the game will fail. Such lack of consent can be felt by an enormous diversity of users. This is not merely a right-holder versus consumer question. Lack of consent and difficulties of access to the game may also be confronted at any point of entry, including production and enforcement.

A Game of Chance

> However unlikely it might seem, no one had tried out before then a general theory of chance. Babylonians are not very speculative. They revere the judgments of fate, they deliver to them their lives, their hopes, their panic, but it does not occur to them to investigate fate's labyrinthine laws nor the gyratory spheres which

---

30   Deleuze & Guattari, *Anti-Oedipus*, 197.
31   Deleuze & Guattari, *Anti-Oedipus*, 197.
32   Wittgenstein, *Blue and Brown Books*, 23.

reveal it ... If the lottery is an intensification of chance, a periodical infusion of chaos in the cosmos, would it not be right for chance to intervene in all stages of the drawing and not in one alone?[33]

The Babylonian Lottery, like Schumpeter's analysis, looks to introduce chance at every stage. The difficulty when it comes to the relationship of indebtedness within the intellectual property system is that the duty of debt seeks to neutralize chance. The debt to the producer limits the actions available (and therefore limits familiar production): 'Intellectual curiosity puts chance beyond my reach. I seek it and it escapes, as if I just missed it.'[34] That is, chance is integral to becoming, to the ethics of expression: 'If it was definitive, chance wouldn't be chance.'[35] The momentum and seduction of familiar production is the intrinsic character of chance that both liberates production from the economic organization of work, notwithstanding the possible relinquishing of that freedom to an extent in the capture of familiar production within the business models of social media: 'Part of human life escapes from work and reaches freedom This is the part of play that is controlled by reason, but, within reason's limits, determines the brief possibilities of a leap beyond those limits. Play, which is as fascinating as catastrophe, allows you to positively glimpse *the giddy seductiveness of chance.*'[36]

Indeed, this seductiveness of chance is relevant not only to revolutionizing research and development models, where the creator is perhaps less 'personal', as it were, but also to art, where the creator's identity, authority and authenticity is foregrounded more legibly. Illustrating this point is the example of architectural theory and accident. In considering architecture and the accidental 'fall' of buildings (through neglect or through natural or human disaster[37]), the architect Lebbeus Woods argues 'Goal-oriented predictability must be expanded to embrace accidents,[38] chance, randomness, unpredictability. This as we know has already happened in physics, biology, cognitive science. It has happened in art to a more limited extent ... But it has not yet happened in architecture and design.'[39] Wood is identifying precisely the opportunities of the Caucus-race that are lost when

---

33   JL Borges, 'The Lottery in Babylon', JM Fein (trans), in JL Borges, *Labyrinths: Selected Stories and Other Writings*, DA Yates & JE Irby (eds), New Directions, New York, [1962]/1964: 30, at 33–34.

34   G Bataille, *Guilty*, B Boone (trans), Lapis P, Venice CA, [1961]/1988: 80.

35   Bataille, *Guilty*, 77.

36   Bataille, *Guilty*, 72.

37   Compare Alice's 'fall' which is precipitated by a decision to follow the rabbit down the rabbit-hole.

38   The concept of 'accident' in relation to risk narratives is considered in more detail in Chapter 5.

39   L Woods, 'The Fall', in P Virilio, *Unknown Quantity*, Thames & Hudson, London, 2002: 150, at 155. See further the discussion of design and professionalizing the designer and design industries later in Chapter 4.

attempting to reproduce a certain predictable and causal 'narrative' in innovation (whether that narrative is characterized by risk mitigation, economic modelling or similar): 'When design aims only at enabling a desired stability – social, economic, psychological – the goals are determined in advance ... But when, as is often the case today, the goal is to enable unpredictability, to give people a high degree of freedom in how and why they need or use space, it is no longer possible to think of function, or purpose, or meaning as we have before.'[40] The *giddy seductiveness of chance* is, as Bataille has noted, the escape from 'work' to reach freedom. And meaning is achieved not through predetermined and purposeful solutions, but through use and chance encounters. Meaning is in the game of chance with the user who thus becomes intrinsic and critically important to the production of meaning. Value in familiar production is therefore generated through use, but not predetermined through utility.

The problem for a game of chance in intellectual property is the distinction between the limits imposed on the innovation concept and the 'use' of innovation: 'The human mind is set up to take no account of chance, except insofar as the calculations that eliminate chance allow you to forget it: that is, *not take it into account.*'[41] The task is to facilitate use without frontiers while at the same time reconciling meaning in the intellectual property system and for the users of that system (whether that is industry relying on it commercially, or users seeking products and so on). Indeed, the very notion of eschewing chance becomes questionable: 'Value not based on chance would be arguable.'[42] How might this relationship between concept and use be articulated in order to achieve legitimacy for all participants in the intellectual property system?

A Long and Sad Tail

> "Mine is a long and a sad tale!" said the Mouse, turning to Alice and sighing.
> "It *is* a long tail, certainly," said Alice, looking down with wonder at the Mouse's tail; "but why do you call it sad?"[43]

It *is* a long tail,[44] an infinite tale perhaps. But as distinct from the infinite debt, what is sought is the endless game of chance: 'Chance is an effect of gambling. This effect can never come to rest. Wagered again and again, chance is a *misunderstanding* of

---

40   Woods, 'Fall', 155.
41   Bataille, *Guilty*, 71.
42   Bataille, *Guilty*, 75.
43   *Wonderland*, 22.
44   This resonates with the long tail of statistical distributions that has gained currency in economics discussions in recent years. See, for instance, C Anderson, *The Long Tail: How Endless Choice is Creating Unlimited Demand*, Random House Business Books, London, 2006.

anguish (to the extent that anguish is a desire for rest, for satisfaction).'[45] That is, a desire for rest and satisfaction is an antidote to innovation, to change, to difference. Chance is thus integral to the becoming of innovation, the unlimited potential of innovation: 'Chance, though, isn't capable of dawdling, and its lightness of foot protects it from this "more." It wants to have its success incomplete and quickly emptied of meaning, one success is soon left behind for another ... Chance wants to be gambled, gambled again, wagered endlessly whenever the cards are dealt in a new game.'[46] Chance is the speed of the Caucus-race.

The story itself gives a 'picture' of the Mouse's tale, 'Fury and the Mouse', which is itself the waves of a long tail, parodying the relationship between recollecting (the tale) and the picture (of the tail).[47] Upon interrupting the Mouse's story, Alice apologizes and suggests where the Mouse's 'place' in the story had been:

"I beg your pardon," said Alice very humbly: "you had got to the fifth bend, I think?"

"I had *not*!" cried the Mouse angrily.

"A knot!" said Alice, always ready to make herself useful, and looking anxiously about her. "Oh, do let me help to undo it!"

"I shall do nothing of the sort," said the Mouse, getting up and walking away. "You insult me by talking such nonsense."

"I didn't mean it!" pleaded poor Alice. "But you're so easily offended, you know!"

The Mouse only growled in reply.

"Please come back and finish your story!" Alice called after it. And the others all joined in chorus, "Yes, please do!" but the Mouse only shook its head impatiently and walked a little quicker.

"What a pity it wouldn't stay!" sighed the Lory, as soon as it was quite out of sight; and an old Crab took the opportunity of saying to her daughter, "Ah, my dear! Let this be a lesson to you never to lose *your* temper!"[48]

We are returned to the potential of the rhizome and the contradiction of its 'knots of arborescence'.[49] This indeed is not a simple dualism of private property and public digital. It seems it will be a very long tale indeed, but never lose *your* temper.

---

45 Bataille, *Guilty*, 75.

46 Bataille, *Guilty*, 75–76.

47 Wittgenstein explains 'What is essential is to see that the same thing can come before our minds when we hear the word and the application still be different. Has it the *same* meaning both times? I think we shall say not.' See Wittgenstein, *Philosophical Investigations*, §140.

48 *Wonderland*, 22.

49 Deleuze & Guattari, *Thousand Plateaus*, 20.

The Turn of Chance

> And if by chance you see a chance beside me, take it!
> It's your chance, not mine.
> No more than I, can you grasp this chance.
> *You'll know nothing about it, though you take a chance on it.*
> In fact who sees it without gambling?[50]

Don't think, but look![51] Who sees it without gambling? But how to introduce chance into intellectual property and the industries and models of innovation based on intellectual property? 'What is more frightening for humankind than play?'[52]

> "Come away, my dears! It's high time you were all in bed!" On various pretexts they all moved off, and Alice was soon left alone.[53]

It's high time.

---

50   Bataille, *On Nietzsche*, 90.
51   Wittgenstein, *Philosophical Investigations*, §66.
52   Bataille, *Guilty* 77.
53   *Wonderland*, 24.

PART II
Of Objects

Chapter 4
Taste

Just a Little Taste

> By this time she had found her way into a tidy little room with a table in the window, and on it (as she had hoped) a fan and two or three pairs of tiny white kid gloves: she took up the fan and a pair of the gloves, and was just going to leave the room, when her eye fell upon a little bottle that stood near the looking-glass. There was no label this time with the words "DRINK ME," but nevertheless she uncorked it and put it to her lips. "I know *something* interesting is sure to happen," she said to herself, "whenever I eat or drink anything; so I'll just see what this bottle does. I do hope it'll make me grow large again, for really I'm quite tired of being such a tiny little thing."[1]

Alice knows something interesting is sure to happen. She is literally remembering the new. Through her memory of Wonderland, Alice is both becoming responsible for her future and innovating upon her past. In a debt economy, Alice has been granted the credit to anticipate her future, and so becomes accountable:

> But how many things this presupposes! To ordain the future in advance in this way, man must first have learned to distinguished necessary events from chance ones, to think causally, to see and anticipate distinct eventualities as if they belong to the present, to decide with certainty what is the goal and what the means to it, and in general be able to calculate and compute. Man himself must first of all have become *calculable, regular, necessary*, even in his own image of himself, if he is to be able to stand security for *his own future*, which is what one who promises does![2]

Notwithstanding this inception of calculability and responsibility, at the same time, and in an important way, Alice is leaving her future to chance, liberated by her use and her performance of the gustatory catalysts of communication in Wonderland. Throughout Alice's journey she is required to eat, to drink, to taste. The very curious charm of her journey is also the mechanism by which Alice makes sense on her journey – a journey of becoming from the body of noise and

---

1   *Wonderland*, 27.
2   Nietzsche, *On the Genealogy of Morals*, 58.

ingestion to the surface of speech,[3] from the depths of the empty well to the surface of unlimited innovation. She is progressing, literally, by 'word of mouth'.

Tasting, sampling, consuming are in this way instrumental to Alice's journey of becoming. To taste is to catalyse change. To select is to tribute. To use is to provoke innovation. This is the very intimacy and investment of self in familiar production.

A Taste for Something

> Very soon the Rabbit noticed Alice, as she went hunting about, and called out to her in an angry tone, "Why, Mary Ann, what *are* you doing out here? Run home this moment, and fetch me a pair of gloves and a fan! Quick, now!" And Alice was so much frightened that she ran off at once in the direction it pointed to, without trying to explain the mistake it had made.
>
> "He took me for his housemaid," she said to herself as she ran. "How surprised he'll be when he finds out who I am! But I'd better take him his fan and gloves – that is, if I can find them." …
>
> "How queer it seems," Alice said to herself, "to be going messages for a rabbit! I suppose Dinah'll be sending me on messages next!"[4] And she began fancying the sort of thing that would happen: "'Miss Alice! Come here directly, and get ready for your walk!' 'Coming in a minute, nurse! But I've got to watch this mouse-hole till Dinah comes back, and see that the mouse doesn't get out.' Only I don't think," Alice went on, "that they'd let Dinah stop in the house if it began ordering people about like that!"[5]

The confusion of class and role is peculiar to Alice and cause for some disquiet and then satisfaction that, no matter what, 'they' would ensure ordinary roles would be resumed. At the time of Alice's adventure, the industrial revolution had been underway for some decades and the relationship between manufacture, class mobility and taste was of foremost interest (and concern).

Indeed, the relationship between taste and intellectual property is delectable. Contemporary with Alice's tale is the role of intellectual property in the machinery of public taste. And more recently, the notion of taste and 'to taste' has become central to the articulation of intellectual property around use, consumption and sampling. Fashion is a very clear example of this construction of a chronological reality for taste in concert with the development of intellectual property in the industrial era. Arguably the intellectual property system continues to maintain a certain rigour with respect to public taste, particularly in reaffirmation of the original through the culture of the copy (copycats and counterfeits) in fashion, and

---

3   Deleuze, *Logic of Sense*, 186–87.

4   Dinah is Alice's pet cat.

5   *Wonderland*, 25–26.

the language of unoriginality and unimaginativeness that attaches to examples of innovation deemed appropriative (or imitative),[6] generating 'classes' of innovation.

However, 'sampling' in intellectual property is almost an inevitable characteristic of the nature of innovative narratives, more noticeably in intellectual property associated with identity (trade marks and brands, music, fashion, literature, film and so on). Similarly, in high technology products, 'identity' is overcoming any trajectory in technology, with markets in new (incremental) technologies being at once arguably markets in brands and their auras. 'Sampling' is considered quite particularly not in terms of an absolute infringement, but in terms of its interference with the integrity, boundaries, discrete identity of the 'original'. That is, sampling,[7] perhaps more than outright copying, is much more troubling to conventional business models of intellectual property, because sampling upsets not only the boundaries of the object but also the boundaries of genre, class and taste. The intellectual property object, despite the assertion by intellectual property rules to the contrary, is not autonomous. Sampling is more than an issue of infringement, it is an issue of cultural legitimacy.

An Ethics of Aesthetics

Aesthetics and taste arguably are motivated by the 'kinship'[8] of meanings and comparison in production: 'The kinship is that of two pictures … The kinship is just as undeniable as the difference.'[9] In other words, in the repetition of the digital there is the production of difference in the 'blurred edges',[10] but nevertheless the persistent kinship of meaning in the proliferation of original events, as distinct from an obligation to the representation of the same, the original: '"Anything – and nothing – is right." – And this is the position you are in if you look for definitions corresponding to our concepts in aesthetics and ethics.'[11]

Therefore, the 'concepts' of taste and aesthetics are instead in the performance[12] as distinct from the definition, and so cooperative with an aesthetic production is the structural character of the industry as well as the architecture and cultural

---

6   For example, transformative or appropriation art, sampling in music and elsewhere, traditional knowledge, incremental innovation.

7   Sampling is relevant not only in music, but also in other copyright industries (including books and fashion), as well as in other areas of aesthetic intellectual property (such as trade marks and parody, design).

8   The notion of kinship is central to the concept of familiar production (introduced in earlier discussions in 'Use' and explored further in Chapter 9).

9   Wittgenstein, *Philosophical Investigations*, §76.

10   Wittgenstein, *Philosophical Investigations*, §71.

11   Wittgenstein, *Philosophical Investigations*, §77.

12   Bourdieu notes, 'It means performing the typically social, and quasi-magical, operation of the encounter between an already objectified discourse and an implicit

production of aesthetic display.[13] In other words, tastes are 'the product of an encounter (a pre-established harmony) between goods and a taste ... Among these goods one must, at the risk of shocking some people, include all the objects of election, of elective affinity, such as the objects of sympathy, friendship or love.'[14] Aesthetics, as an element of ethics, are a taste of things to come.

There are seemingly two competing and, at times, contradictory streams in contemporary fashion and design industries: price and quality. Increasing efficiency of manufacture and price and the outsourcing of the 'product' pose a seemingly insurmountable challenge to market leadership that may be based on 'affective'[15] aspects of quality and aesthetic originality alone. The latter must be supported by an engine of taste and class; that is, a kind of 'affective labour',[16] or labour which is based on the relationship to the product (whether a finished good or a service), which may or may not be understood and legitimated in conjunction with the intellectual property system. While it might be suggested that the relationship between consumers and consumer products is largely built upon connotative meanings outside the immediate function or character of the product, arguably the 'distinction' that characterizes markets for digital intellectual property is largely formulated by 'affective' relationships to the product,[17] the very 'aesthetics' of judgment in the resolution of innovation within capital: 'Why call a judgment of taste *aesthetic*? Because, in order to distinguish whether a thing can be called beautiful, I do not consult the relation of the representation to the *object*, with a

---

expectation, between a language and certain dispositions that only exist in the practical state.' Bourdieu, *Sociology in Question*, 109.

13   Bourdieu and Darbel note, 'This is why, therefore, aesthetics can only be, except in certain cases, a dimension of the ethis (or, better, the ethos) of class. In order to "taste", that is "to differentiate and appreciate" the works on display and in order to understand them and give them value, the uncultivated visitor can only invoke the quality and quantity of the work put into them, with moral respect taking the place of aesthetic admiration.' P Bourdieu & A Darbel, *The Love of Art: European Art Museums and their Public*, C Beattie & N Merriman (trans), Polity P, Cambridge, [1969]/1997: 47.

14   Bourdieu, *Sociology in Question*, 108. Bourdieu notes further: 'Tastes are the product of this encounter between two histories, one existing in the objectified state, the other in the incorporated state, which are objectively attuned to one another' and describes the producer as the 'absent third party' with respect to 'the encounter between a work of art and the consumer' and who transforms taste into an object. See Bourdieu, *Sociology in Question*, 109.

15   The concept of 'affect' was introduced in more detail in the setting out of familiar production provided in 'Use'.

16   Hardt and Negri refer to affective labour as 'labor that produces or manipulates affects such as a feeling of ease, well-being, satisfaction, excitement, or passion'. See Hardt & Negri, *Multitude*, 108.

17   Bourdieu speaks of the love of art as speaking the same language as that of romantic love: '"love at first sight" is the miraculous encounter between an expectation and its realization.' Bourdieu, *Sociology in Question*, 109.

view to knowledge (the judgment of taste does not give us any knowledge) but its relation to the *subject* and to its affect (pleasure or unpleasure).'[18]

Therefore, this moves forward not only a question of aesthetics but also one of contemporary notions of 'taste' as a family of meaning, of coadaptation: 'The philosophical faculty of coadaptation, which also regulates the creation of concepts, is called *taste* ... taste appears as the triple faculty of the still-undetermined concept, of the persona still in limbo, and of the still-transparent plane.'[19] In other words, taste is the very articulation of invention and creativity, innovation and diffusion: 'it is necessary to create, invent, and lay out, while taste is like the rule of correspondence of the three instances that are different in kind.'[20] In a way, taste is an inalienable quality of 'self' that must interact with other aesthetic selves: 'Or what is it like for someone to have no idea how to fathom another's taste?'[21] This exchange, this confrontation between tastes, between producer and consumer, is an exchange of perspective that is intrinsic to the production of meaning and value through use.

A Brief History of Taste

Focusing on the United Kingdom for the moment, the relationship between the law and the construction of taste has persisted arguably throughout the development of intellectual property protection in its design industries. Indeed, taste is as relevant today as part of the 'environment' of intellectual property enforcement, particularly in areas more overtly attuned to an 'aesthetic' and a process of preference,[22] including of course fashion and design.[23] In particular, the concept of 'aesthetics' in the fashion and design industries[24] traces the very strategic relationship between judgment and aesthetics and indeed, ultimately, the relationship between judgment and ethics.

---

18   J Derrida, *The Truth in Painting*, G Bennington & I MacLeod (trans), U of Chicago P, Chicago, [1978]/1987: 44.

19   Deleuze & Guattari, *What is Philosophy?* 77.

20   Deleuze & Guattari, *What is Philosophy?* 77.

21   Wittgenstein, *Culture and Value*, 95e.

22   Bourdieu notes that tastes are 'choices made among practices ... and properties ... through which *taste*, in the sense of the principle underlying these choices, manifests itself'. Bourdieu, *Sociology in Question*, 14.

23   This relationship between intellectual property law and taste is actually very important, but perhaps less well understood, in the efforts to maintain international competitiveness in fashion and design in a market where efficiency of manufacture and price pose a seemingly insurmountable challenge to market leadership on quality and aesthetic originality alone. The latter must be supported by an engine of taste and class, as understood and legitimated within and outside the intellectual property system.

24   Traced here through a brief history of 'aesthetics' in the United Kingdom.

In the early nineteenth century, the burgeoning Britain fashion industry faced remarkably similar challenges of price and quality as introduced above, but from a different, almost reverse, perspective. British manufacture was very fast, cheap and was possible in great quantities. Where it struggled to compete internationally was on aesthetics.[25] This relationship between the law and a design aesthetic is crucial, not only in its social construction, but also in the actual intellectual articulation of these two discourses. Key features in this relationship between the law and taste included informal regulation of the industry both within intellectual property and outside, such as through the professionalization of fashion and design (the creation of training and industry standards). This historical development that aligns professionalism and taste (and protection) resonates with the labour-desert argument for property rights considered earlier.[26] In other words, to merit protection there must be evidence of organized and systematic labour in the production of commodities, and the professionalization of that labour.

In addition to these practices within the industry, growing alongside the intellectual property system was an archive of taste, with a growing nineteenth-century 'industry' of museum culture, fashion and the construction of public taste.[27] This alignment of 'taste' and aesthetics with the professional development of the industry is wholly consistent with applying an economic structure of production to the incarnation of an aesthetic taste.

Towards the early development of the fashion industry, the efforts at producing a national aesthetic, through various influences including the law, offer insight into the efforts for national fashion industries to compete today in an international and digital marketplace. This effort to compete in an increasingly difficult international market, as well as the international context for production and digital manufacturing, has been accompanied by an emergence (or re-emergence) of 'national' aesthetics and their branding as such.

In fact, intellectual property has much to say about the development of a national aesthetic, of national taste. From the perspective of the generation of meaning and value through use, fashion provides a crucial example of an 'analog' version of familiar production, as it were. That is, while digital technologies and environments have transformed the fashion industry, nevertheless the use of finished products (and 'unfinishing' them through use and recombination) provides important insight into the way in which digital tools are merely technological instruments in an otherwise revolutionary period in the cultural life of use and aesthetics. In particular, what insight might the example of the development of fashion and aesthetics provide for the way in which intellectual property laws today deal with the user (as consumer and producer), and with unanticipated and

---

25   B Sherman & L Bently, *The Making of Modern Intellectual Property Law*, Cambridge UP, Cambridge, 1999: 63–64.

26   See the earlier discussion in 'Use'.

27   This relationship between the intellectual property system, memory and the archive is also explored further in chapter 7.

emerging forms of use? That is, what insight into familiar production and digital innovation might be provided by further examination of the relationship between fashion and intellectual property?

A Sense of Taste

> There was no label this time with the words "DRINK ME," but nevertheless she uncorked it and put it to her lips. "I know *something* interesting is sure to happen," she said to herself, "whenever I eat or drink anything; so I'll just see what this bottle does. I do hope it'll make me grow large again, for really I'm quite tired of being such a tiny little thing!"[28]

In their apparent objectivity, intellectual property frameworks appear to mimic a universalist, artistic taste.[29] The rules of intellectual property, in identifying their subject matter, necessarily create an exemplary quality for each inventive, original, distinctive event that is recorded within that framework.[30] So, in other words, where subject matter is accredited with the status of intellectual property, it becomes at the same time part of the narrative of 'quality' that is somewhat implied, presenting a kind of universalist concept of taste in a disinterested and unchanging register: 'Taste makes acceptable.'[31] But this notion of taste as fixed and transcendent in value is arguably fallacious; the register itself is one in which the objects regularly become obsolescent and replaced, classifying and defining the good,[32] as it were: 'The fashion wears out more apparel than the man.'[33] The very nature of the register is such that progression is inevitable and necessary: *I know* something *interesting is sure to happen.* This ensures the appearance of originality and novelty: 'For taste must be an original faculty; whereas one who imitates a model, while showing skill commensurate with his success, only displays taste as

---

28   *Wonderland*, 26–27.

29   See further the discussion of 'taste' in Gibson, *Creating Selves*, 67 and 73–74.

30   This notion of classification and hierarchization as instrumental to the production of 'taste' is conceptually coincident with the processes of authentication, validation and enforcement in intellectual property frameworks. See further the discussion of taste as part of a process of classification in Bourdieu, *Sociology in Question*, 108–16.

31   Wittgenstein, *Culture and Value*, 68e.

32   Bourdieu explains: 'In order for there to be tastes, there have to be goods that are classified, as being in "good" or "bad" taste, "distinguished" or "vulgar" – classified and thereby classifying, hierarchized and hierarchizing – and people endowed with principles of classification, tastes, that enable them to identify, among those goods, those that suit them, that are "to their taste".' Bourdieu, *Sociology in Question*, 108.

33   Shakespeare, *Much Ado About Nothing*, Act III, Scene iii.

himself a critic of this model.'[34] The narrative thus is presented as one premised upon originality, not imitation, although this in itself will be challenged.

In other words, in this understanding, taste is presented as an original faculty and a faculty of originality, but 'Taste rectifies, it doesn't give birth.'[35] Indeed, 'The *most refined* taste has *nothing* to do with creative power'[36], it merely assimilates the register as presented. This introduces a somewhat complicated relationship between taste and the new, or taste and innovation. Is the education of taste intended to sustain the obligation to the same, to the original, or is it in concert in some way with the accreditation of an always already obsolescent and obsolescing register of production? In other words, what is the relationship between fashion and genre (and the innovative tribute or reference to previous eras, designers and so on) and imitation (in the form of slavish or unimaginative copy)?

This understanding of quality and the accreditation of production (through institutionalized judgment, including notions of aesthetics or taste) and innovation in fashion, while at the same time honouring styles and themes that ensure relevance and recognition, provides some insight into the relationship between innovation and tradition (or indeed, the distinction between the momentous event and continuous change, that is, familiar production). Recalling earlier discussions of the translation of traditional knowledge within conventional models of innovation, tradition is but a mechanism among others in order to narrate innovation. It is not an end in itself.[37] The supposed originality credited by intellectual property is premised upon the idea of the new.

In particular, consider the concept of technology itself. The contemporary use of the word in itself ushers in ideas of innovation, of change, of development and suggests civilized (that is, both propertied and deserving of property) accomplishment: 'Remember also that technology always means *new* technology.'[38] In other words, technological progress is itself confounded by the fallacy of representation. Innovation is not a linear, directional narrative capable of representing innovation as finite progressions and starts. And indeed this causality is challenged radically in a digital environment, as well as in traditional models of incremental innovation. This causality is a matter of pragmatics not, quite literally, a matter of law. Technology is always *new*.

So what of fashion as technology, fashion as new? Indeed, what of taste as new? In the coordination of invention, innovation and diffusion, it may be that

---

34   I Kant, *The Critique of Judgement*, JC Meredith (trans), Oxford, Clarendon P, [1790]/1952, Book I: Analytic of the Beautiful.

35   Wittgenstein, *Culture and Value*, 68e.

36   Wittgenstein, *Culture and Value*, 68e.

37   As noted in the earlier discussion in 'Use', with respect to traditional knowledge and familiar production, this misapprehension causes notable problems for the recognition of traditional knowledge within the conceptual framework that is over-governed by intellectual property.

38   J-F Lyotard, *Peregrinations: Law, Form, Event*, Columbia UP, New York, 1988: 28.

the construction and manipulation of taste itself drives the obsolescence of products in an affective market. That is, the 'new' in fashion may precede the product if there is no 'use' for the vehicle in which that new is diffused to the market: 'But there will also be cases where goods do not find the "consumers" who would find them to their taste.'[39] Thus, while it might be said that fashion is apparently always new, creative, innovative, it must also be understood that the fashion system also maintains the archive and interrupts the previous sufficiently to introduce new desires, without entirely unsettling the narrative altogether. That is, fashion maintains its relevance through an intricate family of meanings through tribute, signature and quotation. The cultural technology of fashion is also part of this quest for the new, and yet at the same time, unlike the more literal and linear narrative implied by intellectual property, it is also a strategic and necessary imitation of and tribute to its precursors.

Therefore the case of fashion provides a very specific insight into the way in which innovation is reconciled with a 'product' and market relevance within a much wider range of technologies (and, indeed, intellectual properties): 'This raises the question of whether the goods that precede tastes (apart from the producers' taste, of course) help to make tastes; the question of the symbolic efficacy of the supply of goods or, more precisely, of the effect of the embodiment of a particular taste, that of the artist, in the form of goods.'[40] Indeed, is the concept of 'taste' in fact central not only to the conventional architecture of intellectual property and industry, but also to its reform?

A Product of Taste

As considered in earlier discussions, the quest for newness is so central to the intellectual property narrative that it belies its preoccupation with history and change. In the technology of fashion, this disingenuousness, if you like, is not as readily maintained. Intellectual property, like histories, is edited and selected, heterological, yet it suggests the complete picture. It is presented as thorough, comprehensive and memorable. Similarly, the story of intellectual property is

---

39   Bourdieu, *Sociology in Question*, 108.

40   Bourdieu, *Sociology in Question*, 108. Bourdieu explains further: 'One can imagine a field of production which takes off and "grows" its consumers. That has been true of the field of cultural production, or some sectors of it at least, since the nineteenth century. But it also happened, quite recently, in the religious field. Supply preceded demand; the consumers were not asking for it … That is a case where the logic of the field is operating "in neutral", confirming the central idea that I am putting forward, namely that change does not result from adjustment of the product to the demand. Without forgetting the cases of mismatch, we can say that, in general, the two spaces, the space of production of goods and the space of production of tastes, change at broadly the same rate.' See Bourdieu, *Sociology in Question*, 113.

the very mechanism by which is communicated a sense of completeness and universality with respect to the way innovation, invention, originality and value are judged, even though it is accepted that much is unimaginable and inconceivable within that history, including traditional and indigenous knowledge and, indeed, the cultural technologies of fashion. In this way, indirectly, intellectual property is implicated in the constitution of universal and original taste – creating the appetite, the taste for the new: *I know something interesting is sure to happen.*

But as well as the new, taste is of great interest and significance for this discussion in that taste immediately implicates the consumer, the user. Indeed, taste is also part of the circumstances for use in that it is part of the 'desire' for consumption, the sense of a 'lack' with respect to new consumption, and the way in which we use and interact with new products and new tools. In this way, the coincidence in the development of the concepts of consumer and taste provides some insight.[41]

In its early incarnation, the word taste originally referred to the sense of touch or feeling, a kind of physical consumption, making taste in this respect, literally, the common sense. Common sense is consistent, predictable, plain – it is disinterested, objective and universal – indeed, it is dry: 'Taste, like judgement in general ... gives stability to the ideas, and qualifies them at once for permanent and universal approval, for being followed by others, and for a continually progressive culture.'[42] Therefore, taste is presented in this way as existing in a kind of stable relationship to aesthetics. Taste in this sense, that is touch, fixes sensation, it gives sensation a form: *you can't buy good taste.* However, arguably this notion of stability and endurance in taste is just part of the mythology. Good taste can be bought, good taste can be learnt – that is, one can acquire cultural capital. So taste is learnt through behaviour and experience, that is, through sensation, through relation, through association, through use. The key question then, is the type and quality of access each individual has to the resources in order to complete that transaction.

A Taste of Things to Come

> Bohan, the successor of Dior, talks about his dresses in the language of good taste, discretion, moderation, sobriety, implicitly condemning all the eye-catching provocations of those who are to his 'left' in the field; he speaks of his left as the *Figaro* journalist speaks of *Libération*. As for the avant-garde couturiers, they speak of fashion in the language of politics (our survey was done just after the events of 1968), saying that fashion has to be 'brought on to the streets', and that '*Haute couture* should be within reach of everyone.' It can be seen that there are equivalences between these autonomous fields such

41   R Williams, *Keywords: A Vocabulary of Culture and Society*, Fontana, London, [1976]/1988: 314–15.

42   Kant, *Critique of Judgement*, Book I: Analytic of the Beautiful.

that language can pass from one to the other with apparently identical but really different meanings. This raises the question whether, when people talk about politics in certain relatively autonomous spaces, they are not doing the same as Ungaro talking about Dior.[43]

Intellectual property similarly closes a moment in intellectual progress – the expression binds the copyright text, the patent defines the invention, the trade mark designates the identity, the design perpetrates the individuality. It makes the intangible idea seemingly tangible, touchable, tasty.

However, changes in behaviour precede changes in perception – in other words, changes in use precede changes in products. People cannot simply be instructed to think differently, or presented with a new behaviour in a new product. Instead, through use a different approach is demonstrated, experienced: 'To repeat: don't think, but look!'[44] This presents insight not only for new business models and practices, but also for reform of the law itself. Law will not change behaviour on its own: 'We learn nothing from those who say: "Do as I do". Our only teachers are those who tell us to "do with me", and are able to emit signs to be developed in heterogeneity rather than propose gestures for us to reproduce.'[45] The familiarity and ethics of *do with me* outstrips the moral prescription of the dictate, *do as I do*.

> "It was much pleasanter at home," thought poor Alice, "when one wasn't always growing larger and smaller, and being ordered about by mice and rabbits. I almost wish I hadn't gone down that rabbit-hole – and yet – and yet – it's rather curious, you know, this sort of life! I do wonder what *can* have happened to me! When I used to read fairy-tales, I fancied that kind of thing never happened, and now here I am in the middle of one! There ought to be a book written about me, that there ought! And when I grow up, I'll write one – but I'm grown up now," she added in a sorrowful tone; "at least there's no room to grown up any more *here*."[46]

Despite the pleasant stability of Alice's conventional life, the instability of the rabbit-hole is *rather curious*. This is the rather curious nature of the narrative of fashion, which, problematically for any universalist notion of taste (let alone good taste), resists necessarily any kind of fixation. Indeed, fashion presents the opportunity for mobility beyond a disinterested taste ideal – fashion is a clear

---

43   Bourdieu, *Sociology in Question*, 113.

44   Wittgenstein, *Philosophical Investigations*, §66.

45   Deleuze, *Difference and Repetition*, 23. See further Wittgenstein on ethics as unteachable (unpresentable): 'Teaching this could not be an ethical training. And if you wanted to train anyone ethically & yet teach him like this, you would have to teach the doctrine after the ethical training, and represent it as a sort of incomprehensible mystery.' Wittgenstein, *Culture and Value*, 93e.

46   *Wonderland*, 28.

example of British mobility, as it were: *what can have happened to me?* While there are discrete objects of fashion, the original quality is much less discernible. Familiar production thus defeats the stability of Alice's conventional life and received manners, at the same time enticing her with the chance and opportunity of the curious life of the rabbit-hole, the curious *form of life*.

So, in a way, fashion approximates the kinds of difficulties imposed by the digital environment for commercialization and use, towards a 'universalist' ethic in intellectual property frameworks. As in the digital environment, fashion operates in a world where there is nothing but text. The actual objects of fashion are in many ways at odds with the traditional narrative of intellectual property. They are compatible rather with the mobility of acquired cultural capital, including an education in fashion,[47] thus performing in the round the very exchange of meaning through use that is integral to familiar production. Taste is thus not fixed, pre-existing, but rather it is familiar and in exchange: 'Someone who teaches philosophy nowadays gives his pupil foods, not because they are to his taste, but in order to change his taste.'[48] *Do with me.*

Accounting for Tastes

So how has the fashion industry been co-opted in the development of taste and what role for the law in maintaining that appetite? How does design law in particular cooperate with the measure of distinction? Perhaps an important part of this process has been the creation of a story for fashion itself, in the same way that intellectual property creates a story for innovation: in other words, turning the construction of clothing from a trade into a creative process. An important aspect of this process has been the professionalization of the story of the industry itself. This story was told through three major areas of development in the nineteenth century: first, the establishment of design schools (an important part of the professionalization of the industry and also in developing 'national' aesthetics through particular training backgrounds); secondly, the opening of public museums dedicated to designs and fashion as works of art (the Victoria and Albert), that is, validating the output of the fashion industry; and finally developments in the legal regime itself.[49]

---

47   Bourdieu explains, 'The field of cultural production is the area par excellence of clashes between the dominant fractions of the dominant class, who fight there sometimes in person but more often through producers oriented towards defending their "ideas" and satisfying their "tastes," and the dominated fractions who are totally involved in this struggle.' Bourdieu, Field of Cultural Production, 102.

48   Wittgenstein, Culture and Value, 25e.

49   Sherman & Bently, *The Making of Modern Intellectual Property Law*, 64. This notion of the museum as validating output as artistic shares much with the concept of the curator as author; on this see D von Hantelman, 'Affluence and Choice: The Social

Textile production (particularly cotton) was one of the very significant driving forces of industrialization for Britain, one of the first nations to industrialize.[50] Cotton had become quite fashionable as a material in the eighteenth century and the British cotton industry grew exponentially in the nineteenth century.[51] Nevertheless, despite nineteenth-century Britain enjoying very fast and cheap manufacturing capacity with the ability to produce goods in far greater quantities than many international competitors, these goods did not compete well internationally (such as with France), not because of price but because of aesthetics. Indeed, it was widely considered that the goods themselves were inferior aesthetically. There was thus considered to be an urgent need to improve the state of British design and a national aesthetic, and that the law was the way to do it.[52] This is very striking in the context of current debates in intellectual property and the relationship between cultural aesthetics and the law, with the notion of skill itself as a commodity.[53] Today, competitiveness of the British fashion industry is threatened by the inability to compete on price and quantity.[54] Arguably, further efficiency of manufacturing cannot resolve the current issues for industry. Therefore, it appears that the relationship between a refined aesthetic and competitiveness becomes immediately relevant again today, and part of this environment includes a renewed interest in the role of intellectual property in the fashion industry.

Returning to the developments of the nineteenth century, the law then became directly deployed in the quest to improve the design aesthetic of British industry. From the outset, this included an emphasis on the figure of the designer and property (and the implications of preservation, introduced in earlier discussion),[55]

Significance of the Curatorial', in B von Bismarck et al. (eds), *Cultures of the Curatorial*, Sternberg P, Berlin, 2012: 41–51.

50   P Deane & WA Cole, *British Economic Growth: 1688–1959*, 2nd ed, Cambridge UP, Cambridge, [1962]/1969: 182–213.

51   Deane & Cole, *British Economic Growth*, 184–92.

52   On these points see the discussion in Sherman & Bently, *Making of Modern Intellectual Property Law*, 73–76.

53   For example, the British traditional knowledge and skill in the bespoke tailoring of suits in Savile Row may be understood itself as commodity.

54   See further the discussion in N Owens & A Cannon Jones, *A Comparative Study of the British and Italian Textile and Clothing Industries*, DTI Economics Paper No 2, DTI, April 2003.

55   The first statute dealing with designs protection in the United Kingdom was the Calico Printers' Act of 1787. This was designed to encourage work in designing and printing linens, cottons, calicos and muslins. The idea was that the encouragement of the industry would come about by giving designers, printers and proprietors properties in the designs for a limited period. In other words, the statute contributed early on towards a transformation of the cultural and social conceptualization of the designer itself. Notably, for these early attempts to protect and thus motivate the fashion industry, the Act was modelled on early copyright protection – that is, on principles of authorship and originality. Although it started

immediately introducing notions of quality and aesthetics in conjunction with ownership and selfhood. Further developments included a hierarchy or classification of design itself with respect to suggestions of use (and thus resources), separating design into ornamental and utility,[56] and an implied wasteful uselessness, as it were, of conspicuous ornamental consumption.

A design register was established in 1839,[57] an early example of intellectual property registration that is a principle deployed throughout the intellectual property framework today. And just like today, the register provided evidence of originality as well as a source of information for further innovation and creativity (as well as to ensure that that innovation was original and new). The register contributed to the emerging status of the designer as creator, as author, and to the consideration of design as an act of inspiration and creativity, as well as ensuring the importance of its protection (through intellectual property) as an act of record and archive. The register also served to stabilize and standardize the language and depiction of design – including standards of both written and graphical representation of design – at the same time allowing for certainty in the replication, categorization, measurement and information associated with design. The register thus both educated the emerging professional body of designers in a British design aesthetic and provided historical record and story for the creative process. The introduction of a register in this context was thus part of the very important public remit of the developing intellectual property system as well as a significant record and collective memory of a national aesthetic. Intellectual property thus 'archived' public taste, at the same time operating in a way complicit with a growing bourgeoisie, whose 'indifference tacitly acknowledges the monopoly'.[58]

To reiterate, the changes seen in these laws also usher in the idea of the individual, the designer, as the focus for the organization of the law. This is perhaps especially clear when considering the modelling of these laws on author-based approaches to protection (and the promise of the debt of the author). The

---

out as a temporary measure, the Calico Printers Act was renewed twice and was eventually given permanent effect. See the further discussion in Sherman & Bently, *Making of Modern Intellectual Property Law*.

56   In 1839, the 1787 Calico Printers' Act was repealed and replaced by two new Acts: the Copyright of Designs Act and the Designs Registration Act. These two Acts were soon repealed and replaced by the 1842 Ornamental Designs Act and the 1843 Utility (Non-Ornamental) Designs Act. These Acts extended the subject matter protected under the 1839 Designs Registration Act to include calico (9 months protection). The Acts also separated design into two categories: ornamental design and utility design.

57   Under the Design Registration Act 1839 (2 & 3 Vict c.17). This is also an interesting quality and value built into the system that is perhaps compromised within the modern system itself, in that very few designs are registered.

58   P Bourdieu, *Distinction: A Social Critique of the Judgement of Taste*, R Nice (trans), Routledge, London, [1979]/1986: 62.

basis of protection thus lay in the labour and expense of producing a design.[59] In this way, design is presented as warranting protection by virtue of its status as unique creation, rather than as a reinterpretation of trends. At the same time, what is emerging is the early fashioning of the designer as unique brand, and the contribution of both the museum and the law to the notion of artistic style, and to the signature of the individual design aesthetic.

The designer, with institutional endorsement (through the museum and similar) is established as the 'owner', the creditor of not only the particular contribution, but also of taste. And visitors to the museum are able to 'borrow' from that resource of taste, running up a bill, as it were. The rabbit sends in a little bill, but 'I wouldn't be in Bill's place for a good deal.'[60] Individual subjectivities are thus appropriated and rendered accountable to good taste (that is, granted a 'memory' of taste through the archive 'in those cases where promises are made'[61]), to the 'credit' of design professionals. This proprietary logic of personhood is re-inscribed by the language liabilities of 'beneficiary' and 'consumer' that sustain the user as indebted, obliged, a borrower.[62]

These historical legislative developments were also undertaken in the context of a general acceptance that the appropriate way in which to achieve protection for designs would be to model design legislation on that accorded to literary property. In other words, the value of the fashion industry was elevated as instrumental in the development of a national aesthetic, in accordance with literary and artistic taste. It was a cultural property and designers were creators, not imitators. And fashion, as distinct from high art, was emerging as a seemingly democratized, participatory, individual portal for the general public into that aesthetic.

Artful Life

The legal developments worked in concert with certain cultural developments that followed, linked to the validation of fashion as art. In 1852, after the success of the Great Exhibition in 1851, funds were directed towards setting up a museum of manufacture, which was established in London at Somerset House, soon thereafter moving to the location that would become home to the future Victoria and Albert Museum.[63] The rationale was that a museum would contribute to the

---

59   Recall the labour-desert theory of property discussed in 'Use'.

60   *Wonderland*, 31.

61   Nietzsche, *On the Genealogy of Morals*, 58.

62   See further the discussion in Lazzarato, *Making of the Indebted Man*, 38, where the author examines the subjectivity of the 'indebted man' also in language such as 'worker', 'entrepreneur' and 'unemployed'.

63   B Robertson, 'The South Kensington Museum in Context: An Alternative History', 2(1) *Museum and Society*, March 2004: 1–14. In 1899, Queen Victoria laid the foundation stone for the new grand façade and main entrance of what became renamed as

development of the public taste (and indeed, a national aesthetic and the associated reputational qualities) by making art accessible and available, thus enticing as well as educating the general public in a sense of the aesthetic.[64] Fashion was thus understood and presented as a visual art in everyday life, and the fashion museum became a critical tool in preparing and exposing the general and working public to an education in aesthetics. In other words, the museum took fashion as an art form and thus delivered artistic flavour for consumption by the general public: 'The most notable curatorial interventions have aimed to establish common ground between the registers of fashion and art, possibly reflecting an attempt by the institutions concerned to reach valuable new constituencies of visitors, supposedly put off by the elitist connotations of painting and sculpture more usually associated with the publicly funded museums sector in Britain.'[65]

The museum also recorded and displayed fashion as an accessible, legitimate and participatory history of taste. This was genuine cultural participation, meaning through use, familiar production through participation – the clothes displayed often survived precisely through use, precisely because they had been worn. And the museum itself was also a kind of 'register' of this use, as considered earlier, acting to inform as well as inspire British designers and manufacturers, thus contributing to the development of a national aesthetic within British design.

At the same time, fashion was becoming a means by which individuals could mobilize between classes, where taste and class become inextricably intertwined.[66] Alongside the formal presentation of fashion within the somewhat regulated setting of the museum, public displays of fashion identity became much more common in the nineteenth century, with the emergence of the 'promenade' behaviour and the nineteenth-century dandy.[67] This parading in public became very important in the nineteenth century and part of the early development of the notion of a fashion aesthetic (as distinct from a more uniform approach to dress) and an affective relationship to consumption: 'Empathy with the commodity is fundamentally empathy with exchange value itself. The flâneur is the virtuoso of this empathy.

today's Victoria and Albert Museum. See further A Burton, *Vision & Accident: The Story of the Victoria and Albert Museum*, V&A Publications, London, 1999.

64   Burton, *Vision & Accident*, 108. See further, on the relationship between the museum and the cultivation of taste, Bourdieu & Darbel, *Love of Art*, in particular 110 (on the naturalization of taste in the process of 'common culture').

65   C Breward, *Fashion*, Oxford History of Art, Oxford UP, Oxford, 2003: 12.

66   B English, *A Cultural History of Fashion in the 20th Century*, Berg, Oxford, 2007: 18.

67   M Denny, 'Framing the Victorians: Photography, Fashion, and Identity', in J Potvin (ed), *The Places and Spaces of Fashion 1800–2007*, Routledge, New York, 2009: 34. See further Breward, *Fashion*, 161–67.

He takes the concept of marketability itself for a stroll. Just as his final ambit is the department store, his last incarnation is the sandwich-man.'[68]

The museum thus tells a story – a chronological, historical aesthetic narrative – but it is increasingly a narrative in which the participant, the user, the spectacle itself, may interrupt: the birth of the consumer.[69] The museum gives action to creativity – an identifiable, recognizable, tangible narrative of productivity. This both gives form to creativity as well as stabilizes the story of fashion in a way compatible with the kind of conceptual framework that is also relevant for intellectual property. In other words, the museum is a kind of mixed space of familiar production, both smooth and striated, unfettered innovation and structured organization.

Fan Laws

"I wonder what they *will* do next! If they had any sense, they'd take the roof off."[70]

Indeed, take the roof off, and ventilate the building. Take aesthetics for a stroll.

There is here an interesting resonance with the notion of the fan in today's debates on access to intellectual property material in the cultural industries. The fan is at once a commender and a follower, a leader and a propellant. The fan 'ventilates' the product. Arguably the museum provides reinforcement of fan-based cultures, through providing a link to the artist. Fashion becomes the product of the designer as distinct from slavish imitation. The quality of scarcity, so critical to intellectual property markets, is created not necessarily in the product itself, but in the designer – the brand. In other words, the museum identifies and narrates originality in the creator – the proper name of fashion. The value and artistry of fashion is therefore derived from the approbation of followers as distinct from ring-fenced value in the objects to be merchandised. This is important when we think of the translation of this concept within the modern department store.

The modern department store similarly displays merchandise as a museum, recommending and creating a continuing appetite (that is, taste) for fashion.[71] The museum, as it were, provides a tested business model for department stores and the 'curation' of aesthetic consumption: 'department stores and museums already

---

68   W Benjamin, *The Arcades Project*, H Eiland & K McLaughlin (trans), Belknap-Harvard UP, Cambridge MA, [1988]/2002: 448.

69   Roland Barthes, on the death of the author, explains: 'we know that to give writing its future, it is necessary to overthrow the myth: the birth of the reader must be at the cost of the death of the Author.' Barthes, 'The Death of the Author', 148.

70   *Wonderland*, 33.

71   See the discussion in M Bouquet, 'Thinking and Doing Otherwise: Anthropological Theory in Exhibitionary Practice', in BM Carbonell (ed), *Museum Studies: An Anthology of Contexts*, Blackwell, Malden MA, 2012.

shared many concerns in making the ordered display of valued objects visible to spectators who, as they became part of the spectacle, were also cast as a new self-regulating citizenry.'[72] Where the museum requires superior cultural capital (knowledge) to participate, the department store requires the material resources necessary for participation.[73] Nevertheless, that does not preclude the quiet contemplation of window-shopping; indeed, browsing is an integral part of this model.[74] This model of display, of 'ventilation', this departure from the privacy of elite fashion houses to the spectacle of the department store,[75] is actually a significant manifestation of the relationship between taste and consumption (or more accurately, consumerism) and the affect and engagement of the consumer: 'For the efforts of style-leaders, advertisers, editors, and directors over the past 200 years have all in some way been oriented towards an epiphanic moment of engagement between customer and fashionable product which inevitably happens for the first time within the confines of the retail store, or in related scenarios of spending and consuming.'[76]

Thus, in many respects, this theatricality[77] of shopping as experience is the crucial articulation of the 'use' of consumption and the relationship between the art of production and the artistry of consumption. This has been accompanied by a contemporary consolidation of the relationship between art and fashion in the design and form of the modern store itself as an object of curatorial splendour.[78]

---

72   Bouquet, 'Thinking and Doing Otherwise', 181.

73   Bourdieu, *Distinction*, 273.

74   Christopher Breward notes, 'many of the characteristics that are recognized as indicators of modern systems of selling were well established by the mid-eighteenth century. These included an emphasis on eye-catching window displays, the atmospheric styling of retail interiors with cascading draperies and exotic props, the construction of impressive temporary vistas, seasonal promotions, the arrangement of goods into discrete departments, an explicit encouragement of browsing.' See Breward, *Fashion*, 143.

75   For example, writing on the early twentieth-century design of Paul Poiret, Nancy Troy explains: 'Along the way he developed a spectacular marketing strategy that proved to be enormously effective, although it differed markedly from his customary practice, which was characterised by the privacy and elitism of his couture house in Paris. The high visibility of Poiret's tour was a result of the particular conditions governing the merchandising of French fashions in America, which took place in the public arena dominated by leading department stores, rather than the carefully controlled environment of Poiret's *hotel*, the site of his business headquarters as well as his private home.' See NJ Troy, 'Poiret's Modernism and the Logic Of Fashion', in G Riello & P McNeil (eds), *The Fashion History Reader: Global Perspectives*, Routledge, London, 2010: 455, at 461.

76   Breward, *Fashion*, 143.

77   Christopher Breward describes a 'theatricalization of fashion merchanidizing' extending to the architecture and splendour of consumption and fashion life. See Breward, *Fashion*, 143–46.

78   Gilles Lipovetsky and Veronica Manlow maintain, 'This new relationship between art, fashion and luxury is not merely a result of fashion being recognized as art

Indeed, many scholars of fashion argue that the modern department store has eroded the dichotomy of high art and commercialism, 'the gap between art and business'.[79] This theatricality is not dissimilar from other 'cultural' exhibits, including science and innovation, which attempt to close the gap between high science conservation efforts and commercialism.[80]

As for the promenade, similarly, the modern fashion show is itself a presentation point, a kind of 'museum' presentation, as well as the parading of a particular fashion aesthetic as a significant development in the business model,[81] with 'exhibition' and sharing being part of the protection of the product and consolidation of market share and anticipation prior to commercialization, as distinct from secrecy until market. Alongside this, fashion shows have developed predominantly into a moving display in the runway form, the promenade, as distinct from a static presentation of information. The Arts Festival similarly presents objects; but not as goods, rather almost as props in a larger performance.[82]

---

and showcased in museums and art galleries. Nor is it because artists now sign collections for commercial brands (Takashi Murakami, Stephen Sprouse, Richard Prince), assist in designing fashion shows (Edouard Levé, Haim Steinbach) or make fashion and branding the subject of their work (Olaf Nicolai, Elmgreen & Dragset, Gilles Barbier, Sylvie Fleury and Claude Lévêque). It is the very design of these stores that is becoming an explicitly creative pursuit, in line with today's slogan: "no concept, no business". The stores are now caught up in an endless race to outdo each other in architectural size, innovation and renovation.' See G Lipovetsky & V Manlow, 'The "Artialization" of Luxury Stores', in J Brand et al. (eds), *Fashion and Imagination: About Clothes and Art*, ArtEZ Press, Arnhem, 2009: 154, at 155.

79   English, *Cultural History of Fashion in the 20th Century*, 14–15.

80   For example, see the discussion in G Thomas & T Caulton, 'Communication Strategies in Interactive Space', in SM Pearce (ed), *Exploring Science in Museums*, Athlone P, London, 1996: 107–22. A particularly 'notorious' example is that of the so-called 'Crocodile Hunter', Steve Irwin, where the teachings of conservation failed to an extent in that his 'celebrity' overrode imperatives of conservation when his death from a sting-ray led to a slaughter of the species by 'fans' on Australian beaches: 'Stingrays slaughtered in QLD', *ABC AM*, broadcast 13 September 2006.

81   For instance, the press accompanying the Design Museum exhibit 'Hello, My Name is Paul Smith' includes a reference to 'a catwalk show as part of the process of design rather than the endgame'. I. Fox, 'Paul Smith: a designer in his own fashion', *The Guardian*, 10 June 2013.

82   Further, tastes are not in a sense confined to goods, but rather, to the process of distinction: 'In fact there can be taste without goods (taste in the sense of a principle of classification, a principle of division, a capacity for distinction), and goods without a taste.' Bourdieu, *Sociology in Question*, 108.

No Accounting for Taste

> "A barrowful of *what*?" thought Alice. But she had not long to doubt, for the next moment a shower of little pebbles came rattling in at the window, and some of them hit her in the face. "I'll put a stop to this," she said to herself, and shouted out, "You'd better not do that again!" which produced another dead silence.
>
> Alice noticed with some surprise that the pebbles were all turning into little cakes as they lay on the floor, and a bright idea came into her head. "If I eat one of these cakes," she thought, "it's sure to make *some* change in my size; and, as it can't possibly make me larger, it must make me smaller, I suppose."[83]

So let them eat cake.

What of the current emphasis on articulating the relationship between the fashion industries in the United Kingdom and accessing and utilizing the intellectual property system? This recalls the way in which the law becomes an instrument in the regulation of creativity, in the regulation of taste. Intellectual property laws have become firmly institutionalized as the legitimate way in which industrialized societies recognize, validate, and celebrate innovation, originality and, above all (at least in terms of the current rhetorical flourishes), creativity. In other words, the system itself has become mimetic of creativity in so far as it has adopted 'creativity' as its purpose, its responsibility, and thus its prestige. The emphasis upon 'creativity' affirms, somewhat problematically, a moral narrative for what is increasingly becoming a fundamentally economic system.

Interestingly, the fashion industries appear to be experiencing a renewed self-consciousness regarding the distinction between art and business. Arguably, much of this comes from the inability to compete, this time on price, for which again a comprehensive and aspirational aesthetic is recognized as the appropriate remedy. For this to be achieved, again the law emerges, as in the nineteenth century. The renewed self-consciousness about art and originality in fashion has led to a renewed attention to intellectual property in fashion. However, this attention is not necessarily about stopping copying in fashion. Indeed, imitation is a critical part of the business model of fashion. Without imitation, the original cannot be successful; indeed, this is why intellectual property is important in fashion, but rarely for the purposes of litigation. The imitator, the tribute, the 'sample' is vital to the success of the original, reinforcing the original, marking out its success at identifying the trend. Enforcement is relevant only where it gets beyond the tribute. The role of intellectual property is therefore about reinforcing the tiered structure of fashion; that is, it is about inscribing certain examples as originators and others as generic.[84] It is thus about validating the creativity and originality of fashion: 'Fashion is affirmed by the desire to be the event', however 'As soon

---

83   *Wonderland*, 34.

84   Recall the earlier distinction between ornamental and utility designs, for example.

as it happens, it ceases to be an event, it becomes a piece of information, which circulates and which loses its destabilizing force.'[85]

In a way, intellectual property returns cultural output to a Kantian model of taste, but one for which the motor, as it were, is the process of 'distinction' about which Bourdieu speaks. In the same way that Kant explained taste as a superior capacity for aesthetic experience that detaches art from its earthly bonds and mere gratification, the crisis of legitimacy in the current intellectual property framework (particularly in discussions of copyright) in a way severs cultural products from the user, or from relevant use. The intellectual property system renders cultural resources scarce, rare and elite. It introduces a distinction, limits upon access, creating classes of taste, not only in terms of classes of products but also in terms of classes of access to cultural products. Only those materially able can use (and waste) the products in the senses anticipated by the intellectual property framework. And so, in this way, it appears to write the user out of the narrative of intellectual property. However, this would be a mere disguise. Taste is indeed pretentious – the scarce objects of intellectual property are identified by the system and imposed upon the user.[86] Taste is not only the architecture of convention, but also the appetite for reform.

In other words, at the very same time, this framework of distinction that is imposed by the intellectual property system also provides for mobility, not only with respect to the user (and class mobility) but also with respect to reform of the system itself: 'the idea of taste cannot now be separated from the idea of the consumer.'[87] The intellectual property system arguably contributes to the segregation of society in order to create different audiences and maintain taste – those with legitimate versions, the originals, as distinct from those with the followers, the imitations. In other words, taste in fashion, like taste in art, appears to be persistently class-based, putting the intellectual property system somewhat at odds with everyday consumption. In this way, does fashion actually interfere in choice to the extent that participation in an aesthetic universe is not always unhindered? Paradoxically, does this present an insight into new business models through the way in which a 'product' is revealed and its value is manifest through affect? That is, 'Popularization devalues.'[88]

The relationship between taste and scarcity is crucial to a reimagining of affective value with respect to cultural and intellectual products in the digital sphere. In particular, the digital ignites the relationship between access and aesthetics. Where once, the 'aesthetics' (of scarcity) could be more strictly

---

85   J-F Lyotard, *Political Writings*, B Readings & K Paul (trans), U of Minnesota P, Minneapolis, 1993: 24.

86   Bourdieu, *Sociology in Question*, 135.

87   Williams, *Keywords*, 314–15.

88   Bourdieu, *Sociology in Question*, 114.

maintained through the level of supply,[89] the hierarchy of tastes in the digital is radically unsettled. Although writing long before the era of digital downloads, Bourdieu offers a notable lesson in the generation of value through tribute and expertise, that is, the affective products of experience and the affective labour of the expert in the context of music:[90] 'The rarity of the product and the rarity of the consumer decline in parallel ... The cult of "vintage" records and live recordings can be understood in the same terms. In all these cases, it is a question of bringing back scarcity: nothing is more commonplace than the Strauss waltzes, but what charm they have when conducted by Fürtwängler; and as for Mengelberg's Tchaikovsky!'[91] In the contemporary, digital context, while Mengelberg's Tchaikovsky is more immediately disseminated, deciphering ways in which to recognize and chance upon the rarity and scarcity of the experience and expertise, whether through the social and intellectual life of the store or through the affective experience of live music, arguably will be crucial to maintaining value in the relationship between consumer and product.

The art in fashion, and thus in familiar production, may be perhaps artfully revealed through law. Therefore, this inquiry is not necessarily about abandoning or dismantling the framework, but rather, devising new strategies in the game. The translation of familiar production within legal frameworks is not necessarily about assimilating that production[92] within existing production models (and thus, displacing producers as users within those models), but rather it is a transformation that must account for changes not only in models of production but also in models of distribution and in cultures of use. The question is the way in which to reconceptualize business perspectives in order to account for what is indeed a radical, 'digital' reorientation in the relationship to production, value and use. What the negotiation of that question might involve is a central concern of this inquiry.

---

89   Writing before the era of digital downloads, Bourdieu notes: 'The contribution of change in demand to change in tastes is seen clearly in a case like that of music, where the rise in the level of demand coincides with a lowering of the level of supply, with the *record* (the same thing is seen in the area of reading, with the paperback). The rise in the level of demand includes a translation of the structure of tastes, a hierarchical structure, which runs from the rarest, Berg or Ravel nowadays, to the least rare, Mozart or Beethoven. To put it more simply, all the goods offered tend to lose some of their relative scarcity and their distinctive value as the number of consumers both inclined and able to appropriate them grows.' Bourdieu, *Sociology in Question*, 114.

90   See the discussion of community, information and expertise in G Calamar and P Gallo, *Record Store Days: From Vinyl to Digital and Back Again*, Sterling, New York, 2012. See further, A Glover, 'Independent Record Shops say they are open for business', *BBC News*, 16 January 2013; D Guarini, 'Record Store Day: saving independent music stores since 2008', *Huffington Post*, 20 April 2012.

91   Bourdieu, *Sociology in Question*, 114.

92   Wittgenstein notes, 'Taste is refinement of sensibility; but sensibility does not *act*, it merely assimilates.' Wittgenstein, *Culture and Value*, 68e.

"Oh dear! I'd nearly forgotten that I've got to grow up again! Let me see – how *is* it to be managed? I suppose I ought to eat or drink something or other; but the great question is what?"[93]

The great question *is* what.

---

93   *Wonderland*, 36.

Chapter 5
Risk

Explain Yourself!

> "Who are *you*?" said the Caterpillar.
>
> This was not an encouraging opening for a conversation. Alice replied, rather shyly, "I – I hardly know, sir, just at present – at least I know who I *was* when I got up this morning, but I think I must have been changed several times since then."
>
> "What do you mean by that?" said the Caterpillar sternly. "Explain yourself!"
>
> "I can't explain *myself*, I'm afraid, sir," said Alice, "because I'm not myself, you see."
>
> "I don't see," said the Caterpillar.[1]

The journey thus far has been a treacherous one. From the depths of a spurious and greedy causality to the potential of an unlimited and instantaneous innovation, it seems the relationship between intellectual property and innovation is not the simple mapping between the two that is at times maintained within policy and economic debates.

A particularly significant example of this comes from the high technology industries, although the same might be said across the diverse and incomparable range of innovation and industries. Nevertheless, concentrating for the moment on high technology products, the cultures of research and development in these industries complicate the seemingly linear and causal relationship between problem and technical solution which underpins much of the philosophy of conventional patent protection and patent value. In this way, technological innovation provides much insight into the way in which meaning can be generated within a system of innovation apparently with no end in sight, a field of familiar production; that is, the smooth space of innovation, the innervation of high technology research and development.

How might technological advances actually create new products (devised to account for those advances, after the fact, as it were) and new markets in the high technology sector in particular? The roles of the 'user' and 'producer' and the reception of the meaning of a high technology invention are therefore perhaps more complex and less overt than in the simpler relationships understood as applying to mechanical inventions or even to medicinal products, for example. Approaching this narrative of technological development and new markets through the logic of

---

1   *Wonderland*, 36–7.

sense-making in language provides some insight into the application of modelling in the context of the innovative industries, specifically, the way in which markets and new products are devised.

What Have I Become?

> "I don't see," said the Caterpillar.
>
> "I'm afraid I can't put it more clearly," Alice replied very politely, "for I can't understand it myself to begin with; and being so many different sizes in a day is very confusing."
>
> "It isn't," said the Caterpillar.
>
> "Well, perhaps you haven't found it so yet," said Alice; "but when you have to turn into a chrysalis – you will some day, you know – and then after that into a butterfly, I should think you'll feel it a little queer, won't you?"
>
> "Not a bit," said the Caterpillar.
>
> "Well, perhaps your feelings may be different," said Alice; "all I know is, it would feel very queer to *me*."
>
> "You!" said the Caterpillar contemptuously. "Who are *you*?"[2]

Technological advancement is perhaps more adequately understood as being driven by technological potential itself, that is, about the capacity for change, rather than with a specific product narrative in mind.[3] In a conventional world, a world of representation, the world of intellectual property and trade, and, as in the previous chapter, the world of law and aesthetics, it is the value of identity (of the author, of the product, of the consumer) that is of central importance. Identity thus both defines relationships as well as limits them. In Wonderland, Alice is unable to represent her reality, but encounters fresh and unanticipated presentations, the repetition of difference. Wonderland is a world of becomings, it is the world of pure innovation:

> Becoming is a rhizome, not a classificatory or genealogical tree. Becoming is certainly not imitating, or identifying with something; neither is it regressing-progressing; neither is it corresponding, establishing corresponding relations; neither is it producing, producing a filiation or producing through filiation.

---

2   *Wonderland*, 37.

3   For example, see the discussion in the following: T Kouloupoulos, 'Innovation isn't about new products, it's about changing behavior', *Fast Company*, 31 July 2012; J Dalrymple, 'Delivering innovation vs delivering products', *The Loop Magazine*, 8 January 2013; L Magid, 'Apple's "next big thing" will be an innovation, not an invention', *Forbes*, 26 March 2013.

> Becoming is a verb with a consistency all its own; it does not reduce to, or lead
> to, 'appearing,' 'being,' 'equalling,' or 'producing.'[4]

How then does such innovation arrive at a product? How does it explain itself? For this, innovation relies upon the overcoding by intellectual property orders, encircling and providing the stability of an innovative event for the purposes of market connections. However, it is not a strict overwhelming of the smooth space of innovation. Intellectual property, in its inherent obsolescence, the 'creative destruction' inbuilt in the system, actually restores the smooth space of innovation, innovation as becoming, the logic of chance. So, as we can see in the sea, the intellectual property system is after all one of the most complex examples of the mixture of smooth and striated space.[5]

Again, the problem of representation ensures an infinite regress of explanation: *Who are you?* Any effort to contain entities of intellectual property within its system of accountability requires the comparability of innovation within the conventions of intellectual property: 'A *picture* held us captive. And we could not get outside it, for it lay in our language and seemed to repeat it to us inexorably.'[6] It is impossible to stand outside the system of representation in order to account for itself: 'Representation must encompass an expression which it does not represent, but without which it itself would not be "comprehensive," and would have truth only by chance or from outside.'[7] Therefore, a new product can only be 'invented', 'apprehended', provided its history can be recollected. This is the discomforting circularity or mimicry of that which already exists; this is 'the prerequisite of any possible classification'.[8]

> "You!" said the Caterpillar contemptuously. "Who are *you*?"
> Which brought them back again to the beginning of the conversation. Alice felt a little irritated at the Caterpillar's making such *very* short remarks, and she drew herself up and said, very gravely, "I think you ought to tell me who *you* are, first."
> "Why?" said the Caterpillar.[9]

---

4   Deleuze & Guattari, *Thousand Plateaus*, 239.

5   Deleuze and Guattari explain that 'it is the de jure distinction that determines the forms assumed by a given de facto mix and the direction or meaning of the mix (is a smooth space captured, enveloped by a striated space, or does a striated space dissolve into a smooth space, allow a smooth space to develop?).' Deleuze & Guattari, *Thousand Plateaus*, 475. See further the discussion of the mixture of smooth and striated space in Chapters 2 and 3.

6   Wittgenstein, *Philosophical Investigations*, 115.

7   Deleuze, *Logic of Sense*, 145.

8   Deleuze, *Difference and Repetition*, 247.

9   *Wonderland*, 37–39.

This circularity mimics the drama and agony of representation, and mocks models of linearity, such as those applied not only within the intellectual property rules themselves, but also through the economics and management theory of policy debate. Much of the new 'technology' that Alice is acquiring in Wonderland is bound in similarities and resemblances of her own knowledge and skills. *Keep your temper*: keep your knowledge. Innovation 'objects' are therefore never perfectly discrete, alienable, communicable, abstracted, conveyable. That is, they are never perfectly transferable as new, being always already present in people's knowledge resources. Indeed, this is how the 'innovation' of the product is possible. Thus, linear models of innovation become redundant and misleading, when innovation always brings us back to the beginning of the conversation.

This is also the reality of innovation research and modelling. Linear models are yielding to more dynamic models of networking and cooperation. There has been a shift in emphasis away from the traditional linear models of push (through knowledge, technology and research, with the corporation as innovator[10]) and pull[11] (by identifiable demands)[12] to collaborative social networks[13] such as those described in Actor-Network Theory (ANT)[14] and in the Social Shaping of Technology (SST).[15] Such models of cooperation are networks not only with competitors but also with users themselves; they are incarnations of the community as innovator. In this way, so-called fifth-generation innovation is innovation in *use*; in other words, such innovation is an aspect familiar production.

This dilemma of innovation, of the new, is again manifest through the problem of representation. It is not conceivable to assume a privileged position from which

---

10   See the discussion of the corporation as instrument of innovation and progress in C Freeman & L Soete, *The Economics of Industrial Innovation*, 3rd ed, MIT P, Cambridge MA, 1997.

11   See in particular the discussion of the use of ethnographic research and other customer data in TJ Kelley with J Littman, *The Art of Innovation: Lessons in Creativity from IDEO, America's Leading Design Firm*, Currency, New York, 2001. See also A Ulnwick, *What Customers Want: Using Outcome-Driven Innovation to Create Breakthrough Products and Services*, McGraw-Hill, New York, 2005.

12   On push and pull in innovation, see further CA Voss, 'Technology Push and Need Pull: A New Perspective', 14(3) *R&D Management* 1984: 147–51.

13   R Rothwell, 'Towards the Fifth-Generation Innovation Process', 11(1) *International Marketing Review* 1994: 7–31. See further I Tuomi, *Networks of Innovation: Change and Meaning in the Age of the Internet*, Oxford UP, Oxford, 2002; and more generally S Lindegaard, *The Open Innovation Revolution*, Wiley, Hoboken NJ, 2010. See further in relation to social factors and early industrial innovation, AFC Wallace, *The Social Context of Innovation*, U of Nebraska P, Lincoln, [1982]/2003.

14   See B Latour, *Reassembling the Social: An Introduction to Actor-Network Theory*, Oxford UP, Oxford, 2005; and see further the collection J Law & J Hassard (eds), *Actor Network Theory and After*, Blackwell, Oxford [1999]/2006.

15   R Williams & D Edge, 'The Social Shaping of Technology', 25 *Research Policy* 1996: 856–99.

one can both observe for oneself and take account of oneself in the process of that observation. This problem of representation is one that has preoccupied philosophy since Plato: how to represent the unrepresentable;[16] how to admit the difference. Wittgenstein explains: 'If you observe your own grief, which senses do you use to observe it? A particular sense; one that *feels* grief? Then do you feel it *differently* when you are observing it? And what is the grief that you are observing – is it one which is there only while it is being observed?'[17] It is therefore impossible for Alice to answer the question: Who are *you*? It is impossible to finalize the endless question: What is innovation? It is impossible to represent the unrepresentable: '"Observing" does not produce what is observed. (That is a conceptual statement.)'[18] Rather, there must be something forgotten, a remainder, in order to accept a representation as complete: 'Again: I do not "observe" what only comes into being through observation. The object of observation is something *else*.'[19] The object of observation is always displaced, deferred in any act of representation. The question of change therefore becomes non-sensical, *out-of-the-way*, out of Alice's reach, and yet immediately apparent.

Keep Your Temper

> "Come back!" the Caterpillar called after her. "I've something important to say!"
> This sounded promising, certainly: Alice turned and came back again.
> "Keep your temper," said the Caterpillar.[20]

While it has become a platitude to mention the word 'balance' in relation to intellectual property, arguably this is one of the most important words, but not for the reasons usually provided. It is not so much important in relation to a kind of social contract, or a balance of rights, but in relation to a balance between chance and risk. In the 'calculation' of this balance the concept of risk (as commercial and investment risk) has emerged as pre-eminent. Risk is attributed a value in terms of both the calculation of the value in the specific product as well as the calculation of the broader 'value' of the intellectual property system in respect of the credit (term of protection, strength of rights and so on) to the producer: 'Risking … takes

---

16   The dilemma of representation is traced throughout this inquiry. See further, J-F Lyotard, 'Representation, Presentation, Unrepresentable', in J-F Lyotard, *The Inhuman: Reflections on Time*, G Bennington & R Bowlby (trans), Stanford UP, Stanford, [1988]/1991: 119–28.

17   Wittgenstein, *Philosophical Investigations*, 187.

18   Wittgenstein, *Philosophical Investigations*, 187.

19   Wittgenstein, *Philosophical Investigations*, 187.

20   *Wonderland*, 39.

risking for granted as a value. In the process of risking, value is simply displaced from the object to the risking and to the contestation itself.'[21]

Where the relationship between users and producers of the intellectual property system fails most successfully, as it were, is where the attention to risk is far greater than the appreciation of chance. This failure may be experienced in terms of access (and the 'debt' with respect to the use of the intellectual property object) or indeed in terms of innovation itself (where the value placed on risk and its mitigation exceeds the value placed on chance and the potential for innovation). The intellectual property system thus fails where undue emphasis is placed on minimizing unintended and undesirable consequences (risk) and too little or none on maximizing unanticipated benefits and opportunities (chance).

This preoccupation with 'risk' thus contributes to the indebtedness of users to producers of intellectual property; and the 'promise' of intellectual property carries with it the risk of failure and the memory of success. Intellectual property is a society's memory, its debt, its debt as memory: 'the debtor-creditor relation, which on both sides turns out to be a matter of memory – a memory straining toward the future'.[22] In the subjective debt economy, the user assumes the costs and risks of innovation. Thus, as in other examples (health, transport and so on), the costs and risks of innovation are externalized in such a way that individuals (consumers) carry the risks;[23] and most significant of all is the risk of access.[24] Further, insofar as the structural risks might be seen as necessarily deferred to adjudication in court (for instance, the scope of protection, of works, of claims and so on), the risk for any uncertainty is borne by all users, consumers and right-holders alike, as distinct from the State.

Keep your temper. Temperance is indeed one of the four cardinal virtues, and is characterized by a 'habit', a received 'custom' of rational restraint, of avoidance, of waiting. Keeping one's temper refers not only to passion, but also to opinion, and even to appetite. Temper also gives a notion of what is optimum, proportionate, proper. It is the path of compromise, or perhaps more accurately in the present sense, of balance. This principle of balance is tangible in Wonderland,[25] not merely one of physical or even emotional balance, but one of the balance between chance

21   Bataille, *On Nietzsche*, 126.

22   Deleuze & Guattari, *Anti-Oedipus*, 190.

23   Lazzarato, *Making of the Indebted Man*, 51.

24   For instance, this question of access is experienced acutely in the debate on access to medicines, where access is understood in the International Covenant on Economic, Social and Cultural Rights (ICESCR) as a 'right' within the inclusive right to health (Art 12), and also arguably within the right to the benefits of scientific research (Art 15(b)). See further Gibson, *Intellectual Property, Medicine and Health*, on the detailed discussion of Art 12 and the right to health (chapter 3), and the relationship between health and culture, including access to medicines and the scientific benefit (chapter 4).

25   This resonates also with the loss of balance in *Through the Looking-Glass* in 'Re Use'.

and risk. The Caterpillar advises Alice to keep her temper, indeed to wait for what is *to come*.[26]

"So you think you're changed, do you?"[27]

The Invention of Chance, the Innovation upon Change

The corollary of risk is accident. To invent is to risk: 'To invent something is to invent an accident.'[28] Notably, much of invention might be without use, without need, 'a happy accident'.[29] In this way, technological change must articulate the use (product) for an invention that may have occurred 'independently of any practical need'.[30] That is, integral to the process of technological change is ensuring the proximity (and meaning) to users. And in a way then, finding the product is the innovation in modern and emerging technologies; that is, use is innovation.

In law, accidents have historically been articulated in terms of a relationship between the agent and the victim. Fundamentally, if 'risk imports relation'[31] then risk is also essential to meaning and use. This 'relation' of risk is complicated more recently by the way in which high technology and the digital translate this from an individual to the population as a whole:[32] 'if *interactivity* is to information what *radioactivity* is to energy, then we are confronted with the fearsome emergence of the "Accident to end all accidents", an accident which is no longer *local* and

---

26   This notion of the deferred, the 'to come', is inherent in the problem of representation that is explored throughout this book. Derrida speaks of 'democracy to come' considered here in later discussions. See J Derrida, *Negotiations: Interventions and Interview, 1971–2001*, E Rottenberg (ed, trans), Stanford UP, Stanford, 2002: 178–83; Similarly, Deleuze speaks of the time of Aion (the time of becoming) as 'limitless in either direction. Always already passed and eternally yet to come.' See Deleuze, *Logic of Sense*, 165. See also the imperative to Alice, 'Come on!', discussed in chapter 10.

27   *Wonderland*, 39.

28   P Virilio, 'Virilio: Cyberesistance Fighter: An Interview with Paul Virilio', D Dufresne (interview), J Houis (trans), *Aprés-Coup Psychoanalytic Association*, 1999. Available at www.apres-coup.org/mt/archives/title/2005/01/cyberesistance.html.

29   Schumpeter, *Business Cycles*, 81 n 25.

30   Schumpeter, *Business Cycles*, 81 n 25.

31   *Palsgraf v Long Island R.R.*, 248 N.Y. 339, at 344 (1928).

32   To an extent, similar principles apply in the theories and practices of population-wide studies, including population-wide genomic studies. In this case, the notion of 'risk' and responsibility is crucial to the garnering of support and contribution by participants to such studies. See further the UK Biobank Ethics and Governance Framework, Version 3.0, October 2007. See also the discussion in LØ Ursin et al., 'The Informed Consenters: Governing Biobanks in Scandinavia', in H Gottweis & A Petersen (eds), *Biobanks: Governance in Comparative Perspective*, Routledge, Abingdon, 2008: 177–209. See also Gibson, *Intellectual Property, Medicine and Health*, 126–27, 131–37.

precisely situated, but *global* and generalized. We are faced, in other words, with a phenomenon which may possibly occur everywhere simultaneously.'[33]

In modern renditions of risk, therefore, we have moved from the potential of contingency and chance, to a concept of over-regulation (and thus, loss of opportunity). Risk is attended by a narrative of causality, while chance occupies the smooth space of infinite potential and instantaneity. Risk is the language of effect and materialities. Chance is unregulated, unanticipated, and unseen. To foresee risk is to foresee a loss. To take a chance is to stumble upon a gain.

Indeed, Alice's entire journey commences with an 'intended' accident, as she falls down the rabbit-hole. Arguably, the concept of risk in legal and regulatory environments (risk assessment, risk aversion, risk management) has become something which may be expected, intended. Indeed, it is almost impossible to insure against unexpected events (acts of God). Alice's own journey is one that commences without foresight, and yet with intention; paradoxically therefore without risk, as it were, but full of chance.

Contrast this with the episode in the United States, where in early 2013 the Prince George's County Board of Education proposed for 'Works created by employees and/or students specifically for use by the Prince George's County Public Schools or a specific school or department in PGCPS' to be 'properties of the Board of Education even if created on the employee's or student's time and with the use of their materials'.[34] The Prince George's County Board attempted to minimize its risk to such an extent that it jeopardized the entire learning relation, and as such was met with considerable public outcry on an international scale.[35] In this case, the attempt to mitigate risk was so extreme that use, and thus meaning and legitimacy, was expelled from the system.

The notion of insurance in intellectual property similarly banks upon the 'credit' of intellectual property through the practice of investment in private insurance, as distinct from mutualizing risk through sharing or a kind of simplistic 'crowd' model of risk aversion (that is, dividing up costs among investors). In other words, conventional credit and debt models of intellectual property persist in a private insurance framework, while familiar production maximizes chance, accountability and productivity through the crowd.

Indeed, business models which have attempted to mitigate risk by over-regulating the relationship between user and producer have failed significantly and disastrously, primarily because of this rupture in the meaning-generation of the system. For example, in the music industry, many commentators have suggested

---

33   Virilio, *Information Bomb*, 134.

34   J Collen, 'Maryland Board wants copying homework to be a federal offense', *Forbes*, 22 February 2013.

35   The outcry included the Center for Rights, which responded with the launch of a petition, 'Don't Copyright Me', available at www.dontcopyrightme.com. See further L Clark, 'Don't Copyright Me launches to protect US Schoolkids' homework', *Wired*, 12 February 2013.

that over-zealous policing and enforcement have failed.[36] Arguably this is largely because an engagement with the structural problem fails to redress the conceptual changes and build upon and commercialize the affective labour and experience. Problematic business and social strategy seemingly demonized the industry itself, consequently legitimizing for many the otherwise illegitimate access (through file-sharing and so on). Ultimately such strategies disrupted the affective relationship to the product and severed the privity of fan and artist. In eliminating risk, the opportunities for new business models concentrating on extending and valuing the fan and artist relationship, experience and other areas of value-generation were considerably delayed and in some ways ignored altogether, resulting in not only enormous losses economically, but also considerable (and seemingly irrecoverable) losses normatively and socio-politically. The music industry now appears to be ill-equipped to the extent that business models (not only industry models, but also the disruption from models of illegitimate acquisition and distribution) are also changing the very artistic product itself. Rather than engaging with an artist and the album as a work, music is increasingly consumed as disaffected singles, with no contextual or experiential relationship. Further, in what is also widely recognized as a detriment to the industry itself,[37] music stores are rapidly diminishing in number, compromising the affective value that accompanies the social and intellectual life of the 'expert'. To compromise the 'use' relationship is literally to silence the speaker from whom the system demands consent.

Stable Environments and Chance Relationships

> "Oh, I'm not particular as to size," Alice hastily replied; "only one doesn't like changing so often you know."[38]

The space of innovation has so far been recognized as striated by intellectual property and attending social conventions (taste), all of which provide a certain construction of stability. This persists throughout intellectual property business models which emphasize the mitigation of risk and the 'own' producer. In contrast, through familiar production and in technological innovation for innovation's sake, the fundamental relationship is one of chance. The 'product' is arguably difference itself, the desire created and animated in users to require an unanticipated change. *How did I ever live without it?* The product, as demarcated by intellectual property rights, is an incidental 'invention' of the new technology, the market thus created, rendering the smooth space of innovation striated. However, as has been seen, this

---

36   R Reid, 'What to do when attacked by pirates', *Wall Street Journal*, 1 June 2012.

37   Comments to the author and to delegates from various industry executives at the Association of Independent Music (AIM) Conference, 9 May 2012.

38   *Wonderland*, 42.

relationship between smooth and striated space is mixed, and this is clear in high technology industries in particular.

In the same way, Alice appreciates the importance of some stability in making sense of an otherwise potentially infinite pool of innovation, but is now open to the opportunity of her body's new fluidity in the uncharted space of Wonderland. Use is thus assisted by manners and habit, identified with respect to functions and class, and is itself an acquired aesthetics of taste. But at the same time, like Alice, use is frustrated by construct and artifice. The challenge is to understand the way in which business and economic models in intellectual property might be usefully configured, if at all, upon chance as distinct from risk, and how innovation and creativity might emerge rather than be planned.[39]

Chance Relationship

> "One side will make you grow taller, and the other side will make you grow shorter."
>
> "One side of *what*? The other side of *what*?" thought Alice to herself.
>
> "Of the mushroom," said the Caterpillar, just as if she had asked it aloud; and in another moment it was out of sight.
>
> Alice remained looking thoughtfully at the mushroom for a minute, trying to make out which were the two sides of it; and as it was perfectly round, she found this a very difficult question. However, at last she stretched her arms round it as far as they would go, and broke off a bit of the edge with each hand.[40]

So in respect of Alice's own innovation, the Caterpillar presents her with the tools of chance. At first Alice approaches the dilemma with a conventional risk assessment:

> "And now which is which?" she said to herself, and nibbled a little of the right-hand bit to try the effect: the next moment she felt a violent blow underneath her chin: it had struck her foot!
>
> She was a good deal frightened by this very sudden change, but she felt that there was no time to be lost, as she was shrinking rapidly; so she set to work at once to eat some of the other bit. Her chin was pressed so closely against her

---

39   See, for instance, the discussion of planned and emergent creativity and innovation in C Miller & RN Osborn, 'Innovation as a Contested Terrain: Planned Creativity and Innovation Versus Emergent Creativity and Innovation', in MD Mumford et al. (eds), *Multi-Level Issues in Creativity and Innovation*, JAI P, Bingley, 2008: 169–89. Miller and Osborn note 'we wonder about the applicability of the model to radical innovations calling for architectural changes and innovations involving technological discontinuities.' Miller & Osborn, 'Innovation as a Contested Terrain', 187.

40   *Wonderland*, 44.

foot, that there was hardly room to open her mouth; but she did it at last, and managed to swallow a morsel of the left-hand bit.[41]

Alice literally keeps her temper this time, showing great self-restraint in appetite and moderation of her curiosity by nibbling from each piece, at the same time maximizing her chances of success. Alice at once faces the uncertainty of risk and attempts to rationalize her choices, indeed, her taste, as it were; that is, imposing limits upon the chance and uncertainty before her. However, in so doing she is also taking responsibility for her actions and, to an extent, addressing the logic of Wonderland. The mushroom at once affirms chance and, despite Alice's initial attempts to put limits on that chance, ultimately facilitates her chance-taking. Alice's action therefore suggests not only her engagement but also her relation to the innovative space of Wonderland: 'It's true, the omnipotence of reason limits luck's power ... At the extremes, there's freedom.'[42]

Alice's behaviour at first betrays the rationality that is attributed to risk, but at the same time shows the necessary social aspects of taking a chance. She embraces the options and renders an otherwise causal and abridged version of the Mushroom to a rhizomatic system of growing and shrinking as it suits her. While she merely ran the risk in the Caucus-race, here Alice literally ventures to take a chance and affirms the difference in repetition: 'Turning around the object that interests us is far from useless; it is a task we cannot shirk as long as we remain in the order of signification. We only always perceive one side of this object at a time; it never changes, but if we have gone around properly, what we observe from the newly revealed side introduces us to a new discourse. Different repetition.'[43] And what to do when there is no other side? The other side of what? Where there are all sides at once, and no sides at all?

At the same time Alice engages as user, she commends and participates: 'In order to realize the power to act, we need to believe (trust) in the "moving present," the present as possibility, that is, in the world and the new possibilities of life that it holds.'[44] It appears, therefore, that Alice is beginning to enter the world of play in her new surroundings. She is beginning to see the benefits of participation in the game:

> It is claimed that man does not know how to *play*: this is because, even when he is given a situation of chance or multiplicity, he understands his affirmations as destined to impose limits upon it, his decisions as destined to ward off its effects, his reproductions as destined to bring about the return of the same, given a winning hypothesis. This is precisely a losing game, one in which we risk losing as much as winning because we do not affirm the *all* of chance: the

---

41   *Wonderland*, 44–45.
42   Bataille, *Guilty*, 71.
43   Lyotard, *Discourse, Figure*, 72.
44   Lazzarato, *Making of the Indebted Man*, 70–71.

pre-established character of the rule which fragments has as its correlate the condition by default in the player, who never knows which fragment will emerge.[45]

This notion of game and chance is intriguing in that 'risk' is not ordinarily in the rhetoric of game until it enters into the discourse of the game of work. The conventions of work overwrite the creative potential of the game, re-emphasizing risk in a place where chance prevailed. This freedom and opportunity of chance, within the 'reason' of the game, grants precisely the possibilities of innovation beyond the causal models of risk, and an intellectual property system beyond the debt and credit of risk. Indeed, this is *'the giddy seductiveness of chance'*.[46]

To impose limits is a losing game.[47]

Fair Play

Innovation is thus eschewing mainstream economics of marginalism and is instead inciting the game. However, 'What is more frightening for humankind than play?'[48] So, how might the 'behaviour' of game theory[49] be applied in the win-win game of the Caucus-race? Is the game a pure game and what is the 'strategy' in a game of chance? As it turns out, 'Chance is forever at the mercy of itself. It's always at the mercy of play, always *in* play. If it was definitive, chance wouldn't be chance.'[50]

The most significant aspect of game theory for the present discussion is its relevance to activity over time, most notably the cooperative innovation with the user community in technological development. The concept of extensive form games allows for analysis of interactions over time.[51] This form of the game is most appropriate in considering the kinds of 'identity-based' or reputational drivers of innovation considered in earlier chapters and examples, and crucial to affective value in the 'digital'. Strategic benefits and decisions are understood beyond simple short-term profits, and so interactions may be seen to maximize the 'chances' for benefit (beyond profit transactions and including 'transactions' in reputation, for example).

---

45   Deleuze, *Difference and Repetition*, 115–16.

46   Bataille, *Guilty*, 72.

47   Deleuze's argument here has a seductive resonance with that in popular psychology rhetoric, including P Bronson & A Merryman, *Top Dog: The Science of Winning and Losing*, Hatchette, New York, 2013, where the authors claim that ignoring the odds can mean the chance of greater success.

48   Bataille, *Guilty*, 77.

49   See further J von Neumann & O Morgenstern, *Theory of Games and Economic Behaviour*, Princeton UP, Princeton, 1983.

50   Bataille, *Guilty*, 77.

51   DG Baird, RH Gertner & RC Picker, *Game Theory and the Law*, Harvard UP, Cambridge MA, 1994: 50–52.

To generalize, the original incarnation of game theory assumes and examines rational behaviour (*keep your temper*) to model options (*chance*).[52] In relation to intellectual property discussions, the tragedy of the commons[53] might be presented as an important example of the game, where unmitigated and selfish pursuit of rational self-interest (*lack of temperance*) defeats the game and destroys the resource. Further, while game theory has been applied to some extent in the industrial arena of innovation, less attention has been paid to the creative industries,[54] perhaps in part due to the unconscious preservation of the perceived distinct spheres of industry and creativity, distinct spheres which must be and are disrupted in familiar production: 'the act of being "truly creative" ... can be understood as a strategic response to the environment in which the artist finds himself or herself, and is not something that needs to be perpetually shrouded in mystery.'[55]

Thus, in the present discussion, recognizing the broader remit of the term 'innovation', it is perhaps useful to consider this methodology as an opportunity to understand further this balance between risk and chance, and to provide resources for modelling environments for innovation more generally. However, at the same time, it is somewhat counter-intuitive to accept that the unrepresentable innovation (and justice) can be calculated (and thus pre-determined and subjected

---

52   While game theory utilizes mathematical modelling, sociology generally utilizes game theory for analysis of institutional and social structures, without necessarily attempting mathematical analysis, the latter being questioned in sociology as possibly oversimplifying very complex social situations. For more on the uneasy position of game theory in sociology, see further R Swedberg, *Principles of Economic Sociology*, Princeton UP, Princeton, 2009, in particular 289–90; P Abell, 'Sociological Theory: What Has Gone Wrong and How to Put It Right, a View from Britain', in J Hage (ed), *Formal Theory in Sociology: Opportunity or Pitfall*, SUNY P, Albany NY, 1994: 105, at 110.

53   The concept of 'The Tragedy of the Commons' comes from a 1968 essay by ecologist, Garrett Hardin, in which the author explains the way in which certain social and cultural problems and behaviours in association with resources cannot be resolved by technical means: G Hardin, 'The Tragedy of the Commons', 162 *Science* 1968: 1243–48. The concept has been subsequently adopted (and at times curiously adapted) in the context of intellectual property. For an insightful analysis of Hardin's exploration of morality as a solution, as distinct from technical measures, in the context of intellectual property, see S Ghosh, 'The Fable of the Commons: Exclusivity and the Construction of Intellectual Property Markets', 40 *UC Davis Law Review* 2007: 855.

54   WDA Bryant & D Throsby, 'Creativity and the Behavior of Artists', in VA Ginsburgh & D Throsby (eds) *Handbook of the Economics of Art and Culture*, North-Holland, Amsterdam, 2006, 507–29, at 516.

55   Bryant & Throsby, 'Creativity and the Behavior of Artists', 517.

to judgment).[56] This would indeed run counter to an ethical approach to innovation. Indeed, it simply is not clear that innovation can count on the numbers alone.[57]

Imbalance of Power

Nevertheless, consider Alice's journey and her adaptation from what appears at first to be rational and predictable behaviour, but gradually habituates and accustoms herself in her new environment and new players, finding her participating in games for which she knows no rules, or indeed for which there are no rules. Taken as the pure game of chance, this is perhaps the ultimate opportunity of innovation. Recognizing this adaptability is the reflexivity of game theory and economics: 'Reflexivity is no longer a stranger even to economic theory.'[58] Game theory agitates the linear model of equilibrium theory, as a fallacious preconception of markets, and challenges it with nonlinearity,[59] a pure becoming of the system. This indeed is the pure becoming, the paradoxically nonlinear process of innovation. While technology appears to provide a beginning, the innovation in the product is as if 'by chance', through that technology's interaction with the user, the public – that is, through the acquisition of meaning.

Game theory illustrates the difference between rational behaviour in a transaction and the ethics of relation in chance.[60] Identity relations of reputation, loyalty and so on may indeed infuse knowledge economies with an ethics of mutual benefit, over and above pure transactional economics. The disruption of the overcoding of knowledge transactions by the digital is informative here. In some extreme examples it has been taken to undermine the rule of law,[61] but in other aspects of community as innovator it is important not to underestimate or

---

56   For instance, see the discussion of the limitations of the 'racing models' based on innovation and the patent system in Stoneman, *Soft Innovation*, 141–42.

57   For instance, Dasgupta and Stiglitz argue that intellectual property rights (like patents) lead to a racing model with overinvestment in R&D in order to win the race: P Dasgupta & J Stiglitz, 'Industrial Structure and the Nature of Innovative Activity', 90.358 *Economic Journal*, 1980: 266–93. See further the discussion of this in Stoneman, *Soft Innovation*, 251.

58   G Soros, *Open Society: Reforming Global Capitalism*, Little, Brown, London, 2000: 14.

59   Soros, *Open Society*, 88–89.

60   See the discussion of the replacement of relationships by transactions in Soros, *Open Society*, 113–14. See further the observations of Mark Buchanan, who notes that 'The trouble with this mathematical sophistication is that it is psychologically naïve ... the mere existence of an equilibrium says little about whether any real economy might actually stumble into such a condition.' See M Buchanan, *Forecast: What Physics, Meteorology and the Natural Sciences Can Teach Us About Economics*, Bloomsbury, London, 2013: 101.

61   See further the phenomenon of virtual 'crowds' in the context of social media platforms such as Twitter and Reddit, discussed in Chapter 11.

overlook the possible potential of the social context for innovation in the digital environment. This is becoming rapidly a space of anonymous relation, a lottery of communication, where chance is seemingly infinitely replicated in a non-linear system. Social media is at once both a tool of globalization[62] and a remedy, deterritorializing the transactional quality of international trade and knowledge exchange, and reterritorializing it with relationships based not on proximity but on other markers of community, including shared interests, values, beliefs.

A crucial example of this mixture in tensions is provided by the momentum towards harmonization in intellectual property. The momentum for harmonization appears to suggest that the 'crimes' of intellectual property (such as unequal trading environments, unequal enforcement environments and so on) are such that it is warranted to appeal to an order over and above the sovereignty of individual states.

The very dynamics of chance in familiar production defeat the notion of balance and equilibrium: 'No one, after all, not even a grand master champion, plays chess by working out the Nash equilibrium perfect strategy – first, because no human being can possibly calculate what it is, and second, because there's no guarantee your opponent will play that way either.'[63] It is not possible nor is it advantageous to ascribe to innovation the notion of a purposeful, directed, solution approach to innovation. Rather, it is the pure chance and intimacy of associational, aggregated intercourse in which 'value' is generated in familiar production. That is, value and meaning is generated by virtue of the very participation in the production itself; that is, in use. Indeed, this is the 'familiar economy' of production in the digital.

Chance in the Crowd

"Come, my head's free at last!" said Alice in a tone of delight, which changed into alarm in another moment, when she found that her shoulders were nowhere to be found: all she could see, when she looked down, was an immense length of neck, which seemed to rise like a stalk out of a sea of green leaves which lay far below her.[64]

Alice is refuting the crowd of the forest and the profundity of its foliage, its surface. The forest is ordinarily '*higher* than man ... its real density, that which makes it

---

62   Soros notes that 'Globalization works in the same direction by increasing the scope for transactions and diminishing the dependence on relationships.' Soros, *Open Society*, 114.

63   Buchanan, *Forecast*, 102. See further the unpredictability and 'chance' nature of the game: 'But it becomes even clearer in experiments with computers playing games. In experiments it's possible to study what happens gradually as games get harder, so that rational behavior becomes ever more impossible. The outcome is that the games begin to look a lot like markets.' Buchanan, *Forecast*, 102.

64   *Wonderland*, 45.

a forest, is its foliage; and this is overhead. It is the foliage of single trees linked together which forms a continuous roof; it is the foliage which shuts out the light and throws a universal shadow.'[65] But Alice has grown so tall, to such heights, that she can no longer see what is at the surface, she can no longer recognize the leaves themselves: '"What *can* all that green stuff be?"'[66] And at the same time, she is no longer able to plunge to the depths of Wonderland.

All at once, in departing from the familiarity of the forest, Alice enters into a battle: 'a sharp hiss made her draw back in a hurry: a large pigeon had flown into her face, and was beating her violently with its wings.'[67] As soon as Alice attempts to ignore the details at the surface, at the border, she is violently reminded of her foundations in the familiar. She must take her chances in the crowd, the territory of familiar production, 'which allows itself to be cut down to the last man before it gives a foot of ground'.[68]

Trying to Invent Something

"Serpent!" screamed the Pigeon.

"I'm *not* a serpent!" said Alice indignantly. "Let me alone!"

"Serpent, I say again!" repeated the Pigeon, but in a more subdued tone, and added with a kind of sob, "I've tried every way, and nothing seems to suit them!"

…

"And just as I'd taken the highest tree in the wood," continued the Pigeon, raising its voice to a shriek, "and just as I was thinking I should be free of them at last, they must needs come wriggling down from the sky! Ugh, Serpent!"

"But I'm *not* a serpent, I tell you!" said Alice. "I'm a – I'm a –"

"Well! *What* are you?" said the Pigeon. "I can see you're trying to invent something!"[69]

To innovate, to invent, is to invent the accident, to disrupt, to intervene. To create or invent a product, it would seem, is to imagine and invent a use. In other words, the problem presents itself once the solution has been invented. Inventive activity is by definition a departure from the norm and, thus, an impropriety as it were: 'it inserts a disorder into the peaceful ordering of things, it disregards the proprieties.'[70] Alice has indeed struggled with expected roles and departures and has learnt to reinvent herself for each new aspect of the journey.

---

65   Canetti, *Crowds and Power*, 84.
66   *Wonderland*, 45.
67   *Wonderland*, 46.
68   Canetti, *Crowds and Power*, 85.
69   *Wonderland*, 46–47.
70   Derrida, *Acts of Literature*, 312.

"Come, there's half my plan done now! How puzzling all these changes are! I'm never sure what I'm going to be, from one minute to another! However, I've got back to my right size: the next thing is to get into that beautiful garden – how *is* that to be done, I wonder?"[71]

How *is* that to be done? I wonder.

---

71   *Wonderland*, 49.

Chapter 6

Change

A Great Letter

> The Fish-Footman began by producing from under his arm a great letter, nearly as large as himself, and this he handed over to the other, saying, in a solemn tone, "For the Duchess. An invitation from the Queen to play croquet." The Frog-Footman repeated, in the same solemn tone, only changing the order of the words a little, "From the Queen. An invitation for the Duchess to play croquet."[1]

The letter is delivered, the critical event of its 'use'. However, neither the sender (the Queen) nor the intended recipient (the Duchess) is present at the moment where the message is communicated. The intended meaning appears to have been delivered effectively in its reiteration of the message by the Queen's footman, and then the repetition of the message (if not the words) by the Duchess's footman, but that meaning is somehow changing. This episode similarly reiterates the problem of representation throughout the tale (and indeed throughout the intellectual property system), the problem of the 'third man', the problem of representation: 'Representation must encompass an expression which it does not represent, but without which it itself would not be "comprehensive," and would have truth only by chance or from outside.'[2]

Notably, in the iterability of the footmen's messages, reading the actual content of the letter appears to be redundant, it is the activity of its delivery that is significant. Indeed, the content of the letter remains necessarily hidden, literally enveloped within the letter:

> Representation attains this topical ideal only by means of the hidden expression which it encompasses, that is, by means of the event it envelops. There is a 'use' of representation, without which representation would remain lifeless and senseless. Wittgenstein and his disciples are right to define meaning by means of use. But such use is not defined through a function of representation in relation to the represented, nor even through representativeness as the form of possibility ... use is in the relation between representation and something extra-representative, a nonrepresented and merely expressed entity. Representation envelops the event in another nature, it envelops it at its borders, it stretches until this point, and it brings about this lining and hem. This is the operation which

---

1   *Wonderland*, 49.
2   Deleuze, *Logic of Sense*, 145.

> defines living usage, to the extent that representation, when it does not reach this
> point, remains only a dead letter confronting that which it represents, and stupid
> in its representiveness.[3]

Further, its intended destination (the Duchess) is exceeded in that Alice too, it comes to pass, is inadvertently 'invited' to the game of croquet. In her act of observing the message, Alice somewhat changes its intention, and changes herself: *so you think you're changed, do you*? The greater the spectacle of the object, the greater the transformation achieved in the game (without necessarily attributing a value to that transformation at this stage). This is the impossibility of discounting or discrediting that which is necessary for representation, the 'third man', Alice, the user. In a literal respect, as it were, use takes up space. This is the domain of familiar production.

Occupy Movement, Occupy Use

Use is the occupational risk of the exchange in a digital environment. Nevertheless, risk imports relation, the use of the exchange. Use in a digital environment, as discussed in previous chapters, is not so much about risk management, however, as it is about chance relations. At the moment of delivery neither the sender nor the receiver need be present, either virtually or most certainly actually. In various forms of social media, despite the rhetoric of 'conversation' more usually the risk is a posting to no-one, a deferred receipt if received at all, reflected back in an unnecessary introspection. Nevertheless, this experience is being cast as the new social, and so meaning through use becomes all the more crucial and complex, and particularly difficult to hear over the drumming rhetoric and the dead letters.

Crucial to the quality and sustainability of familiar production in the digital is thus the notion of the familiar itself as place, as abode, as 'occupation'. This affective labour of the familiar can be traced in the 'cultural' models of social media as well as translated into effective business models.[4] Earlier discussions have noted that this is more complicated than simply production which is communal or collaborative.[5] In more 'traditional' social media formats, such as Facebook, the conventional markers of producer and user can nevertheless be made out. Facebook preserves the habitual, the commonplace, and the analog identity of the producer, the 'self' as brand. Therefore, although there is considerable content generated and shared through Facebook, arguably the construction of identity is remarkably similar to conventional, linear narratives of subjectivity. This also plays out in the discussions of use and commercial applications of Facebook as a branding exercise, where Facebook easily assimilates conventions of self and property,

---

3   Deleuze, *Logic of Sense*, 146.
4   Considered in more detail in '*Re* Use'.
5   See the earlier discussion in 'Use'.

of 'own' identity and even notions of intimacy and privacy. The page itself is a banner of commonplace marks of subjectivity and information, and identities are readily marked and traced through the acts of 'adding friends', 'likes' of pages and 'shares'.

Where social media becomes more interesting in terms of familiar production is through innovations upon these earlier social media models. Other forms of social media, such as Twitter, although allowing for conventional markers of identity, offer regimes that are far less organized and structured with respect to user information. The use of Twitter, for example, is thus more incidental, exponential and proliferative. Community is more dialogical and accidental, by chance and wager, inconstant and in common. The private is nevertheless still there for all to see, but the use of Twitter and other platforms is not in respect of collating that information, in the way that Facebook and early platforms are. Therefore, paradoxically, although Facebook is arguably more concerned with conventions and customs of privacy and identity, these are more 'at risk' than in the field of production in Twitter and other platforms. In other words, where Facebook maintains a banal taxonomy of users, Twitter proliferates a community of use.

Twitter and more recent social media platforms are without real identity – it is a conversation that is fundamentally text-based. If a user is not available at the moment the tweet arrives, it is unlikely the user will see it later unless specifically searching another user. In a very interesting way, therefore, Twitter imports a sense of 'relationship' and proximity into the digital environment, with an emphasis on conversation, as distinct from celebrating one's own identity or personal brand (as in Facebook). That is not to deny the branding capacity of Twitter, but to note that it operates in a particularly different way, and perhaps one that is more compatible with progressions in the way in which people use social media.[6] While Facebook, on the other hand, relies on a more 'analog' version of identity, Twitter arguably produces a mutually constitutive and 'digital' version of identity, an identity through 'use', an identity through familiar production.

The counterpoint of this is that in commercial business models for the raising of revenue through advertising, and utilized in large digital platforms such as Facebook and Google, relation imports risk. Deploying the information on users and traffic in order to offer advertisers targeted advertising, companies have found their advertisements appearing alongside undesirable content (in particular, images depicting and advocating sexual violence against women[7]). Major advertisers

---

6   See the findings in the study by M Madden et al., *Teens, Social Media, and Privacy*, Pew Research Center, Washington DC, 21 May 2013. See further the 'digital' discussions in K Belgrave, 'Why I'm done with Facebook (and moving on to Twitter)', *Tarrytown-SleepyHolly*, 21 May 2013; R Robinette, 'Twitter outlasts Facebook as a "cool" site for young people', *Trinity Tripod*, 22 April 2013; C Warzel, 'Teens explain why they don't care about Facebook anymore', *BuzzFeed*, 21 May 2013.

7   R Cellan-Jones, 'Sexism campaign: Facebook learns a lesson', *BBC News*, 29 May 2013.

have indeed withdrawn commercials, some temporarily,[8] leading to radical rethinking of advertising business models in social media platforms. This is the quintessential example of the problem of trying to predict community and custom in familiar production, and in social media users, through the use of keywords and other indices of identity and taste. Advertising business models thus far deployed in social media have nevertheless still relied upon the conventional notion of products and audiences (passive users, passive receivers), quite contrary to their own models of use and community. It is a problem of not only the intention of the message, but also the proliferation of meaning in its receipt: 'In the best case, it is theory as a message in a bottle.'[9]

Before the Lore

Alice went timidly up to the door, and knocked.

"There's no sort of use in knocking," said the Footman, "and that for two reasons. First, because I'm on the same side of the door as you are; secondly, because they're making such a noise inside, no one could possibly hear you." And certainly there was a most extraordinary noise going on within – a constant howling and sneezing, and every now and then a great crash, as if a dish or kettle had been broken to pieces.

"Please, then," said Alice, "how am I to get in?"

"There might be some sense in your knocking," the Footman went on without attending to her, "if we had the door between us. For instance, if you were *inside*, you might knock, and I could let you out, you know." He was looking up into the sky all the time he was speaking, and this Alice thought decidedly uncivil. "But perhaps he can't help it," she said to herself; "his eyes are so *very* nearly at the top of his head. But at any rate he might answer questions. – How am I to get in?" she repeated, aloud.

"I shall sit here," the Footman remarked, "till to-morrow –"

At this moment the door of the house opened, and a large plate came skimming out, straight at the Footman's head: it just grazed his nose, and broke to pieces against one of the trees behind him.

---

8   A subsequent withdrawal of advertising came from Nissan and several other commercial entities: 'Companies pull Facebook ads over violent content', *CBS News*, 29 May 2013; R Carroll, 'Facebook gives way to campaign against hate speech on its pages', *The Guardian*, 29 May 2013; D Lee, 'Facebook bows to campaign groups over "hate speech"', *BBC News*, 29 May 2013. On consumer and civil society action, see N Kemp, 'Consumers urge brands to boycott Facebook over domestic violence', *Marketing Magazine*, 22 May 2013. A large number of civil society groups for women and consumers composed an 'Open letter to Facebook', published in *The Huffington Post*, 21 May 2013.

9   Theodor Adorno in T Adorno & M Horkheimer, *Towards a New Manifesto*, R Livingstone (trans), Verso, London, [1989]/2011: 100.

"– or next day, maybe," the Footman continued in the same tone, exactly as if nothing had happened.

"How am I to get in?" asked Alice again, in a louder tone.

"*Are* you to get in at all?" said the Footman "That's the first question, you know."[10]

As distinct from the rhizomatic passage that greeted Alice when she first tumbled to the bottom of the rabbit-hole, Alice is now faced with one door and utter uncertainty as to entry. This is strikingly similar to Kafka's story 'Before the Law':

> Before the Law stands a doorkeeper. To this doorkeeper there comes a man from the country and prays for admittance to the Law. But the doorkeeper says that he cannot grant admittance at the moment. The man thinks it over and then asks if he will be allowed in later. 'It is possible,' says the doorkeeper, 'but not at the moment.' Since the gate stands open, as usual, and the doorkeeper steps to one side, the man stoops to peer through the gateway into the interior. Observing that, the doorkeeper laughs and says: 'If you are so drawn to it, just try to go in despite my veto. But take note: I am powerful. And I am only the least of the doorkeepers. From hall to hall there is one doorkeeper after another, each more powerful than the last. The third doorkeeper is already so terrible than I cannot bear to look at him.' These are difficulties the man from the country has not expected; the Law, he thinks, should surely be accessible at all times and to everyone, but as he now takes a closer look at the doorkeeper in his fur coat, with his big sharp nose and long, thin, black Tartar beard, he decides that it is better to wait until he gets permission to enter.[11]

The man sits for days waiting and then gives the doorkeeper everything he has to try and bribe him, but the doorkeeper merely says 'I am only taking it to keep you from thinking you have omitted anything.' Finally, after many years of waiting and at the end of his life, the man thinks to ask a question he has not yet asked:

> 'Everyone strives to reach the Law,' says the man, 'so how does it happen that for all these many years no one but myself has ever begged for admittance?' The doorkeeper recognizes that the man has reached his end, and, to let his failing sense catch the words, roars in his ear: 'No one else could ever be admitted here, since this gate was made only for you. I am now going to shut it.'[12]

In order to enjoy a privileged position of representation, the law must forget or deny its context, the infinite regress of representation: 'To be invested with its

---

10   *Wonderland*, 50.

11   F Kafka, 'Before the Law', W Muir & E Muir (trans), in *The Complete Short Stories*, NH Glatzer (ed), Minerva, London, 1992: 3–4, at 3.

12   Kafka, 'Before the Law', 4.

categorical authority, the law must be without history, genesis, or any possible derivation. That would be *the law of the law*. Pure morality has no history.'[13] An important aspect of Kafka's story is that the door remains open. The 'openness' indeed immobilizes the man for there is nothing to open, nothing to reform, nothing with which to engage in order to enter into play.[14] In other words, access to the law brings with it the counterpoint of immobilization of one's position with respect to the law, to be banned and thus abandoned, banished: 'The imperative is always imperious, whatever its kind. Otherwise, it is not an imperative but advice – an exhortation or a mere recommendation.'[15]

The imperative of the law thus removes the ethics of decision thus paradoxically rendering it impassable: 'The man from the country is delivered over to the potentiality of law because law demands nothing of him and commands nothing other than its own openness.'[16] In its very acceptance the law does not invite the user into play, does not entertain choice and chance, but rather, interpellates and positions the user within the law. Thus, the user, in 'access' to the law, is fixed within the law, not in play but in debt: 'law applies to him in no longer applying, and holds him in its ban in abandoning him outside itself. The open door destined only for him includes him in excluding him and excludes him in including him. And this is precisely the summit and the root of every law.'[17] Contrary to popular refrains regarding the failures of the law, intellectual property law does indeed ensure 'access' for the user, but it does so by writing the user outside the credit of knowledge. That is, the law does not limit access; rather, it negates play, and thus use. Access to the law is not absent, but meaningful access (that is, meaning through use) is beyond the law.

Ultimately this compromises the legitimacy of the law; it is a genuine perturbation in the very processes and customs of legitimation[18] in intellectual property. Thus, the law, in its application, subjugates the user to an extent that compliance does not engage the will, simply the performance of instructions in the

---

13   Derrida, *Acts of Literature*, 191.

14   See in particular the discussion in G Agamben, *Homo Sacer: Sovereign Power and Bare Life*, D Heller-Roazen (trans), Stanford UP, Stanford, [1995]/1998: 49.

15   J-L Nancy, 'From the Imperative to Law', in BC Hutchens (ed), *Jean-Luc Nancy: Justice, Legality and World*, Continuum, London, 2012: 11–18, at 11.

16   Agamben, *Homo Sacer*, 50.

17   Agamben, *Homo Sacer*, 50.

18   Lyotard defines legitimation as 'the process by which a "legislator" dealing with scientific discourse is authorized to prescribe the stated conditions'. Lyotard, *Postmodern Condition*, 8. Jürgen Habermas describes a crisis in the acceptance and legitimation of the law in his concept of 'legitimation crisis', which may be understood as not simply a collapse, but rather a rejuvenescence of emergent social potential. J Habermas, *Legitimation Crisis*, T McCarthy (trans), Beacon P, Boston, 1976. See further a similar notion of crisis in legitimacy in the concept of '*Geltung ohne Bedeutung*' or 'Being in force without significance': Agamben, *Homo Sacer*, 51.

execution of an infinite debt. The ethics of intellectual property must address the notion of a law that is without any significance for its users, for consumers. This is acute in the case of copyright, where the user is not only deferred but entirely displaced by the law: 'For life under a law that is in force without signifying resembles life in the state of exception, in which the most innocent gesture or the smallest forgetfulness can have most extreme consequences.'[19] In this respect, the relationship between use, users and producers can present as almost arbitrary in that it is without belief and relevance, without consent and legitimacy: 'law is all the more pervasive for its total lack of content, and in which a distracted knock on the door can mark the start of uncontrollable trials. Just as for Kant the purely formal character of the moral law founds its claim of universal practical applicability in every circumstance, so in Kafka's village the empty potentiality of law is so much in force as to become indistinguishable from life.'[20] In the context of familiar production, particularly in the context of social media and consequences for incidental inclusion, or streaming and the uncertain relationships of culpability,[21] this 'distracted knock on the door' becomes almost inevitable.

While behind Alice's door is 'such a noise inside, no one could possibly hear you',[22] behind Kafka's door is silence: 'the Sirens have a still more fatal weapon than their song, namely their silence.'[23] Rather than the 'infinite negotiations with the doorkeeper' that characterize access 'before the law', and the impassability of a judgment without decision, Alice embraces chance and enters the language-game: 'Following a rule is analogous to obeying an order. We are trained to do so; we react to an order in a particular way. But what if one person reacts in one way and another in another to the order and the training? Which one is right?'[24]

An impassable law of silence is intolerable and impossible for Alice, for the user. She is getting closer to her entry into the game, she is approaching meaning. As distinct from the outset of her journey, where she sought to reconcile the novel by referencing the lessons of old, Alice is now starting to disrupt class boundaries, limits of identity, function and habit. If the footman cannot open the door she will open it herself. Observing social order will merely condemn her to one side of the

---

19   Agamben, *Homo Sacer*, 52.

20   Agamben, *Homo Sacer*, 52–53.

21   This uncertainty is demonstrated in the EU context by the referral from the UK Supreme Court to the Court of Justice in *PRCA v Newspaper Licensing Agency* [2013] UKSC 18. The Court found that reading a pirated webpage should be treated in the same way as reading a pirated book (a freedom to read, as it were), yet it felt that, despite its firm view, the opinion of the Court of Justice was necessary. Uncertainty pervades at least for now.

22   *Wonderland*, 50.

23   F Kafka, 'The Silence of the Sirens', W Muir & E Muir (trans), in F Kafka, *The Complete Short Stories*, NH Glatzer (ed), Minerva, London, 1992: 430, at 431.

24   Wittgenstein, *Philosophical Investigations*, §206.

door. She makes a decision, assumes the risk, takes her chances, opens the door, and enters the game.

Becoming

> "If everybody minded their own business," the Duchess said in a hoarse growl, "the world would go round a deal faster than it does."
>
> "Which would *not* be an advantage," said Alice, who felt very glad to get an opportunity of showing off a little of her knowledge. "Just think what work it would make with the day and night! You see the earth takes twenty-four hours to turn round on its axis —"[25]

Recalling the discussion of becoming in the previous chapter, 'Becoming concerns speed'.[26] This is the instantaneity of familiar production and justice.[27] If everybody minds their own business, the world will go faster. Business implies competition, that is, in the sense of participation: beware competition and mind your own business. Contrary to the flow of becoming, 'businesses are constantly introducing an inexorable rivalry presented as healthy competition, a wonderful motivation that sets individuals against one another and sets itself up in each of them, dividing each within itself.'[28] In order to attain the speed of becoming, one must resist turning everything into a business, and simply mind one's own. This is the smooth space of familiar production, the smooth space of becoming.

What might this introduce regarding new business models in the innovative and creative industries? As considered earlier, in some ways intellectual property laws are regarded as an anathema to competition,[29] however in many ways they might be said to produce a specific set of circumstances for a particular kind of competition, based upon the speed of innovation. But this is still a causal, linear model of change as distinct from the unlimited potential of innovation. Further, innovation is distributive and distributed, not only in terms of its sites of innovation, but also in terms of the very products themselves: 'capitalism in its present form is no longer directed towards production, which is often transferred to remote parts of the Third World … It's directed toward metaproduction.'[30] In a digital world of metaproduction the 'product' is no longer the good, but the service. Contemporary intellectual property industries such as music, publishing and the like are facing revitalization around the subject matter of experience and availability, as distinct from the good and access. This is the time of innovation

---

25   *Wonderland*, 52
26   Massumi, *User's Guide*, 104.
27   See the discussion of speed and justice in Chapter 11.
28   Deleuze, *Negotiations*, 179.
29   See the earlier discussions in 'Use' and Chapter 2.
30   Deleuze, *Negotiations*, 181.

in a capitalist economy: 'What it seeks to sell is services, and what it seeks to buy is activities. It's a capitalism no longer directed toward production but toward products, that is, toward sales or markets. Thus it's essentially dispersive, with factories giving way to businesses.'[31] The important question remains as to how business might set out to overcome the banality of a competition model in a market economy. Is it possible? What is it to become?

> The baby grunted again, and Alice looked very anxiously into its face to see what was the matter with it. There could be no doubt that it had a *very* turn-up nose, much more like a snout than a real nose; also its eyes were getting extremely small for a baby: altogether Alice did not like the look of the thing at all. "But perhaps it was only sobbing," she thought, and looked into its eyes again, to see if there were any tears.
>
> No, there were no tears. "If you're going to turn into a pig, my dear," said Alice, seriously, "I'll have nothing more to do with you. Mind now!" The poor little thing sobbed again (or grunted, it was impossible to say which), and they went on for some while in silence.
>
> Alice was just beginning to think to herself, "Now, what am I to do with this creature when I get it home?" when it grunted again, so violently, that she looked down into its face in some alarm. This time there could be *no* mistake about it: it was neither more nor less than a pig, and she felt that it would be quite absurd for her to carry it any further.
>
> So she set the little creature down, and felt quite relieved to see it trot away quietly into the wood. "If it had grown up," she said to herself, "it would have made a dreadfully ugly child: but it makes rather a handsome pig, I think."[32]

The baby embodies the enormous potential of Wonderland, the production of a new assemblage of affects and creative potential, of which Alice is a part.[33] The baby literally becomes animal, becomes pig, challenging the fixation on identity that has preoccupied the manners, habit and custom from which Alice has emerged: 'becoming is not an evolution, at least not an evolution by descent and filiation. Becoming produces nothing by filiation; all filiation is imaginary. Becoming is always of a different order than filiation. It concerns alliance. If evolution includes any veritable becomings, it is in the domain of *symbioses* that bring into play beings of totally different scales and kingdoms, with no possible filiation.'[34] This fixation and limitation of identity not only restrains new business models in existing industries, but also has failed in the context of social media and emergent

---

31   Deleuze, *Negotiations*, 181.

32   *Wonderland*, 54.

33   Wonderland is full of becomings, translated in *Through the Looking-Glass* in respect of the refrain on innovation itself. See '*Re* Use' and the kitten as Queen.

34   Deleuze & Guattari, *Thousand Plateaus*, 238.

modes of production.[35] There is a critical opportunity to understand new business models in the context of the creative and innovative transformation of social and political life in the digital. The creative potential of the war machine[36] of familiar production is becoming-animal,[37] a genuine symbiotic assemblage of production that is dispersed and distinct, continuous and in common. Understanding the way in which the law and new business models engage with use as distinct from users, becoming as distinct from being, is arguably the fundamental challenge for reform of intellectual property law and practice.

Which Way?

> "Cheshire Puss," she began, rather timidly, as she did not at all know whether it would like the name: however, it only grinned a little wider. "Come, it's pleased so far," thought Alice, and she went on. "Would you tell me, please, which way I ought to go from here?"
>
>   "That depends a good deal on where you want to get to," said the Cat.
>
>   "I don't much care where –" said Alice.
>
>   "Then it doesn't matter which way you go," said the Cat.
>
>   "– so long as I get *somewhere*," Alice added as an explanation.
>
>   "Oh, you're sure to do that," said the Cat, "if you only walk long enough."[38]

The pure becoming of innovation is directionless and unlimited. While Alice is concerned to overcode this with her attachment to causality and everyday direction, the Cat assures her that no matter which way Alice chooses, she will always get *somewhere*.

> "By the by, what became of the baby?" said the Cat. "I'd nearly forgotten to ask."
>
>   "It turned into a pig," Alice quietly said, just as if it had come back in a natural way.
>
>   "I thought it would," said the Cat, and vanished again.[39]

---

35   For example, the difficulties with advertising models in social media, as seen in the Facebook example discussed earlier in this chapter.

36   This concept was introduced in Chapter 2.

37   Deleuze and Guattari note: 'The hunting machine, the war machine, the crime machine all entail all kinds of becomings-animal that are not articulated in myth, *still less in totemism* … The war machine is always exterior to the State, even when the State uses it, appropriates it. The man of war has an entire becoming that implies multiplicity, celerity, ubiquity, metamorphosis and treason, the power of affect.' Deleuze & Guattari, *Thousand Plateaus*, 242–43.

38   *Wonderland*, 55.

39   *Wonderland*, 56.

Alice is left quite literally with an interpretation hanging in the air: '"But how can a rule shew me what I have to do at *this* point? Whatever I do is, on some interpretation, in accord with the rule." – This is not what we ought to say, but rather: any interpretation still hangs in the air along with what it interprets, and cannot give it any support. Interpretations by themselves do not determine meaning.'[40] In reforming the law, interpretation of intellectual property by itself will not determine meaning. There must be use, hanging in the air along with its interpretation: '"Then can whatever I do be brought into accord with the rule?" – Let me ask this: what has the expression of a rule – say a sign-post – got to do with my actions? What sort of connexion is there here? – Well, perhaps this one: I have been trained to react to this sign in a particular way, and now I do so react to it.'[41] What such interpretation offers to a discourse of reform is merely the passage, not the meaning: 'But that is only to give a causal connexion; to tell how it has come about that we now go by the sign-post; not what this going-by-the-sign really consists in. On the contrary; I have further indicated that a person goes by a sign-post only in so far as there exists a regular use of sign-posts, a custom.'[42]

The custom of intellectual property is therefore not in and of itself interchangeable with innovation. It is simply a custom in relation to innovation and creativity: 'To obey a rule, to make a report, to give an order, to play a game of chess, are *customs* (uses, institutions).'[43] This is to limit interpretation to the imperative as distinct from the potential of the law through its meaning: 'To understand a sentence means to understand a language. To understand a language means to be master of a technique.'[44] Rule-following may in some senses appear to be arbitrary, and if devoid of content (and consent) will fail for legitimacy. Genuine and effective reform is not merely a consultation but a dialogical and iterative production of consent to certain linguistic and regulatory communities; it is consent to a form of life: 'So you are saying that human agreement decides what is true and what is false? – It is what human beings *say* that is true and false; and they agree in the *language* they use. That is not agreement in opinions but in form of life.'[45] The challenge for intellectual property in the digital is the inoperability of consensus, as it were, not through disagreement but through marginalization of the user from the game itself. Where the user is disenfranchised by the system (whether they participate in the conventions of production or consumption) and the teaching, then the instruction fails. And as noted in earlier discussion, this is perhaps exhibited most extremely in copyright, where the user is almost wholly absent.

---

40 Wittgenstein, *Philosophical Investigations*, §198.
41 Wittgenstein, *Philosophical Investigations*, §198.
42 Wittgenstein, *Philosophical Investigations*, §198.
43 Wittgenstein, *Philosophical Investigations*, §199.
44 Wittgenstein, *Philosophical Investigations*, §199.
45 Wittgenstein, *Philosophical Investigations*, §241.

Meaning cannot be found at the bottom of the well for Alice. It flourishes through dialogue, through use.

A Grin without a Cat

> Alice waited a little, half expecting to see it again, but it did not appear, and after a minute or two she walked on in the direction in which the March Hare was said to live. "I've seen hatters before," she said to herself; "the March Hare will be much the most interesting, and perhaps, as this is May, it won't be raving mad – at least not so mad as it was in March." As she said this, she looked up, and there was the Cat again, sitting on a branch of a tree.
>
> "Did you say pig, or fig?" said the Cat.
>
> "I said pig," replied Alice; "and I wish you wouldn't keep appearing and vanishing so suddenly: you make one quite giddy."
>
> "All right," said the Cat; and this time it vanished quite slowly, beginning with the end of the tail, and ending with the grin, which remained some time after the rest of it had gone.
>
> "Well! I've often seen a cat without a grin," thought Alice; "but a grin without a cat! It's the most curious thing I ever saw in all my life."[46]

Alice is left, literally, with an interpretation in the air: 'any interpretation still hangs in the air along with what it interprets, and cannot give it any support'.[47]

> "I almost wish I'd gone to see the Hatter instead!"[48]

---

46   *Wonderland*, 56.

47   Wittgenstein, *Philosophical Investigations*, §198.

48   *Wonderland*, 57.

Chapter 7
Time

Plenty of Room, But No Time

> The table was a large one, but the three were all crowded together at one corner
> of it. "No room! No room!" they cried out when they saw Alice coming. "There's
> *plenty* of room!" said Alice indignantly, and she sat down in a large arm-chair at
> one end of the table.[1]

While there is plenty of room at the table, it is time that in fact creates the
'scarcity' of resources here. While the intellectual property system is accredited
with imposing the rights controlling *access* (as the quality governing scarcity),
arguably the scarcity integral to the modern intellectual property system is *actually*
one of time.

The time of the Mad Tea-Party is in stark contrast to the linear, causal, arboreal
time of the rabbit-hole. It is the time of the rhizome: 'any point of a rhizome can
be connected to any other, and must be. This is very different from the tree or root,
which plots a point, fixes an order.'[2] In contrast to the rhizome, the taxonomical
and arboreal metaphor dominates conventions of representation in knowledge, and
indeed in innovation: 'Now, there is no doubt that trees are planted in our heads:
the tree of life, the tree of knowledge, etc. The whole world demands roots. Power
is always arborescent. There are few disciplines which do not go through schemes
of arborescence ... And yet, nothing goes through there, even in these disciplines.'[3]
Here, instead of the 'tree of life', of the 'tree of knowledge', everything is *now
here*: '"You can think now of *this* now of *this* as you look at it, can regard it now
as *this* now as *this*, and then you will see it now *this* way, now *this*" – *What* way?
There *is* no further qualification.'[4]

Compared to the 'tree-like' genealogy[5] of conventional narratives of innovation,
the digital presents a rhizomatic potential (and 'catastrophe') for the language-
game: 'A method of the rhizome type ... can analyze language only by decentering
it onto other dimensions and other registers. A language is never closed upon itself,

---

1   *Wonderland*, 57.

2   Deleuze & Guattari, *Thousand Plateaus*, 7.

3   G Deleuze & C Parnet, *Dialogues*, H Tomlinson & B Habberjam (trans), Columbia
UP, New York, [1977]/1987: 25–26.

4   Wittgenstein, *Philosophical Investigations*, 200e.

5   Indeed, 'There is always something genealogical about a tree.' Deleuze & Guattari,
Thousand Plateaus, 8.

except as a function of impotence.'[6] As distinct from the everyday time of causality, linearity, time of innovation is the virtual time of becoming, 'the indefinite time of the event'.[7] This distinction between ordinary linear time and the virtual time of innovation can be explained by Deleuze's concepts of Chronos (everyday, causal time) and Aion (the virtual time of becoming): 'Whereas Chronos was limited and infinite, Aion is unlimited, the way that future and past are unlimited, and finite like the instant. Whereas Chronos was inseparable from circularity and its accidents ... Aion stretches out in a straight line, limitless in either direction.'[8] Innovation is essentially endless, but it is also necessarily directionless. This is indeed the habitual abode of familiar production.

Saying

The Hatter's riddle is a curious episode in the logic of representation. The Hatter asks a question, a puzzle, to which Alice is certain there must be an answer. However, she is left unsatisfied as the riddle's puzzle is to query the difference between knowledge and understanding. While Alice believes she knows the answer, the Hatter suggests she means she could find the answer (could come to know):

> "Your hair wants cutting," said the Hatter. He had been looking at Alice for some time with great curiosity, and this was his first speech.
>
> "You shouldn't make personal remarks," Alice said with some severity; "it's very rude."
>
> The Hatter opened his eyes very wide on hearing this; but all he *said* was, "Why is a raven like a writing-desk?"
>
> "Come, we shall have some fun now!" thought Alice. "I'm glad they've begun asking riddles. – I believe I can guess that," she added aloud.
>
> "Do you mean that you think you can find out the answer to it?" said the March Hare.
>
> "Exactly so," said Alice.
>
> "Then you should say what you mean," the March Hare went on.
>
> "I do," Alice hastily replied; "at least – at least I mean what I say – that's the same thing, you know."
>
> "Not the same thing a bit!" said the Hatter. "You might just as well say that 'I see what I eat' is the same thing as 'I eat what I see'!"
>
> "You might just as well say," added the March Hare, "that 'I like what I get' is the same thing as 'I get what I like'!"[9]

---

6   Deleuze & Guattari, *Thousand Plateaus*, 8.
7   Deleuze & Guattari, *Thousand Plateaus*, 262.
8   Deleuze, *Logic of Sense*, 165.
9   *Wonderland*, 57–58.

The exchange provides an intriguing demonstration of the principle of saying versus showing, to show what is missing in order to say, to pretend at the full picture. The Hatter has literally made a riddle, pierced with holes the argument, and exposed the timely presumption of representation. The Hatter's riddle is a riddle without an answer, the impossibility of representing the unrepresentable. To understand the logic of language, therefore, is to understand the limits of saying, of meaning, of sense: 'What *can* be shown, *cannot* be said.'[10] Again, this is an episode in the iterations upon representation throughout this tale: 'Philosophy cannot escape from the limits of philosophy, of language, that is.'[11]

The Hatter's riddle thus confounds the usual pastime (and past time) of the riddle, that is, of the purpose of the answer. A riddle is literally a question, the purpose of which is surely the answer. However, in the case of the Hatter's riddle, the usual intentionality of the teller is absent. The riddle has no answer.[12] This too is a riddle. But, just like the '"answerless riddles" of *Alice*',[13] whether it is a riddle without an answer or an answer that poses more questions is the paradox of this inquiry itself: 'There is therefore an aspect in which problems remain without a solution, and the question without an answer.'[14] In many respects, therefore, the riddle is also a 'riddle' of time. This game of question and answer, traditionally to pass the time, is defeated, put out of court and beyond the law by the 'instant' and 'endless' time of the Hatter's tea-party.

Personal Time

Alice does not participate in the language-game and instead continues to search for an answer to the riddle itself, missing what the riddle shows but cannot say. To reiterate, an ordinary riddle is usually something for which there is an answer, an endpoint. That is, the usual purposeful and teleological nature of riddles is at odds with the instantaneity and endlessness of the riddles in Wonderland.[15] And again,

---

10   Wittgenstein, *Tractatus Logico-Philosophicus*, 4.1212.

11   G Bataille, *Erotism: Death and Sensuality*, M Dalwood (trans), City Lights Books, San Francisco, [1957]/1986: 274.

12   The intrigue and mystification of readers at the Hatter's riddle was addressed by Carroll in his preface to the 1896 edition: 'Enquiries have so often been addressed to me, as to whether any answer to the Hatter's Riddle can be imagined, that I may as well put on record here what seems to me to be a fairly appropriate answer, viz. "Because it can produce a few notes, though they are *very* flat; and it is never put with the wrong end in front!" This, however, is merely an afterthought; the Riddle, as originally invented, had no answer at all.' *Preface*, 356.

13   Deleuze, *Logic of Sense*, 56.

14   Deleuze, *Logic of Sense*, 56.

15   Another well-known Carroll tale, 'What the Tortoise said to Achilles', similarly is articulated around a riddle with no answer. See *Tortoise to Achilles*, 455–56.

the riddle is a game, a pastime and is introduced at the Mad Tea-Party as such; that is, literally to pass time, to undertake the impossible. At the Mad Tea-Party, time never passes, and riddles have no answers:

> "Have you guessed the riddle yet?" the Hatter said, turning to Alice again.
> "No, I give up," Alice replied: "what's the answer?"
> "I haven't the slightest idea," said the Hatter.
> "Nor I," said the March Hare.[16]

This question of purpose and answer, the problem of representation itself, not only recurs throughout this journey and this inquiry, but also throughout the dogma of intellectual property: 'Most men are indifferent to this problem. It is not necessary to answer the riddle of existence; it is not even necessary to ask it. But the fact that a man may possibly neither answer it nor even ask it does not eliminate the riddle.'[17] On the one hand is the moral certainty of an incontrovertible purpose and truth, and on the other, the ethics of decision and uncertainty.

> Alice sighed wearily. "I think you might do something better with the time," she said, "than waste it asking riddles with no answers."
> "If you knew Time as well as I do," said the Hatter "you wouldn't talk about wasting *it*. It's *him*."
> "I don't know what you mean," said Alice.
> "Of course you don't!" the Hatter said, tossing his head contemptuously. "I dare say you never even spoke to Time!"
> "Perhaps not," Alice cautiously replied: "but I know I have to beat time when I learn music."
> "Ah! that accounts for it," said the Hatter. "He won't stand beating. Now, if you only kept on good terms with him, he'd do almost anything you liked with the clock."[18]

This personification of time and other concepts (which indeed occurs throughout Alice's journey) is wholly consistent with her journey of becoming: "Climate, wind, season, hour are not of another nature than the things, animals, or people that populate them, follow them, sleep and awaken within them."[19] It is prudent to keep on good terms.

---

16   *Wonderland*, 59.
17   Bataille, *Erotism*, 274.
18   *Wonderland*, 59–60.
19   Deleuze & Guattari, *Thousand Plateaus*, 263. Deleuze and Guattari explain the instantaneity of becoming as follows: 'Five o'clock is this animal! This animal is this place! "The thin dog is running in the road, this dog is the road," cries Virginia Woolf. That is how we need to feel. Spatiotemporal relations, determinations, are not predicates of the thing but dimensions of multiplicities. The street is as much a part of the omnibus-horse

Instant Messaging

"Well, I'd hardly finished the first verse," said the Hatter, "when the Queen jumped up and bawled out, 'He's murdering the time! Off with his head!'"

"How dreadfully savage!" exclaimed Alice.

"And ever since that," the Hatter went on in a mournful tone, "he won't do a thing I ask! It's always six o'clock now."[20]

Unlike the circularity of the conversation with the Caterpillar, with consequences of actions and an adherence to bodily changes, Alice now encounters the time of innovation, the time of the instant, the time of six o'clock.[21] The time of innovation is 'unlimited, the way that future and past are unlimited, and finite like the instant'.[22] Time is freedom. The time is now: 'it's always tea-time.'[23] This is the smooth space-time of familiar production, a time 'without counting' as distinct from striated space-time where 'one counts in order to occupy'.[24] In other words, smooth space-time, the smooth space of familiar production, is occupied through use, rather than through measure. The difference and mixture of time in smooth and striated space 'makes palpable or perceptible the difference between nonmetric and metric multiplicities, directional and dimensional spaces'.[25]

To recall earlier discussions,[26] the incorporation of the social in production has transformed the time of production: 'Consider, for example, the transformation of the working day in the immaterial paradigm, that is, the increasingly indefinite division between work time and leisure time.'[27] In social media platforms, and the obligatory participation in the digital marketplace, the user is, to an extent, 'employed' in the production of surplus value through social media content and the collection of social data within those platforms.[28] Production is not only outside the factory, but also outside time: 'work time tends to expand to the entire time of

---

assemblage as the Hans assemblage the becoming-horse of which it initiates. We are all five o'clock in the evening, or another hour, or rather two hours simultaneously, the optimal and the pessimal, noon-midnight, but distributed in a variable fashion.' Deleuze & Guattari, *Thousand Plateaus*, 263. This is the logic of Wonderland and the 'model' of business that must be deciphered.

20   *Wonderland*, 61.

21   'We are all five o'clock in the evening, or another hour, or rather two hours simultaneously, the optimal and the pessimal, noon-midnight, but distributed in a variable fashion.' Deleuze & Guattari, *Thousand Plateaus*, 263.

22   Deleuze, *Logic of Sense*, 165.

23   *Wonderland*, 61.

24   Deleuze & Guattari, *Thousand Plateaus*, 477.

25   Deleuze & Guattari, *Thousand Plateaus*, 477.

26   In particular, see the earlier discussion in 'Use'.

27   Hardt & Negri, *Multitude*, 111.

28   This is discussed further in Chapter 2.

life. An idea or an image comes to you not only in the office but also in the shower or in your dreams.'[29] This is the infinity and instant of debt: 'Sovereignty died of the maneuvers that brought about the general submission to the concern for the future: Nietzsche alone restored it to the reign of the moment.'[30] Thus, the future, an ethics *to come*, becomes accountable to current power relations.[31]

Time-Barred Debt

In the debt economy, the future is frozen, the debt is infinite: 'For debt simply neutralizes time, time as the creation of new possibilities, that is to say, the raw material of all political, social, or esthetic change. Debt harnesses and exercises the power of destruction/creation, the power of choice and decision.'[32] This is not to describe debt as the time of the instant; on the contrary, it is to understand the way in which debt is time-barred altogether. The consumer is thus rendered both without choice (without taste) and without chance: 'The debt economy ... has deprived them of the future, that is, of time, time as decision-making, choice, and possibility.'[33]

Modern technology is always already in credit: 'objectivising time, possessing it in advance, ... subordinating all possibility of choice and decision which the future holds to the reproduction of capitalist power relations'.[34] In other words, the consumer is entirely accountable within the system of debt: 'debt appropriates not only the present labor time of wage-earners and the population in general, it also pre-empts non-chronological time, each person's future as well as the future of society as a whole. The principal explanation for the strange sensation of living in a society without time, without possibility, without foreseeable rupture, is debt.'[35]

This relationship is explicit and literal in the case of high technology products, where the product is always already obsolete, holding the consumer is without time and in an interminable debt relationship to the producer (and to consumption), either directly (through functionality and operability) or indirectly (through the tie-in

---

29   Hardt & Negri, *Multitude*, 111–12.

30   G Bataille, *The Accursed Share: An Essay on General Economy, Volumes II and III*, R Hurley (trans), Zone, New York, [1976]/1991: 379.

31   Lazzarato, *Making of the Indebted Man*, 46.

32   Lazzarato, *Making of the Indebted Man*, 49.

33   Lazzarato, *Making of the Indebted Man*, 8. Lazzarato is speaking specifically of Europeans and the European debt crisis, however, the principles of the debt economy remain unchanged. As Deleuze and Guattari have explained previously in *Anti-Oedipus*, debt is infinite, 'The infinite creditor and infinite credit have replaced the blocks of mobile and finite debts ... the debt becomes a *debt of existence*' (197) and the law '*is the juridical form assumed by the infinite debt*' (213).

34   Lazzarato, *Making of the Indebted Man*, 46.

35   Lazzarato, *Making of the Indebted Man*, 45–46.

to new use). On the other hand, the producer is in a position to prescribe and sell 'time' (as use, as innovation), as it were, through evergreening technology[36] and similar innovation strategies to maintain market share. Such innovation strategies highlight the key contrast in a culture of risk (central to the debt economy) and a culture of chance (and associated choice). In a debt and credit relationship of use and production respectively, there is a subordination of choice and decision – this is the dictate, the prescript, the imperative of production. Where new business models similarly subjugate use to production, and limit choice and decision, the legitimacy and efficacy of those business models will be in question. An example of one such problem in the translation from policy to use, from the social to the technical,[37] is the Digital Copyright Exchange (DCE).

Proposed in the Hargreaves Review,[38] the DCE[39] and its model to contain all copyright material for the purposes of licensing was in many respects unworkable in practice,[40] largely for want of consensus with respect to use. In order to facilitate the recommendation, following a feasibility study in 2012,[41] it was modified for launch as a hub for information concerning rights ownership and copyright licences.[42] In its original inception, the individual accessing the DCE would not interact; that is, the individual would not use, it would simply follow. In many respects this disrupts the relationship between access and use, and so the 'meaning' generated between issuer and user: 'the presence of antiproduction within production itself'.[43] The user is again displaced and disenfranchised from

---

36   A range of activities can come within the term 'evergreening', to include any legal or business strategies with respect to patents nearing expiration. This can include incremental or associated innovation upon the invention, which may make it possible to obtain a new patent that will also restrict access to the technology covered by the existing patent. In addition, evergreening can refer to other indirect activities with respect to management and competition. The practice of evergreening is especially controversial in the area of pharmaceutical patents and medicinal products.

37   Recall the discussion of the 'Tragedy of the Commons' in Chapter 5.

38   I Hargreaves, *Digital Opportunity: A Review of Intellectual Property and Growth*, May 2011. See further supporting information from the UK Intellectual Property Office (IPO) at www.ipo.gov.uk/ipreview.htm.

39   Hargreaves, *Digital Opportunity*, 28–38.

40   A Orlowski, 'Hargreaves' Digital Copyright Exchange will never happen', *The Register*, 23 May 2011; T Ingham, 'Digital Copyright Exchange: who's going to pay for it?' *Music Week*, 31 July 2012.

41   R Hooper & R Lynch, *Copyright Works: Streamlining Copyright Licensing for the Digital Age*, Intellectual Property Office, Cardiff, July 2012. The report recommends the adaptation of the DCE recommendations to that of an information hub: Annex E, 51.

42   T Ingham, 'Digital Copyright Exchange: UK Music, PPL, PRS applaud Hooper', *Music Week*, 31 July 2012. See also the analysis on behalf of the Authors' Licensing and Collecting Society, R Combes, 'The copyright hub', *ALCS News*, 20 May 2013.

43   Deleuze & Guattari, *Anti-Oedipus*, 235. See further: 'On the one hand, it alone is capable of realizing capitalism's supreme goal, which is to produce lack in the large

the system, so as to render access to the system less accurate and less meaningful: 'The digital language of control is made up of codes indicating whether access to some information should be allowed or denied. We're no longer dealing with a duality of mass and individual. Individuals become *"dividuals,"* and masses become samples, data, markets, or *"banks."*'[44]

Author Series

> "Suppose we change the subject," the March Hare interrupted, yawning. "I'm getting tired of this. I vote the young lady tells us a story."[45]

Given the opportunity of authorship, Alice is at once agitated and disconcerted.

> "I'm afraid I don't know one," said Alice, rather alarmed at the proposal.
> "Then the Dormouse shall!" they both cried. "Wake up, Dormouse!" And they pinched it on both sides at once.[46]

However, it is in the iterative dialogue of story that Alice is both motivated and disruptive in the social life of the narrative, repeatedly derailing and distracting the story into unanticipated routes and territories through a conscientious questioning of the Dormouse:

> "Why did they live at the bottom of a well?"
> The Dormouse again took a minute or two to think about it, and then said, "It was a treacle-well."
> "There's no such thing!" Alice was beginning very angrily, but the Hatter and the March Hare went "Sh! sh!" and the Dormouse sulkily remarked, "If you can't be civil, you'd better finish the story for yourself."
> "No, please go on!" Alice said. "I won't interrupt again. I dare say there may be *one*."[47]

---

aggregates, to introduce lack where there is always too much, by effecting the absorption of overabundant resources. On the other hand, it alone doubles the capital and the flow of knowledge with a capital and an equivalent flow of *stupidity* that also effects an absorption and a realization, and that ensures the integration of groups and individuals into the system. Not only lack amid overabundance, but stupidity in the midst of knowledge and science.' Deleuze & Guattari, *Anti-Oedipus*, 235–36.

44   Deleuze, *Negotiations*, 180. Recall the introduction to this notion of subjugation and submission, and abdication of responsibility (and decision) in the concept of 'machinic enslavement' discussed in 'Use'.

45   *Wonderland*, 61.

46   *Wonderland*, 61.

47   *Wonderland*, 62.

Alice attempts to disrupt the story, to engage as author: 'The author's gesture is attested to as a strange and incongruous presence in the work it has brought to life, in exactly the same way that – according to the theorists of the commedia dell-arte – the Harlequin's *lazzo* incessantly interrupts the story unfolding on the stage and continually unravels the plot.'[48] Indeed, this 'interruption' becomes the site of meaning in the story, the moments of explanation, of departure, of further detail. As the recipient of the story, Alice renders herself responsible, calculable, accountable. Alice, the user, is a worthy guarantor of the story and its meaning.[49]

Meaning is thus a negotiation between the story, the Dormouse and Alice. That is, meaning is a negotiation between text, author and reader. Alice is beginning to dispute the credit attributed to the story-teller and is looking to 'put herself into play' as author.[50] This introduces the critical aspect of knowledge negotiations in contemporary interactions between text, user and producer within the framework of intellectual property debt. Meaning is achieved through 'the gesture through which the author and reader put themselves into play in the text and, at the same time, are infinitely withdrawn from it'.[51]

Memory Games

> "They were learning to draw," the Dormouse went on, yawning and rubbing its eyes, for it was getting very sleepy; "and they drew all manner of things – everything that begins with an M –"
>
> "Why with an M?" said Alice.
>
> "Why not?" said the March Hare.
>
> Alice was silent.
>
> The Dormouse had closed its eyes by this time, and was going off into a doze; but, on being pinched by the Hatter, it woke up again with a little shriek, and went on: "– that begins with an M, such as mouse-traps, and the moon, and memory, and muchness – you know you say things are 'much of a muchness' – did you ever see such a thing as a drawing of a muchness?"[52]

Alice was silent.

While Alice's memory is challenged throughout her adventures in Wonderland, her attempts to repeat and recollect do not anchor her back to her past self.

---

48 G Agamben, *Profanations*, J Fort (trans), Zone, New York, [2005]/2007: 70.

49 On the notion of responsibility and debt, see the discussion in Chapter 2. See further Nietzsche, *On the Genealogy of Morals*, 58; and Deleuze & Guattari, *Anti-Oedipus*, 190–92.

50 Agamben, *Profanations*, 61–72.

51 Agamben, *Profanations*, 71. See further Chapter 8 and the discussion of games, and Chapter 11 and the discussion of accountability to the 'expression'.

52 *Wonderland*, 63.

Instead, they make a difference, a creative difference. Rather than predict her future (necessarily subjugating it to the prescriptives of the 'past' imperatives), her recollections are almost subject to chance: 'Chance commingles with a feeling of déjà vu.'[53] Lessons like her Multiplication Tables, ordinarily classified and ordered in her hierarchical 'tree of knowledge', are muddled and destabilized in her journey of becoming in Wonderland. Her erroneous recitations of lessons throughout her journey resonate here in the Dormouse's story of the three little sisters who live at the bottom of a well. The three sisters learn to draw everything that begins with an M. In addition to the story itself containing a language and memory game (everything that begins with 'M'), the sisters are supposed to be drawing 'memory', as well as 'muchness'. They are not creating new knowledge, but recollecting memory in order to recognize difference (in all its 'muchness'). Memory is therefore not so much a representation, but a performance, a new presentation, an experience. Memory is the déjà vu of chance, the instant of innovation.

This recalls earlier discussions of the archive function of the intellectual property system.[54] Perhaps it is correct after all to attribute an incentive-function to intellectual property, but perhaps for reasons other than those usually offered. In classifying innovation in relation to an archive, a prior art, in providing innovation with a history, the repetition of the system necessarily drives the new. The 'past' constitutes a new present in innovation. It is not necessarily the promise of a right that constitutes a convincing incentive, but rather an experience of the archive, a 'debt' to the archive. The archive thus possesses a certain 'justice' with respect to the collective memory, legacy and heritage. Therefore, while articulating the link between memory and justice, the archive also implements the duty of memory, the 'heritage-debt'.[55] This notion of heritage as debt draws in aspects of succession and a proprietary relationship to heritage that binds notions of self within the language of property and the culpability of debt. Thus, the attachment to the evidence of the register within the juridical narrative of intellectual property is wholly compatible with the notion of heritage in innovation and creativity, and the infinite debt to the producer.[56]

The fact that such an archive is achieved in part within the functioning of the intellectual property system is therefore perhaps more than just a convenience; rather, it is a structural requirement in order to ensure the repeatability and certainty of the system. Whether or not this is translated within other types of 'libraries' is of particular interest to the record of familiar production. The crucial question is how, if at all, familiar and continuous production might be translated into a timely, commercial, economic market; that is, whether it is possible to conduct business

---

53   Bataille, *On Nietzsche*, 73.

54   See in particular the discussion in 'Use' and in Chapter 4.

55   P Ricoeur, *Memory, History, Forgetting*, K Blamey & D Pellauer (trans), U of Chicago P, Chicago, [2004]/2006: in particular, 363–64.

56   This is considered in detail in Chapter 2.

in familiar production. Indeed, is the most important question to ask that of the question of responsibility to the archive rather than incentives for the new? Or is it necessary to consider the 'ethics' of memory in the context of familiar production, and the tantalizing opportunity of ambiguity and chance in the record itself?

Could perhaps other 'libraries' drive the new in a very similar process, whether these are physical libraries, internet archives,[57] personal records,[58] conversations,[59] memory? The haunted absorption and obsessive curation of the internet pretends at the map of all maps, the encyclopaedia of all encyclopaedia, the life of all life.[60] But that archive is indexical, or merely indicative perhaps, always deferred and displaced from life itself. That is, it becomes merely a witness, irrevocably altering life but at the same time utterly unable to account for itself as part of life.[61] In the very fallibility and failure of the archive, a discourse and dialogue of familiarity is 'invented'. This is the success of forgetting.[62]

In other words, while the archive presents memory as a kind of time-bound repository (whether this is achieved through a register of innovation such as that

---

57 For example, the internet archive Wayback Machine is a non-profit organization founded in San Francisco in 1996 in order to construct a library of the internet. The Internet Archive is available at http://archive.org/. See further, on the transition to the internet of things, the discussion in B Stiegler, 'The Indexing of Things', in U Ekman (ed), *Throughout: Art and Culture Emerging with Ubiquitous Computing*, MIT P, Cambridge MA, 2013: 493–501.

58 Such as Facebook, for example.

59 Such as in Twitter, for example.

60 This illogicality of the representation, the necessary forgetting of the archive, is perfected by Borges in the short parable, 'Of Exactitude in Science': ' In that Empire, the craft of Cartography attained such Perfection that the Map of a Single province covered the space of an entire City, and the Map of the Empire itself an entire Province. In the course of Time, these Extensive maps were found somehow wanting, and so the College of Cartographers evolved a Map of the Empire that was of the same Scale as the Empire and that coincided with it point for point.' JL Borges, 'Of Exactitude in Science', in JL Borges, *A Universal History of Infamy*, NT di Giovanni (trans), Penguin, London, [1954]/1975: 131.

61 Bernard Stiegler notes: 'It is only because an object is a memory support in a system of memory with other objects that the psyche of an individual for whom it is the object can project and place itself outside, where its traces remain. Because objects are such supports and such witnesses, the archaeologist and the prehistorian can through these objects reconstitute modes of life without recurring to writing, the main source of information for the historian.' Stiegler, 'Indexing of Things', 495.

62 In the short story 'The Aleph' Borges confronts the impossible experience of everything all at once and not at all, encompassed in the Aleph, 'the place where, without any possible confusion, all the places in the world are found, seen from every angle'. After encountering the Aleph himself, Borges explains, 'I feared that not a single thing was left to cause me surprise; I was afraid I would never be quite of the impression that I had "returned." Happily, at the end of a few nights of insomnia, forgetfulness worked in on me again.' JL Borges, 'The Aleph', A Kerrigan (trans), in JL Borges, *A Personal Anthology*, A Kerrigan (ed), Grove P, New York, [1961]/1967: 138, at 147 and 152 respectively.

provided by intellectual property or through a curation of the internet and so on), familiar production defeats the notion of a linear and logical, clear and complete record. Every utterance, every production, is always already in circulation. The memory of familiar production is therefore continuous, productive, repeated and different, resonating with the 'novelistic manner in which little modifications are torn from the brute and mechanical repetitions of habit, which in turn nourish repetitions of memory'.[63] The archive of familiar production presents not the obligation to the same, but the cacophony and freedom of the repetition of difference: *much of a muchness*.

The Time of Silence

Alice was silent.

The library, the museum, the gallery, the internet and memory itself are all silent: 'don't think, but look!'[64] Bataille explains: 'The supreme moment is indeed a silent one, and in the silence our consciousness fails us.'[65] Again, the problem of representation is presented: 'Words paint and sing, but only at the limit of the path they trace through their divisions and combinations. Words create silence.'[66] This silence is intriguing, in that the archive is at once the diagnostic tool of civilization, its therapy, and its speech, and its silence and preservation. While the archive is sometimes described as a protection against use as waste,[67] and commercialization as a guard against waste of innovation and a direction into use, the paradox is that in both respects, each can be understood as use and as waste, depending on the meaning. In other words, this persistent over-simplification of the role of the archive, and of its supposed counterpoint in industry, must be overcome in the interests of transformative business models for familiar production. Industry itself can operate as an archive, a repository, and in silence. And similarly, the museum can extend use, communication and production, and indeed itself be a site of innovation.

It is thus necessary to understand further the way in which a legal framework can perform an archival function. As noted earlier, the intellectual property system, with perhaps the exception of copyright, is built upon obsolescence, redundancy, and so it at once remembers all innovation and forgets it, effaces it. In this regard copyright is set apart from the rest of intellectual property in that it excludes both the user and its memory. Without register, without archive, without use, copyright is simply assumed, leading to a fundamentally different relationship with respect

---

63   Deleuze, *Difference and Repetition*, 294.

64   Wittgenstein, *Philosophical Investigations*, §66.

65   Bataille, *Erotism*, 276.

66   Deleuze, *Essays Clinical and Critical*, 113.

67   Recall also the notion of preservation and street art in the case of Banksy, discussed in 'Use'.

to value, use and, indeed, to infringement. The irony, as it were, of the rest of the intellectual property system is its 'death drive', its rampant destruction of memory through the obsolescence of product, the obsolescence of effort. The archive of the intellectual property system, its 'wider' world of prior art, actually supports the forgetting in favour of the new. It is archived so that it can be forgotten. It is institutionalized so that it need not be mentioned again: 'The dwelling, this place where they dwell permanently, marks this institutional passage from the private to the public, which does not always mean from the secret to the nonsecret.'[68] This is the silence of the archive – the 'house arrest' of memory.[69] The archive preserves all that is forgotten, erased as timeless (literally without time), 'only kept and classified under the title of the archive by virtue of a privileged *topology*'.[70]

The relationship between the intellectual property system and the archive is therefore primarily one of citation, with the archive being the rich display of resources, the lore of causality, reaffirming the progress of innovation. For contemporary innovation, every archive has 'at once an institutive and conservative function',[71] and for the contemporary archive, with its almost simultaneous doubling of innovation, forgetting is being achieved more and more quickly. In many respects, this makes the system more and more efficient. But what of innovation and the archive *to come*?[72]

Is this indeed the answer to the Hatter's riddle? That is, perhaps the answer is to puncture, to riddle with holes, the front of language in order to reveal meaning, '"boring holes" in the surface of language so that "what lurks behind it" might at last appear ... to allow for the emergence of the void or the visible in itself, the silence or the audible in itself'.[73]

Message in a Bottle

In the best case, it is theory as a message in a bottle.[74]

---

68   J Derrida, *Archive Fever: A Freudian Impression*, E Prenowitz (trans), U of Chicago P, Chicago, [1995]/1996: 2–3.

69   Derrida, *Archive Fever*, 2.

70   Derrida, *Archive Fever*, 3.

71   Derrida, *Archive Fever*, 7.

72   As noted in Chapter 5, Derrida speaks of 'democracy to come': Derrida, *Negotiations*, 178–83; see also the discussion of the 'yet-to-come' or *avenir* in J Derrida, 'Force of Law: The "Mystical Foundation of Authority"', in D Cornell et al. (eds), *Deconstruction and the Possibility of Justice*, Routledge, New York, [1990]/1992: 3–67, at 36. See also the discussion throughout Part III.

73   Deleuze, *Essays Critical and Clinical*, 173–74.

74   Theodor Adorno in Adorno & Horkheimer, *Towards a New Manifesto*, 100.

The riddle of familiar innovation is the presence of the user: 'Sociologists, however, ponder the grimly comic riddle: where is the proletariat?'[75] Where is the user?

> "At any rate I'll never go *there* again!" said Alice as she picked her way through the wood. "It's the stupidest tea-party I ever was at in all my life!"[76]

This marks a turning point in her adventure, Alice is about to enter the game. At any rate.

> Once more she found herself in the long hall, and close to the little glass table. "Now, I'll manage better this time," she said to herself, and began by taking the little golden key, and unlocking the door that led into the garden. Then she set to work nibbling at the mushroom (she had kept a piece of it in her pocket) till she was about a foot high: then she walked down the little passage: and *then* – she found herself at last in the beautiful garden, among the bright flower-beds and the cool fountains.[77]

At last in the beautiful garden … At last.

---

75   T Adorno, *Minima Moralia: Reflections from Damaged Life*, EFN Jephcott (trans), Verso, London, [1951]/1974: 194.

76   *Wonderland*, 64.

77   *Wonderland*, 64.

PART III
Of Games

Chapter 8
Rule

Everything now returns to the surface.[1]

Paint the Roses Red

> A large rose-tree stood near the entrance to the garden: the roses growing on it
> were white, but there were three gardeners at it, busily painting them red.[2]

Alice, the user, is restored to the game at the surface, a scrutiny of the borders,
where there is everything to play for: 'It is not therefore a question of *the
adventures* of Alice, but of Alice's *adventure:* her climb to the surface, her
disavowal of false depth and her discovery that everything happens at the border.'[3]
It is in this negotiation of false depth and curious surfaces that meaning emerges
and one can begin to know one's way about.[4] The roses are literally inscribed
with the intention of the sovereign, rendering the roses compatible with social and
cultural requirements. The guilt is literally inscribed on the bodies of the indebted,
the affect overrun and striated by the sovereign, all in the good name of depth:
'Ah, reason, seriousness, mastery over the affects, the whole somber thing called
reflection, all these prerogatives and showpieces of man: how dearly they have
been bought! how much blood and cruelty lie at the bottom of all "good things"!'[5]

However, Alice is immediately disconcerted by what is going on at the surface:
'Alice thought this a very curious thing, and she went nearer to watch them.'[6]
Throughout the journey, guided by Alice, the genuinely radical intervention into
the property and innovation debates is to go nearer, to recognize the immediacy of

---

1   Deleuze, *Logic of Sense*, 7.

2   *Wonderland*, 64.

3   Deleuze, *Logic of Sense*, 9. Deleuze notes further, 'Paradox appears as a dismissal
of depth, a display of events at the surface, and a deployment of language along this limit.
Humor is the art of the surface, which is opposed to the old irony, the art of depths and
heights.' Deleuze, *Logic of Sense*, 9.

4   In *Philosophical Investigations* §664, Wittgenstein notes, 'In the use of words one
might distinguish "surface grammar" from "depth grammar". What immediately impresses
itself upon us about the use of a word is the way it is used in the construction of the sentence,
the part of its use – one might say – that can be taken in by the ear. – And now compare the
depth grammar, say of the word "to mean", with what its surface grammar would lead us to
suspect. No wonder we find it difficult to know our way about.'

5   Nietzsche, *On the Genealogy of Morals*, 62.

6   *Wonderland*, 64.

the surface of innovation and use (including consumers and producers), as distinct from the premise of presumptive calculability of depth. Whether that depth is abbreviated and asserted through incentives or other manipulations of economic policy narrative, it necessarily (notwithstanding that this may occur unwittingly) defers the surface of meaning through use, risking the masking of the chatter of use and the play of language through law, that languishes 'hidden behind riddles and iridescent uncertainties'.[7] In fact, after all that, the riddles have no answers.

To articulate fully the opportunity of familiar production, 'What is required for that is to stop courageously at the surface, the fold, the skin, to adore appearance, to believe in forms, tones, words, in the whole Olympus of appearance.'[8] It is time to be 'superficial – *out of profundity*'.[9]

On the Face of Things

> Alice was rather doubtful whether she ought not to lie down on her face like the three gardeners, but she could not remember ever having heard of such a rule at processions; "and besides, what would be the use of a procession," thought she, "if people had all to lie down upon their faces, so that they couldn't see it?" So she stood still where she was and waited.[10]

How can one watch a procession with one's eyes to the ground? How can one understand language without the opportunity to survey its use, *to adore appearance*? This is the quality of the language-game, 'the whole, consisting of language and the actions into which it is woven',[11] a context-driven analysis of the relationship between language and representation: 'we look at games and language under the guise of a game played according to rules. That is, we are always *comparing* language with a procedure of that kind.'[12]

Similarly, intellectual property rules (the law, the practice, the normative value) adhere to a language-game, with specific language, grammar as well as non-linguistic social, political, cultural, economic and commercial elements all contributing to use in connection with the innovation about which the language-game speaks. Part of that language-game is the way in which intellectual property rights are used to 'point' to innovation: *this is innovation, and that is not*. However, this is not the meaning of the intellectual property system, nor is it representative of innovation. Rather, meaning is explained by the rules and grammar of intellectual

---

7   F Nietzsche, *The Gay Science*, W Kaufmann (trans), Vintage-Random House, New York, [1887]/1974: 38.

8   Nietzsche, *Gay Science*, 38.

9   Nietzsche, *Gay Science*, 38.

10   *Wonderland*, 66.

11   Wittgenstein, *Philosophical Investigations*, §7.

12   Wittgenstein, *Philosophical Grammar*, 63, proposition 26.

property, the quality or object of innovation, the artefacts themselves and the non-linguistic 'gestures' of the system itself: 'That philosophical concept of meaning has its place in a primitive idea of the way language functions.'[13]

Therefore, overseeing the entire game and in order to keep innovation in play, 'rules' for use must be agreed, and it is this question of 'use' that gives meaning to the entire system. In every sense the potential here is rhizomatic, with the language-game motivated by 'a complicated network of similarities overlapping and criss-crossing: sometimes overall similarities, sometimes similarities of detail'.[14] Recalling the earlier discussion of family resemblances and the coadaptation of meaning in familiar production, together with the activity of 'recognizing' and perpetuating innovation within the rules and language-game of intellectual property, it becomes clear that '"games" form a family'.[15] Games are intrinsic to familiar production. The very play and innovation of the game affirms the difference in repetition and 'establishes a world of nomadic distributions and crowned anarchies ... as a joyful and positive event, as un-founding'.[16] The game of familiar production is 'the joy of the diverse; and the practical critique of all mystifications'[17] in a manifest challenge to the dilemma of representation: 'we misunderstand the role of the ideal in our language. That is to say: we too should call it a game, only we are dazzled by the ideal and therefore fail to see the actual use of the "game" clearly.'[18] Once again, the architecture of representation is exposed in the language-game: 'For if you look at them you will not see something that is common to *all*, but similarities, relationships, and a whole series of them at that. To repeat: don't think, but look!'[19]

Returning in its refrain is the problem of representation: *don't think, but look!* It is not possible to step outside the act of representation, outside language, in order to to provide a complete and perfect explanation. This problem of representation is central to the articulation between intellectual property, innovation and use, and is a crucial limitation of overarching economic policy narrative. In the same way that the meaning of a word is learnt through its use, so too will the meaning (and so the relevance and legitimacy) of the language and grammar of intellectual property be learnt through use. Where there is unreasonable obstruction to genuine participation in the game, for whatever player, meaning will be compromised and the legitimacy of the system undermined. It is participation and use that both contribute to legitimacy and are crucial for an effective and flourishing game

---

13   Wittgenstein, *Philosophical Investigations*, §2.

14   Wittgenstein, *Philosophical Investigations*, §66.

15   Wittgenstein, *Philosophical Investigations*, §67.

16   Deleuze, *Logic of Sense*, 263.

17   Deleuze, *Logic of Sense*, 279. See further the discussion of the mythology of the original in Chapter 9 in the present text.

18   Wittgenstein, *Philosophical Investigations*, §100.

19   Wittgenstein, *Philosophical Investigations*, §66.

(including entry to the system, user principles, enforcement and so on): 'Compare: inventing a game – inventing language – inventing a machine.'[20]

Further, the system is coherent and autonomous where the grammar and language of the game is understood and effective. In other words, a meaningful intellectual property system does not require external corroboration if effective. However, this is not the same as saying it is a 'complete' system: 'the term "language-*game*" is meant to bring into prominence the fact that the *speaking* of language is part of an activity, or of a form of life.'[21] This comes back to the difference between the meaning of a word and the explanation of the meaning: 'we talk about it as we do about the pieces in chess when we are stating the rules of the game, not describing their physical properties.'[22] Put succinctly: 'The question "What is a word really?" is analogous to "What is a piece in chess?"'[23] Language cannot be merely denotative, according to rules; it is meaning in practice, in use: 'a person goes by a sign-post only in so far as there exists a regular use of sign-posts, a custom.'[24] That is, meaning is a public affair.

Similarly the language-game of intellectual property cannot be merely a system of designation; justice within the intellectual property system cannot simply be a foregone conclusion of the rules, despite efforts to model mathematically the criterial evidence of intellectual property (criteria of patentability, for example). This would reduce justice to a mere execution of the calculable, a mere following of rules. It would reduce justice to the coercion of moral certainty and moral judgment, as distinct from the collective use of ethics: 'A decision that didn't go through the ordeal of the undecidable would not be a free decision, it would only be the programmable application or unfolding of a calculable process.'[25] That is, such a decision would only be a form of automated submission to what is otherwise a fixed set of instructions.

Surface Ethics

The difference between the execution of the calculable and prescriptive judgment on the one hand, and the play of decision on the other, is the important distinction between 'morality' and 'ethics'.[26] Morality prescribes the certainty and constraint

20   L Wittgenstein, *Zettel*, GEM Anscombe & GH von Wright (eds), GEM Anscombe (trans), U of California P, Berkeley, 1967: §327.

21   Wittgenstein, *Philosophical Investigations*, §23.

22   Wittgenstein, *Philosophical Investigations*, §108.

23   Wittgenstein, *Philosophical Investigations*, §108.

24   Wittgenstein, *Philosophical Investigations*, §198.

25   Derrida, 'Force of Law', 24.

26   On the difference between an ethics and morality, see the detailed discussion in Deleuze, *Spinoza*, 17–29. See further the comments in Deleuze, *Negotiations*, 100–101. On the ethics of inquiry and the juridical concept of decision, see further the discussion in

of fixed rules, such that judgment according to certain overarching and transcendent values is applied, while at the same time the subject is denied the capacity for choice, the capacity to exercise taste, undertake decisions, or take risks.[27] It is forever six o'clock,[28] and yet the Hatter still has everything to play for, negotiating the present of the tea-party with the endless negotiation of the riddle. While the imperative of moral certainty subjugates the individual to a separate, 'machinic' process of instructions as distinct from choices, the smooth space-time of ethics provides the negotiation and 'play' within its rules; that is, there is capacity for choice, decision and undertaking in order to judge our actions and behaviours according to and in coordination and cooperation with others. This is the ethics of the game.

With respect to policy debates and reform, this distinction between morality (guilt and blame) and ethics (accountability and cooperation) can be understood to explain the difference between the moralizing language of the intractable positions often assumed in intellectual property debates (by various perspectives), as distinct from the ethical and dialectical potential for reform. Moral rhetoric is thus entirely coincident with relationships of guilt and debt, whereas ethics may be understood as constituted by cooperation (duties as well as dues). Further, inherent in such characterization and positioning in the debates is the reverence for 'all or nothing' and 'either/or', but this is entirely contrary to the potential for ethical choice.[29]

It is this community of use that is crucial to the iterative process of the law; that is, justice *to come*. This is not in the sense of an unattainable infinitude, but rather, is to give life to the ethics of chance, 'an unlimited finity'.[30] Logic has no place in the language-game: 'We want to say that there can't be any vagueness in logic. The idea now absorbs us, that the ideal "*must*" be found in reality. Meanwhile we do not as yet see *how* it occurs there, nor do we understand the nature of this

---

A Negri, *Reflections on Empire*, E Emery (trans), Polity P, Cambridge, [2003]/2008: 169–72. The relationship of decision procedure and decision-making to ethics is considered further here in Chapter 11.

27   Additionally, this distinction between morality and ethics brings with it notions of taste and adaptation explored in earlier discussions, where taste is not a measure but 'The philosophical faculty of coadaptation, which also regulates the creation of concepts.' Deleuze & Guattari, *What is Philosophy?* 77.

28   Note also the further discussion of the instant of innovation and the endless potential 'to come' in ethics and justice in later chapters and, in particular, in the concept of *now, here* in '*Re* Use'. See also Derrida, *Negotiations*, 178–83. See also the earlier discussion of the smooth space-time of innovation in Chapter 7.

29   See the discussion in Massumi, *User's Guide*, 112.

30   G Deleuze, *Foucault*, S Hand (trans and ed), U of Minnesota P, Minneapolis, [1986]/1988: 131.

"must". We think it must be in reality; for we think we already see it there.'[31] The logic of innovation is always and evermore a waiting game.

The Queen! The Queen!

> When the procession came opposite to Alice, they all stopped and looked at her, and the Queen said severely, "Who is this?" She said it to the Knave of Hearts, who only bowed and smiled in reply.
>
> "Idiot!" said the Queen, tossing her head impatiently; and, turning to Alice, she went on, "What's your name, child?"
>
> "My name is Alice, so please your Majesty," said Alice very politely; but she added, to herself, "Why, they're only a pack of cards, after all. I needn't be afraid of them!"
>
> "And who are *these*?" said the Queen, pointing to the three gardeners who were lying round the rose-tree; for, you see, as they were lying on their faces, and the pattern on their backs was the same as the rest of the pack, she could not tell whether they were gardeners, or soldiers, or courtiers, or three of her own children.
>
> "How should *I* know?" said Alice, surprised at her own courage. "It's no business of *mine*."
>
> The Queen turned crimson with fury, and, after glaring at her for a moment like a wild beast, screamed, "Off with her head! Off –"
>
> "Nonsense!" said Alice, very loudly and decidedly, and the Queen was silent.[32]

Now that Alice has emerged on the surface of things, all is still nonsense but nevertheless different: 'It is not that surface has less nonsense than does depth. But it is not the same nonsense.'[33] Alice's behaviour demonstrates the necessity of consent and legitimacy for the game to function, in the face of the unlimited potential of chance, in the face of utter nonsense.[34] To obey the Queen's command (to follow the rule), Alice must intend and be able to explain this action: 'What

---

31   Wittgenstein, *Philosophical Investigations*, §101. For Wittgenstein, the pre-determined, calculable regime of logic is damaging to understanding in language. See further Wittgenstein, *Philosophical Investigations*, §81.

32   *Wonderland*, 66–68.

33   Deleuze, *Essays Critical and Clinical*, 22.

34   Deleuze notes 'Carroll's uniqueness is to have allowed nothing to play through sense, but to have played out everything in nonsense, since the diversity of nonsenses is enough to give an account of the entire universe, its terrors as well as its glories: the depth, the surface, and the volume or rolled surface.' Deleuze, *Essays Critical and Clinical*, 22.

this shews is that there is a way of grasping a rule which is *not* an *interpretation*, but which is exhibited in what we call "obeying the rule" and "going against it.'"[35]

Indeed, Alice cannot remember any rules of deference by lying down, in the face of the procession, and so she chooses simply to remain standing: '"obeying a rule" is a practice'.[36] This raises the role of purpose in the game,[37] that is, the way in which the game is defined by the rules, the meaning of which emerges in light of their 'purposeful' nature, their legitimacy. Similarly, the way in which relationships to things might be defined by the rules of intellectual property will show through use those which are considered ineffective (or inessential) to meaning (that is, those rules which lack legitimacy for participants). Further, the way in which intellectual property developments (and users, both consumers and producers) are perhaps considered by users to be rendered subordinate to the economics of contemporary policy discourse might similarly be in disarray for want of legitimacy (both in terms of the law and in terms of business and use). Otherwise inessential rules might give rise to speculation or objection[38] and ultimately to the conclusion of inconsequence in the context of the game: 'If I understand the character of the game aright – I might say – then this isn't an essential part of it. ((Meaning is a physiognomy.))'[39] On the face of things, what is the use of the preoccupation with the economicization of intellectual activity and use, if such analysis forces users to lie face down?

> Following a rule is analogous to obeying an order. We are trained to do so; we react to an order in a particular way. But what if one person reacts in one way and another in another to the order and the training? Which one is right?
>
> Suppose you came as an explorer into an unknown country with a language quite strange to you. In what circumstances would you say that the people there gave orders, understood them, obeyed them, rebelled against them, and so on?[40]

The ambiguity in the game and in this very inquiry is whether the faceless and therefore 'meaningless' gardeners are correct in lying down, or whether Alice the explorer is correct in remaining standing and forcing the question of use to the surface. The object appears clear. Have the courage to explore the surface. Know your place!

---

35   Wittgenstein, *Philosophical Investigations*, §201.
36   Wittgenstein, *Philosophical Investigations*, §202.
37   This is discussed in more detail in Chapter 10.
38   Wittgenstein, *Philosophical Investigations*, §566–§567.
39   Wittgenstein, *Philosophical Investigations*, §568.
40   Wittgenstein, *Philosophical Investigations*, §206.

To Face Down

> "Get to your places!" shouted the Queen in a voice of thunder, and people began
> running about in all directions, tumbling up against each other; however, they
> got settled down in a minute or two, and the game began.[41]

In this new game, Alice is faced with the problem of naming: 'One has already
to know (or be able to do) something in order to be capable of asking a thing's
name. But what does one have to know?'[42] For the Queen, the Sovereign, the act
of naming is of paramount importance, but the meaning remains elusive: 'That a
name can receive any number of meanings introduces a certain contingency into
the relations between field and world.'[43] And so, for Alice, it is certainly none of
her business: 'For naming and describing do not stand on the same level: naming
is a preparation for description. Naming is so far not a move in the language-
game – any more than putting a piece in its place on the board is a move in chess.
We may say: *nothing* has so far been done, when a thing has been named. It has
not even *got* a name except in the language-game.'[44]

This is the dilemma of the rules of intellectual property and the significant
difficulties in implementation, adherence, practice, enforcement and so on (not only
domestically in some jurisdictions more than others, but also cross-border, when
trading in a jurisdiction and language other than one's own): 'The fundamental
fact here is that we lay down rules, a technique, for a game, and that then when we
follow the rules, things do not turn out as we had assumed. That we are therefore
as it were entangled in our own rules.'[45] This experience of 'entanglement', a
'tangled tale', is the explanation of meaning: 'This entanglement in our rules is
what we want to understand (i.e., get a clear view of).'[46] *Know your place.*

Therefore, in devising the names or characters of the game, a decision must
be made. In this way, naming within the game (through use, through interaction,
through community) might be understood to be part and parcel of the ethical
dimension and potential for intellectual property, that is, the act of decision:
'Naming must therefore be a collective and common process ... naming is perhaps
the only process through which a form of decision can be imagined.'[47]

---

41   *Wonderland*, 70.

42   Wittgenstein, *Philosophical Investigations*, §30.

43   G Van Den Abbeele, 'Communism, the Proper Name', in Miami Theory Collective
(ed), *Community at Loose Ends*, U of Minnesota P, Minneapolis, 1991: 30–41, at 31.

44   Wittgenstein, *Philosophical Investigations*, §49.

45   Wittgenstein, *Philosophical Investigations*, §125.

46   Wittgenstein, *Philosophical Investigations*, §125.

47   Negri, *Negri on Negri*, 120.

Put a Face to the Name

What we are faced with at this stage of the journey is, quite literally, the physiognomy of meaning. The physiognomy of meaning, the surface of meaning, highlights the departure from the search for depth explored previously; the fallacy of linear depth as opposed to the rhizomatic potential of unpredictable chance. Meaning is use, on the face of things. It is the 'ridges and furrows' of the Queen's croquet-ground, with living balls, mallets and arches. Indeed, everything is living in the Queen's croquet game, 'the balls were live hedgehogs, the mallets live flamingos, and the soldiers had to double themselves up to stand upon their hands and feet, to make the arches'.[48] This is the intensive becoming in the game of chance where all the characters 'cease to be subjects to become events'.[49] This is crucial to understanding the transformation of the creative subject in the digital, and the transition from exchanges of goods to mutualism in experiences through familiar production. In many ways, the face of this transformation was foreseen, as it were, in the rhetoric surrounding the digital, in the intimacy of familiar production, and the suggestion of dealing with information as itself living.[50] Perhaps information really does want to be free. And we can rest assured, we will know it when we see it.

And so Alice enters the game, that is, she attempts to put herself into play – to realize her function as an actor, an author.

A Very Difficult Game Indeed

> Alice soon came to the conclusion that it was a very difficult game indeed.
> The players all played at once without waiting for turns, quarrelling all the while, and fighting for the hedgehogs; and in a very short time the Queen was in a furious passion, and went stamping about, and shouting, "Off with his head!" or "Off with her head!" about once in a minute.[51]

---

48   *Wonderland*, 70.

49   Deleuze & Guattari, *Thousand Plateaus*, 263. Deleuze and Guattari explain further: 'It is the entire assemblage in its individuated aggregate that is a haecceity; it is this assemblage that is defined by a longitude and a latitude, by speeds and affects, independently of forms and subjects, which belong to another plane. It is the wolf itself, and the horse, and the child, that cease to be subjects to become events, in assemblages that are inseparable from an hour, a season, an atmosphere, an air, a life.' Deleuze & Guattari, *Thousand Plateaus*, 263. See further the more detailed discussion at 260–65.

50   For instance, Brouwer & Mulder draw links between the digital and traditional oral cultures, noting that 'Digital archives are unstable, plastic, living entities, as stories and rituals were in oral cultures.' See J Brouwer & A Mulder, 'Information is Alive', in Brouwer et al. (eds), *Information is Alive*, V2_/Nai, Rotterdam, 2003: 4–6, at 5.

51   *Wonderland*, 71.

The Queen's croquet game appears to decide the fate of all those who play it, their guilt, their culpability, their debt: *Off with his head! Off with her head!* Again, the will of the State and the debt of the user will be literally inscribed on its body. The game of croquet appears to remove the application of rational decision-making processes in order to subject the players to indisputable and incommunicable moral certainties. Alice too is apparently marginalized from the rationality of the game. She risks, literally, losing her head. She cannot calculate with respect to the game. She cannot discern equilibrium. Indeed, she is anticipating the very 'debt' that is being interminably thrust upon her.

The critical question is how Alice, the user, might not only be allowed to be put into play, but also to be necessary to the will of the game. That is, for intellectual property, how might the user enter the game and discharge its debt? Admitting the user to play and gambling the value of familiar production liberates innovative potential that would go some way towards an ethics of intellectual property: 'A life is ethical not when it simply submits to moral laws but when it accepts putting itself into play in its gestures, irrevocably and without reserve – even at the risk that its happiness or its disgrace will be decided once and for all.'[52]

In terms of the debt to production, and the accountability of the user for the risk, admitting the user to the language-game opens the dialogue of chance and familiar production: 'I want only chance ... which is my goal, my only goal, and my sole means.'[53] By entering the game, the generation of meaning through chance (through use) is a strategic and intentional exercise aimed at mutual solutions. As in the Caucus-race, meaning is generated through a cooperative and coordinated game of chance.[54] In other words, the 'taking place' of innovation occurs not in the product, nor in the producer, but in the 'play' between the expression, the 'author' and the 'reader'.[55]

This coordination of results is precisely the concept of the just: 'justice exists solely in relation to the other'.[56] This coordination of mutual benefit is crucial to the just relation of rights and obligations, the relationship between the right and conversely the 'dues' (or means) in order to realize the right.[57] Article 15 of the International Covenant on Economic, Social and Cultural Rights (ICESCR)

---

52   Agamben, *Profanations*, 69.

53   Bataille, *On Nietzsche*, 143.

54   See the earlier discussion in Chapter 3 of the pure game of chance in the Caucus-race.

55   Agamben, *Profanations*, 71.

56   J-L Nancy, *God, Justice, Love, Beauty: Four Little Dialogues*, S Clift (trans), Fordham UP, New York, [2009]/2011: 43.

57   Jean-Luc Nancy explains in 'The Just' the two parts of the definition of justice – the right (what is due) and the means by which to realize that right: 'But then – this is the second part of the definition – what is due to someone? We're not posing the question here of how to give or render to each person what he or she is due. But one can easily distinguish some elements of what is owed to everyone: everyone has the right to live, so that means

articulates not only the right to benefit from one's expression[58] (arguably part of the basis for intellectual property rights[59]), but also the right to the means, that is the right to benefit from the expressions of others (participation in cultural life[60] and the benefits of scientific progress[61]). When considering intellectual property, and its secure basis in the 'right to benefit', arguably we can understand the scope of that right as ensuring also that everyone is also owed the means by which to secure the right to benefit. In order to 'benefit' from scientific progress one must necessarily be able not only to access the law but also to enter (into play). That is, the law must ensure use. And through that play, that struggle, that negation, comes the necessarily dynamic nature of the law itself, bringing us 'to the side of the law and to what will always be in need of change, reform, and modification … The law doesn't change every day, but there are always good reasons to consider transforming it or to consider creating new laws, so that society can become more just.'[62]

This is the infinite nature of the ethical pursuit of justice: 'Impossible not to yield to this truth, that my life implies a beyond of light, a beyond of the chance I love.'[63]

Some Way of Escape

> She was looking about for some way of escape, and wondering whether she could get away without being seen, when she noticed a curious appearance in the air: it puzzled her very much at first, but, after watching it a minute or two, she made it out to be a grin, and she said to herself, "It's the Cheshire Cat: now I shall have somebody to talk to."
>
> "How are you getting on?" said the Cat, as soon as there was mouth enough for it to speak with.
>
> Alice waited until the eyes appeared, and then nodded. "It's no use speaking to it," she thought, "till its ears have come, or at least one of them." In another minute the whole head appeared, and then Alice put down her flamingo, and began an account of the game, feeling very glad she had someone to listen to her. The cat seemed to think there was enough of it now in sight, and no more of it appeared.

---

that everyone is owed the means to live, to feed himself or to protect herself from the elements.' Nancy, *God, Justice, Love, Beauty*, 44.

58   Article 15.1(c).
59   See in particular the discussion in Gibson, *Creating Selves*, 1–3.
60   Article 15.1(a).
61   Article 15.1(b).
62   Nancy, *God, Justice, Love, Beauty*, 45.
63   Bataille, *On Nietzsche*, 109.

"I don't think they play at all fairly," Alice began, in rather a complaining tone, "and they all quarrel so dreadfully one can't hear oneself speak – and they don't seem to have any rules in particular; at least, if there are, nobody attends to them – and you've no idea how confusing it is all the things being alive; for instance, there's the arch I've got to go through next walking about at the other end of the ground – and I should have croqueted the Queen's hedgehog just now, only it ran away when it saw mine coming!"

"How do you like the Queen?" said the Cat in a low voice.

"Not at all," said Alice: "she's so extremely –" Just then she noticed that the Queen was close behind her listening: so she went on, "– likely to win, that it's hardly worth while finishing the game."[64]

The Queen, the sovereign, wins every time. The Queen is always already in credit, and the fate of the indebted, the guilty, is decided through her game. It would seem that the rules are such that the debt, as it were, is rendered infinite.[65] While the player, the user is available to participate, nevertheless the outcome of the game is predetermined and pre-exists in an infinite debt to the sovereign. There are no legitimate rules or reasons contributing to a decision on each player's fate, no rational process of decision-making, no justice. The game loses all legitimacy with Alice, indeed, 'it's hardly worth while finishing the game'.

While the game may not appear to be worth playing, nevertheless this problem of legitimacy is the source of its instability and vulnerability. Just as for contemporary crises for intellectual property (whether in terms of access to the system, access to product, value through enforcement and infringement and so on), legitimacy facilitates the game. The failure to adapt to familiar production and accounting for consensus both in the governing framework and as a tool for innovation in its own right consequently disenfranchises not only users (including producers), but also whole processes of innovation and production (including familiar production itself). Where the intellectual property system comes to be viewed as at best irrelevant and at worst in conflict with the creative and innovative process, the instability of the system may be irreparable.

The question therefore is how to escape and transform the game. In other words, how is the user to connect and use to evolve, rather than persist in a game that is static and unproductive? Once again, the war machine of familiar production is an assemblage of creative and innovative force, facilitating a 'line of flight',[66] a way of escape, through which Alice, the user, will achieve 'actual' connection. Understood in this way, the assemblage of familiar production provides both the

---

64   *Wonderland*, 71–72.

65   Deleuze & Guattari, *Anti-Oedipus*, 197–98.

66   Deleuze and Guattari explain the line of flight as taking a necessary risk. In a way, this is chance versus risk, in that it is action that carries on regardless, despite an unaccountable risk: 'the warrior arises in the infinity of a line of flight'. Deleuze & Guattari, *Thousand Plateaus*, 277.

materiality and immateriality of community, of enunciative and creative forces, as well as the shifting stability of territory (in a way, the 'in common' of familiar production, the 'abode' of ethics) and at the same time the innovative disruption to intellectual property, agitating for reform and revolution in the way in which innovation might not only be recognized, but also be produced.[67] This is the flexibility, the adaptability of the system as a field of chance, as distinct from a regime of risk.[68] And in actualizing connections, familiar production is a 'physical' capacity of the digital, as it were, the territory of social life. The digital has a place. *Get to your places.*

This brings to the surface the experimental and experiential nature of familiar production (chance) as distinct from an otherwise planned, institutionalized and organized innovation of the firm: 'Lodge yourself onto a stratum, experiment with the opportunities it offers, find an advantageous place on it, find potential movements of deterritorialization, possible lines of flight, experience them, produce flow conjunctions here and there, try out continuums of intensities segment by segment, have a small plot of new land at all times.'[69] Importantly, this experimental nature of familiar production is not simply a counter to organized innovation, it is a whole new form of life. That is, it is precisely the qualities of experiment and chance, innovation without purpose and use without question (as distinct from risk and planning), that motivate familiar production. Chance and experiment in the action of the assemblage generate knowledge and innovation that 'pass beyond their own axiomatics, generating increasingly deterritorialized signs, figures-schizzes that are no longer either figurative or structured, and reproduce or produce an interplay of phenomena without aim or end; science as experimentation'.[70]

Familiar production is thus a wholly transformed and transformative relationship to the personal, to 'authorship' and 'ownership', to benefit and purpose, to use and desire. Familiar production is not a reaction, it is a whole new line of innovation, a whole new line of flight. Familiar production unseats the primacy of the original in a way disconcerting to the rhythm of conventional representation and authenticity provided by the intellectual property system. In the game of innovation, familiar production brings the profundity of the superficial to the fore: 'All identities are

---

67   Deleuze and Guattari explain the assemblage as follows: 'On the one hand it is a *machinic assemblage* of bodies, of actions and passions, an intermingling of bodies reacting to one another; on the other hand it is a *collective assemblage of enunciation*, of acts and statements, of incorporeal transformations attributed to bodies.' Further 'the assemblage has both *territorial sides*, or reterritorialized sides, which stabilize it, and *cutting edges of deterritorialization*, which carry it away'. Deleuze & Guattari, *Thousand Plateaus*, 88.

68   Deleuze and Guattari explain: 'a *line of flight* must be preserved to enable the animal to regain its associated milieu when danger appears'. Deleuze & Guattari, *Thousand Plateaus*, 55.

69   Deleuze & Guattari, *Thousand Plateaus*, 161.

70   Deleuze & Guattari, *Anti-Oedipus*, 371.

only simulated, produced as an optical "effect" by the more profound game of difference and repetition.'[71]

Alice looks about for a way to escape, a line of flight, her freedom. Once again, she is facing an interpretation hanging in the air – the Cheshire Cat:

> The Cat's head began fading away the moment he was gone, and, by the time he had come back with the Duchess, it had entirely disappeared; so the King and the executioner ran wildly up and down looking for it, while the rest of the party went back to the game.[72]

Back to the game.

---

71   Deleuze, *Difference and Repetition*, xix.
72   *Wonderland*, 74.

Chapter 9
Blame

Moralities Play

> "You're thinking about something, my dear, and that makes you forget to talk. I can't tell you just now what the moral of that is, but I shall remember it in a bit."
>
> "Perhaps it hasn't one," Alice ventured to remark.
>
> "Tut, tut, child!" said the Duchess. "Everything's got a moral, if only you can find it."[1]

Everything has a moral, says the Duchess, but the challenge is to find it. To refer to the moral, therefore, is not in the sense of moral certitude or appealing to something which exceeds the reason of ethics. Indeed, the situation for the Duchess is precisely the contrary, in that one must look for the moral; that is, one must engage with the ethical dimension of arriving at meaning. Don't think, but look![2]

The ethical character of the intellectual property system is therefore not a determination at the level of the invention, but rather, the dialogue through use and the mutual responsibilities of the 'speakers' (the users, understood in the broadest sense to include consumers, producers and any other dimension of use). This is the 'balance' of the system, not in terms of an equilibrium or indeed social contract, but in terms of a reciprocity. More accurately therefore, as discussed throughout this inquiry, it may be understood as an imbalance.

The market (in the broadest sense as the meeting place, the moment of use, regulation and meaning), as distinct from the product, is the forum for the relationship of 'ethics' in the contemporary intellectual property system and the industries reliant upon that system. However, as considered in earlier chapters, this social function of meaning has been somewhat displaced by a regulation of use through reference to a debtor–creditor relationship, a propertyless–propertied disequilibrium.[3] Therefore, the challenge is in terms of expanding the understanding of the market and the activities within as a social forum for meaning, largely through a reconsideration of the referents for meaning (that is, the objects or the

---

1 *Wonderland*, 75.

2 Wittgenstein, *Philosophical Investigations*, §66.

3 These concepts are discussed throughout, but in particular, see the extensive discussion of the notion of property and the user as 'peasant', as well as the introduction to the rhetoric of 'social contract' and the circulation of credit and debt, in 'Use'. See further the expansion of the discussion of debtor–creditor relations in Chapter 2.

'focus' of intellectual property themselves). In exploring the move from goods to experience in the digital environment, the question remains: how might the rules of intellectual property be understood as encompassing the necessary 'incentives' for markets, as distinct from the usual justification on incentives for products? This marks what might be considered a significant departure from arguments focused on economic rights in property, towards a cultural and social narrative of knowledge-driven markets. The critical policy question is the translation of that cultural experience into economic growth.

And the moral of *that* is …

Moral Values

So in this exchange, moral is used in the sense of 'meaning' and is taken to be understood as a practical lesson or teaching in conduct. Literature, fables, tales are thus cultural repositories of teaching in meaning, literally moral or practical lessons,[4] and in this way will be said to 'point to' the moral of the words. However, the exchange between the Duchess and Alice shows the very misery of attempting to generalize morals in the face of conditions otherwise not at play. How does one reconcile the concept of the irrefutable 'moral' with the language-game of ethics? In other words, the problem of language and representation is again present in the relationship between morals (and morality) and ethics in the pronouncements and developments of any policy discussion. Morals deal with language and 'the moral of what is said', while ethics might be said to deal with bodies and connections.[5] Morals can merely point to meaning: 'What is this pointing for, and what are these words and what else may accompany them for?'[6] Morals literally guide the user nowhere.

An entertaining moral account of the difference between naming and use is provided in an interlocutory in the comedy movie *Wanderlust*.[7] The moral play occurs between Alan Alda's character (Carvin), the 'eccentric' owner of a farm, the Elysium, used as an 'intentional community' (or commune), and Paul Rudd's character (George), a recently redundant and homeless New York City executive. George and his wife Linda, having just lost their new apartment in the West Village, are forced to put all their possessions in their car and drive to Atlanta to live with and work for George's arrogant and offensive brother, a modern-day member of the bourgeoisie, living with all the trappings, glitz and unhappiness money can seemingly buy. On the way, they stay overnight at the commune's

---

4   These ideas were introduced in the Preamble.

5   Deleuze distinguishes between a morality of words and the ethics of bodies throughout his work. See, in particular, Deleuze, *Negotiations*, 114–15 and Deleuze, *Logic of Sense*, 142–47.

6   Wittgenstein, *Philosophical Investigations*, §670.

7   *Wanderlust* (2012). Dir: D Wain; SP: D Wain and K Marino.

bed and breakfast hotel and, at their departure the next morning, George has the following dialogue with Carvin:

Carvin:	Just remember, money buys nothing.
George:	Well, nothing important.
Carvin:	No, no. Money *literally* buys nothings.
George:	I think you mean metaphorically.
Carvin:	No, literally. Nothing.
George:	Literally money buys most things.
Carvin:	No, nothing. Right, wait, are you saying that …
George:	Well, I'm saying that literally …
Carvin:	No, but I'm saying *literally* money buys nothing. I don't know what you're …
Linda:	… It buys nothing.
George:	You're right. Money pays for nothing.
Linda:	That's right.
[PAUSE]	
George:	But not literally.

But Carvin is right. Money literally buys nothing. In other words, the debt is infinite. Money buys nothing and 'literally' renders the debt infinite. It 'literally' buys nothing: 'In a word, money – the circulation of money – *is the means for rendering the debt infinite* … the abolition of debts or their accountable transformation initiates the duty of an interminable service to the State that subordinates all the primitive alliances to itself (the problem of debts).'[8] Money is thus not about exchange, but about control; that is, the debtor's interpellation by the State apparatus: 'A time will come when the creditor has not yet lent while the debtor never quits repaying, for repaying is a duty but lending is an option.'[9]

---

8   Deleuze & Guattari, *Anti-Oedipus*, 197.
9   Deleuze & Guattari, *Anti-Oedipus*, 197–98.

This interlocutory also performs the very real problem of representation, the problem of language, and the crucial dilemma of the severance in meaning in relationships between users (and classes) of the intellectual property system. Consider the following exchange between Wittgenstein and his 'interlocutor' in *Philosophical Investigations*:

> In giving explanations I already have to use language full-blown (not some sort of preparatory, provisional one); this by itself shews that I can adduce only exterior facts about language.
>
> Yes, but then how can these explanations satisfy us? – Well, your very questions were framed in this language; they had to be expressed in this language, if there was anything to ask!
>
> And your scruples are misunderstandings.
>
> Your questions refer to words; so I have to talk about words.
>
> You say: the point isn't the word, but its meaning, and you think of the meaning as a thing of the same kind as the word, though also different from the word. Here the word, there the meaning. The money, and the cow that you can buy with it. (But contrast: money, and its use).[10]

Moral lessons of the nineteenth century and the time of *Alice* were also particularly concerned with expectations and certainties of behaviour attached to class and roles in Victorian society.[11] In this way moral lessons were supposed not only to guide conduct but also to show certain behaviour and custom (that is, use), and therefore circumscribe particular relations of power. Arguably, as discussed earlier,[12] much of the rhetoric surrounding intellectual property use and misuse is similarly articulated around the creation of 'classes' of participant in the system, creating particular social groups or identities within the intellectual property narrative. Further, moral lessons paradoxically do not coincide with acts of judgment and discretion (moral judgment), precisely for the reason of their very reference to behaviour beyond the will and decision-making of the individual: 'morality presents us with a set of constraining rules of a special sort, ones that judge actions and intentions by considering them in relation to transcendent values (this is good, that's bad …).'[13]

Ethics, on the other hand, will cooperate with greater autonomy, decision, freedom and opportunity: 'ethics is a set of optional rules that assess what we

---

10   Wittgenstein, *Philosophical Investigations*, §120.

11   See the discussion in J Schroeder, 'Education: Introduction', in ME Leighton & L Surridge (eds), *The Broadview Anthology of Victorian Prose, 1832–1902*, Broadview P, Ontario, 2012: 149–55.

12   See the earlier discussion in 'Use'.

13   Deleuze, *Negotiations*, 100.

do, what we say, in relation to the ways of existing involved'.[14] That is, ethics is in relation to situations relevant and in play. Therefore, the moralizing discourse on intellectual property policy (and politics) not only presumes and expects certain behaviours, but also assigns behaviours to particular classes (business, rights-holders, consumers and so on), compromising not only immediate conditions but also processes of implementation and legitimacy. Moralizing discourse therefore reaffirms the hierarchical relationships to use and the circulation of credit and debt. As a rhetorical and discursive strategy, this re-inscribes the 'moral certainty' of the intellectual property system, rendering its defensive prescription something to which all solutions must aspire. However, this is conceivably part of the problem of legitimacy in present conditions of implementation, enforcement and reform. As a guide to conduct, the Duchess's morals fail. The game is underway and moralities play.

The contradiction between moral (and morality) and ethics is therefore significant and relevant to policy reform in intellectual property and innovation, not only to the deciphering of the subjectivities created and circulated within intellectual property discourse and debate, but also in terms of genuine policy responses and reform. Arguably, reform of intellectual property (legislative as well as normative) should encompass an ethics of the 'bodies' within the system, the momentum of the system itself; that is, an 'ought to' of bodies rather than a morality of words based on anticipated conditions. This again emphasizes the question of evidence and decisions on the probability of behaviour, as well as the relevance of 'chance' to the value and efficacy of the system. Judgment arguably emphasizes the 'risk' qualities in the equation and the means by which to mitigate the risk of certain outcomes. In other words, it attempts to remove the 'chance' value from the situation. A move from moral judgment to an ethics of chance and justice is necessitated by the social and political difficulties currently agitating the intellectual property game. An ethics of chance would motivate participation in the game not by prescription but by maximizing the potential benefit to all participants.

> "The game seems to be going on rather better now," she said.
>
> "'Tis so," said the Duchess: "and the moral of it is – 'Oh, 'tis love, 'tis love, that makes the world go round!'"
>
> "Somebody said," whispered Alice, "that it's done by everybody minding their own business!"
>
> "Ah, well! It means much the same thing," said the Duchess, digging her sharp little chin into Alice's shoulder as she added, "and the moral of *that* is – 'Take care of the sense, and the sounds will take care of themselves.'"
>
> "How fond she is of finding morals in things!" Alice thought to herself.[15]

---

14   Deleuze, *Negotiations*, 100. This, of course, does not apply to 'professional ethics' in the sense of a code of conduct or rules for practice.

15   *Wonderland*, 75.

Indeed, take care of the sense, the content, the affect; the words will take care of themselves. Take care of the experience. Take care of the culture and the economics will take care of itself.

Choice Words

> "I dare say you're wondering why I don't put my arm around your waist," the Duchess said after a pause: "the reason is, that I'm doubtful about the temper of your flamingo. Shall I try the experiment?"
>
> "He might bite," Alice cautiously replied, not feeling at all anxious to have the experiment tried.
>
> "Very true," said the Duchess: "flamingos and mustard both bite. And the moral of that is – 'Birds of a feather flock together.'"
>
> "Only mustard isn't a bird," Alice remarked.
>
> "Right, as usual," said the Duchess: "what a clear way you have of putting things!"
>
> "It's a mineral, I *think*," said Alice.
>
> "Of course it is," said the Duchess, who seemed ready to agree to everything that Alice said; "there's a large mustard-mine near here. And the moral of that is – 'The more there is of mine, the less there is of yours.'"[16]

This is the classic version of the scarcity model. Making information, ordinarily indivisible and inexhaustible, conform to a system of rivalry and competition. In this case, intellectual property rules systematize deprivation, rendering knowledge in such a way that it cannot co-exist, as it were. *One cannot have one's cake and eat it too.*[17] This recalls earlier discussions of the relationship between use and purpose, where that purpose is for profit for the beneficiary, the counterpoint being the debt of the user and the depletion and obsolescence of resources.[18] The glory of consumption in this sense is thus expenditure (whether that is in terms of financial or other resources), rendering the user or consumer in debt ('the less there is of yours'). Debt is thus instrumental to the construction of subjectivity in conventional models of capital and production, the 'depth' of measured and measurable striated space.

> "Oh I know!" exclaimed Alice, who had not attended to this last remark. "It's a vegetable. It doesn't look like one, but it is."
>
> "I quite agree with you," said the Duchess; "and the moral of that is – 'Be what you would seem to be' – or if you'd like it put more simply – 'Never

---

16   *Wonderland*, 75–76.

17   See in contrast the Looking-glass cake in '*Re* Use'.

18   See the earlier discussion of use and purpose in connection with waste and the generation of 'lack' in consumers in 'Use'.

imagine yourself not to be otherwise than what it might appear to others that what you were or might have been was not otherwise than what you had been would have appeared to them to be otherwise.'"

"I think I should understand that better," Alice said very politely, "if I had it written down: but I'm afraid I can't quite follow it as you say it."

"That's nothing to what I could say if I chose," the Duchess replied in a pleased tone.[19]

The Duchess thus recuperates to the surface the smooth space of innovation, restoring Alice from the seriousness of measure and debt and the cruelty that lies 'at the bottom', the smooth space of innovation, and the profundity of the superficial.[20] In other words, rather than the ordered, hierarchical and measurable world of representation, where things never appear other than as they are, the Duchess identifies the threat to those conventions, that things might be other than they are, that meaning itself is found in the difference in repetition. Participation in the language-game thus becomes not a burden or debt, but a gift: "'Oh, don't talk about trouble!" said the Duchess. "I make you a present of everything I've said as yet."'[21] As distinct from the prescriptive certitude of morals, use emancipates the ethics of decision and chance:

"Now, I give you fair warning," shouted the Queen, stamping on the ground as she spoke; "either you or your head must be off, and that in about half no time! Take your choice!"

The Duchess took her choice, and was gone in a moment.[22]

Take your choice!

Come On!

Then the Queen left off, quite out of breath, and said to Alice, "Have you seen the Mock Turtle yet?"

"No," said Alice. "I don't even know what a Mock Turtle is."

"It's the thing Mock Turtle Soup is made from," said the Queen.

"I never saw one, or heard of one," said Alice.

"Come on, then," said the Queen, "and he shall tell you his history."[23]

---

19   *Wonderland*, 76.
20   Nietzsche, *On the Genealogy of Morals*, 62.
21   *Wonderland*, 76.
22   *Wonderland*, 77.
23   *Wonderland*, 77.

Come on! is still an order of the game, but a condition of participation not of competition. In other words, it is not to follow a model, but rather, to embrace the *giddy seductiveness of chance*. It presents a sense of collaboration, of familiarity, an assemblage of becoming. Come on! is indeed coincident with principles of emergent and radical innovation, as distinct from planned innovation and creativity: 'innovation itself is being reduced to routine. Technological progress is increasingly becoming the business of teams of trained specialists who turn out what is required and make it work in predictable ways. The romance of earlier commercial adventure is rapidly wearing away, because so many more things can be strictly calculated that had of old to be visualized in a flash of genius.'[24] Come on!

> "Everybody says 'come on!' here," thought Alice, as she went slowly after it: "I never was so ordered about in all my life, never!"[25]

Throughout Wonderland, Alice is asking, which way, which way, and others are commanding her, come on, come on. Importantly, Alice questions and chooses but never properly reaches a final conclusion. This thread of ethical questioning shares much with a Deleuzian ontology of ethics, which is opposed to the transcendence assumed by morality. On the dialogue between Alice and the Duchess, Deleuze explains that 'the possibility of a profound link between the logic of sense and ethics, morals and morality, is confirmed'.[26] The moral provides a rule (the grammar, as it were) while the ethics of Alice's journey is expressed by her experiences and actions in Wonderland: *which way, which way*. Indeed, as is now paradoxically (and familiarly) clear, 'A philosophical problem has the form: "I don't know my way about."'[27]

The moral of intellectual property has been to configure a system such that knowledge can become competitive, objects cannot co-exist, but are rendered sequential and ordered. As it stands, intellectual property has achieved a convincing striation of the smooth space of innovation which includes, most importantly, a pronouncement of knowledge-based communities, transactions and enterprises as quintessentially competitive. The conventional remit of successful competition is the assessment and management of risk, minimizing chance.

The ethics of chance would be to reinvigorate the intellectual property system by creating value through chance. The object of transaction might be one of abundant opportunity and experience, as distinct from objects rendered as scarce and competitive resources both by rights and by manufacturing resources. It is having one's cake and eating it too. The mantra of intellectual property should not be go on, it should be come on.

---

24   Schumpeter, *Capitalism, Socialism and Democracy*, 132.
25   *Wonderland*, 78.
26   Deleuze, *Logic of Sense*, 31.
27   Wittgenstein, *Philosophical Investigations*, §123.

Copy That

> So they went up to the Mock Turtle, who looked at them with large eyes full of tears, but said nothing.
>
> "This here young lady," said the Gryphon, "she wants for to know your history, she do."
>
> "I'll tell it her," said the Mock Turtle in a deep, hollow tone: "sit down, both of you, and don't speak a word till I've finished."
>
> So they sat down, and nobody spoke for some minutes. Alice thought to herself, "I don't see how he can *ever* finish, if he doesn't begin." But she waited patiently.[28]

To observe is, in effect, to double, to copy, to repeat. In other words, to look is to find meaning, and that meaning is a combination of perception and experience: 'a well-defined observer extracts everything that it can, everything that can be extracted in the corresponding system. In short, the role of a partial observer is *to perceive* and *to experience*.'[29] *Meaning is use, on the face of things.* Meaning is experience in observation, but observation which is necessarily always ambiguous, in action, in the game. This is a crucial transformation perhaps in the way in which to orientate oneself with respect to intellectual resources in the digital, and to begin to understand the 'product' of experience. Experience is indeed integral to making meaning out of representation.[30]

The mock turtle is thus not so much a legitimate copy of the original, authentic turtle, or a false version of the true turtle, but rather it is a repetition and difference, the simulacrum: 'The primacy of identity, however conceived, defines the world of representation. But modern thought is born of the failure of representation, of the loss of identities, and of the discovery of all the forces that act under the representation of the identical. The modern world is one of simulacra.'[31] What is intriguing about the mock turtle is that his very existence compromises the authority and primacy of the original.

The mock turtle is thus a startling rendition of the disquiet in contemporary circumstances of innovation, particularly in the context of familiar production, and the original–copy rationale that underpins intellectual property. Indeed, this

---

28   *Wonderland*, 79.

29   Deleuze & Guattari, *What is Philosophy?* 130.

30   This is the problem of representation, that of accounting for the very process itself, accounting for relativism in approximating meaning: 'Perspectivism, or scientific relativism, is never relative to a subject: it constitutes not a relativity of truth but, on the contrary, a truth of the relative, that is to say, of variables whose cases it orders according to the values it extracts from them in its system of coordinates.' Deleuze & Guattari, *What is Philosophy?* 130. See further the discussion in Deleuze & Guattari, *What is Philosophy?* 117–33 (chapter 5, 'Functives and Concepts').

31   Deleuze, *Difference and Repetition*, xix.

relationship between the original and the copy, between repetition and difference, is arguably the primary relationship of uneasiness between familiar production and intellectual property. Recalling the emphasis on origin and causality in the way in which the intellectual property system is formed: 'origins are assigned only in a world which challenges the original as much as the copy, and an origin assigns a ground only in a world already precipitated into universal *ungrounding*.'[32] That is, the primacy of the original is itself an artifice, a pre-emptive strike, not a fundamental value in terms of the production of meaning: 'Everything has become simulacrum, for by simulacrum we should not understand a simple imitation but rather the act by which the very idea of a model or privileged position is challenged and overturned. The simulacrum is the instance which includes a difference within itself.'[33]

It is not as banal as a question of technologies of reproduction now allowing for a dismantling of the authority of the copy. Indeed, the proper names of intellectual property are problematically imposed as originals of the impressions of the subject matter itself. For instance, the patent, if taken as original, is an imprint of the invention. The invention thus may be replicated to the extent to which resources might allow. Similarly, the register, presented as the original, merely indicates the trade mark or the design, an analog which may be reproduced. But what resembles the 'original' in copyright? In original works of art it is the work itself that appears to enjoy a unique material existence, but the expression and experience nevertheless still changes with every use, every repetition, and indeed also every reproduction (including decontextualized reproductions in merchandising). Intellectual property confounds the very notion of original and copy, somewhat dramatically in the case of copyright, even before the advent of a digital conceptualization of information. Increasingly, intellectual property rules now govern the 'recipe' for that replication, as distinct from the good, the unit. The original in intellectual property is a mockery: 'The simulacrum is never what hides the truth – it is truth that hides the fact that there is none.'[34] The mock turtle is true.[35]

In a digital environment, therefore, an attachment to the unit as the repository of value has been confronted by a world of simulacra. Not only business models but also perhaps the regulatory frameworks themselves may be required and indeed obligated to refine the subject matter of their application. Rather than the fixation upon the identity of things, seen elsewhere as the weakness of business models

---

32   Deleuze, *Difference and Repetition*, 202.

33   Deleuze, *Difference and Repetition*, 69.

34   J Baudrillard, *Simulacra and Simulation*, SF Glaser (trans), U of Michigan P, Ann Arbor, [1981]/1994: 1.

35   Baudrillard notes, 'The simulacrum is true.' Baudrillard, *Simulacra and Simulation*, 1.

in the digital,[36] the digital is motivated by a work of appearances and the play of difference. This is the sphere of affect. This is where attention must be directed.

Again, the ability to represent existence (and the moral certainty of representation) is confronted by the aesthetics of difference, by a world of pure simulacra. Meaning and value in the simulacrum is not derived from standing in for an absent original. Rather, the simulacrum itself is incensed with creative potential, that is, the repetition of difference. This is the important 'kinship', as it were, between originals and 'copies' in familiar production: 'If someone were to draw a sharp boundary I could not acknowledge it as the one that I too always wanted to draw, or had drawn in my mind. For I did not want to draw one at all. His concept can then be said to be not the same as mine, but akin to it. The kinship is that of two pictures ... The kinship is just as undeniable as the difference.'[37] Still further, the presumption of a referent, an absent original giving meaning to the copy, is undermined with the necessary blurring and play of vagueness, the ethics of the language-game, the repetition and comparability that is paradoxically only possible because of difference: 'And if we carry this comparison still further it is clear that the degree to which the sharp picture *can* resemble the blurred one depends on the latter's degree of vagueness.'[38]

Importantly, this is not to undermine the 'reality' or 'actuality' of innovation and artistic and inventive process. The legal discourse of certainty and stability is not 'simulated' and unravelled by copies; rather, its presumptions and imperatives are instead revealed. This world of simulacra does not refer to anything 'behind' the everyday; simulacra *are* the everyday. This is the surface effect, the physiognomy of meaning in a digital world. The challenge is to translate a preference and emphasis in the law from transactions in goods to that of regulation of the relationships between.

The surface of the digital immediately reveals the fissures in the moral certainty of copies. Conventional approaches to enforcement of intellectual property in a digital environment attempt to define the copy as referring to something which it is not, a model that is otherwise not present, an absent and unassailable original. However, not only is this difficult to sustain in a digital environment and illogical to the many users in that environment, resulting in the loss of legitimacy of the discourse of intellectual property, but also it is somewhat an artificial and contrived construction in the analog and mechanical world of intellectual property as well. That is, in a digital environment, the repetition of knowledge artefacts is not merely a representation of an original; indeed, there is frequently no such original to which a transaction may be anchored. While the very landscape of the digital itself challenges and proliferates innovative models without reference to a

---

36 For instance, recall the discussion in Chapter 6 of advertising business models in Facebook and other social media platforms.

37 Wittgenstein, *Philosophical Investigations*, §76.

38 Wittgenstein, *Philosophical Investigations*, §77.

physical good, in some cases there is never any interaction with a physical good or a stable identity, merely the simulacrum and familiar production.

In the digital then, each repetition is different in itself. In other words, each 'use' makes a difference, and that difference has a value: 'symbols or simulacra are the letter of repetition itself. Difference is included in repetition by way of disguise and by the order of the symbol. This is why the variations do not come from without ... The variations express, rather, the differential mechanisms which belong to the essence and origin of that which is repeated.'[39] This confounds the very notion of the original that underpins not only the structural logic of intellectual property, but also the socio-political context: 'For this reason, it may be that the end of the *Sophist* contains the most extraordinary adventure of Platonism: as a consequence of searching in the direction of the simulacrum and of leaning over its abyss, Plato discovers, in the flash of an instant, that the simulacrum is not merely a false copy, but that it places in question the very notions of copy and model.'[40] Quite literally, 'imitation self-destructs'.[41]

The presumptions in the relationship between original, copy and the world of simulacra played out in an intriguing way through an exhibition of counterfeit goods at the Bucharest Palace of Justice, to mark World Intellectual Property Day 2013. The contributions to this 'aesthetic' of the counterfeit thus came from State institutions, including the National Customs Authority, the Border Police, the Romanian Office for Intellectual Property (ORDA), the State Office for Inventions and Trademarks (OSIM) and the National Consumer Protection Authority (ANPC).[42] In celebration of World Intellectual Property Day 2013, Bucharest held an exhibit not of intellectual property, but of counterfeit goods. The exhibition was not of the value of the affect and experience (such as with examples of innovation, invention and creativity), but rather the fixation upon the original, through exhibits of counterfeit objects. Users were nowhere to be seen (neither producers nor consumers). The sensation, the aesthetic of the exhibition, is in the appearance of difference – difference between the original and the counterfeit. However, where is that difference in mass-produced objects, or in household consumer goods? It is not a question of distinguishing between skill and craft (or imperfection in an original) and the counterfeit, but in fact of responding to the image of resemblance.[43]

---

39   Deleuze, *Difference and Repetition*, 17.

40   Deleuze, *Logic of Sense*, 256.

41   Deleuze & Guattari, *Thousand Plateaus*, 304.

42   I Popescu, 'Counterfeit goods exhibition at Bucharest Palace of Justice marks World Intellectual Property Day', *Romania Insider*, 26 April 2013.

43   Baudrillard notes 'A kind of unintentional parody hovers over everything, a tactical simulation, a consummate aesthetic enjoyment [*jouissance*], is attached to the indefinable play of reading and the rules of the game. Travelling signs, media, fashion and models, the blind but brilliant ambience of simulacra.' J Baudrillard, *Symbolic Exchange and Death*, IH Grant (trans), M Gane (intro), Sage, London, [1976]/1993: 75.

This makes it clear that it is no longer merely a question of the implementation of effective business models to protect those goods (or those traces of goods). Instead, it has become the need for a complete re-conceptualization of models, copies and identities, not only with respect to those of producers and consumers, but also those models configuring the identity of 'knowledge' itself. This is particularly critical to the management of familiar production and the proliferation of innovation and value in the digital. If the crucial value is that of time, how can time and transactions in experience be valued? How is the game to be played? 'Just as a move in chess doesn't consist simply in moving a piece in such-and-such a way on the board – nor yet in one's thoughts and feelings as one makes the move: but in the circumstances that we call "playing a game of chess", "solving a chess problem", and so on.'[44]

Interlocutory of Experience

> "Well, there was Mystery," the Mock Turtle replied, counting off the subjects on his flappers, "– Mystery, ancient and modern, with Seaography: then Drawling – the Drawling-master was an old conger-eel, that used to come once a week: *he* taught us Drawling, Stretching, and Fainting in Coils."
>
> "What was *that* like?" said Alice.
>
> "Well, I can't show it you myself," the Mock Turtle said: "I'm too stiff. And the Gryphon never learnt it."[45]

Recalling the introduction of experiment and experience in the previous chapter,[46] the Mock Turtle illustrates again the distinction between the representational knowledge (conventional intellectual property models) and the experimentation of familiar production. The play of simulacra in the digital space of familiar production is inconceivable within the system if what is maintained is a continued and persistent deference (both in custom and in time) to the original, the model, the moral judgment that dominates conventional approaches to innovation and intellectual property.[47] While the world of representation 'verifies' the original, the experiment and chance of familiar production maximizes the connections and combinations of innovation and use. In other words, familiar production does not interpret a purpose (and product or imagined model) but rather an unpredictable connection between use and users.

---

44   Wittgenstein, *Philosophical Investigations*, §33.

45   *Wonderland*, 81.

46   See the discussion in Chapter 8.

47   Nevertheless, as will be seen in '*Re* Use', the existing intellectual property system is wholly capable of accounting for familiar production, albeit through a revolution of business models and perspectives.

In many ways, therefore, familiar production is also a living and continuous archive, in that its memory and learning of 'Drawling, Stretching, and Fainting in Coils' must be performed, flexibility permitting, not explained: 'It is difficult to respond to those who wish to be satisfied with words, things, images and ideas. For we may not even say that sense exists either in things or in the mind; it has neither physical nor mental existence.'[48] Nothing in familiar production *necessarily* resembles a product, and yet it is always already archived in the language and connection of the digital. Indeed, this is precisely the challenge presented by familiar production, the challenge of the experience and affect over and above the satisfaction of goods and materiality. It is a solution without a product, a use without a purpose, an idea without a target. Familiar production is, literally, the archive of experience.[49] It is the hunting of the Snark.[50]

> "That's enough about lessons," the Gryphon interrupted in a very decided tone: "tell her something about the games now."[51]

Onward to the games.

---

48   Deleuze, *Logic of Sense*, 20.

49   Brian Massumi explains: 'Relations are not made of zeroes and ones; they are qualities of experience. They are archivable, but only in reactivatable trace-form. It is important not to mistake the inert form of the race for the archive of experience that it really isn't, but that it can rebecome, in the analog repetition of its event.' B Massumi, 'The Archive of Experience', in J Brouwer et al. (eds), *Information is Alive*, V2_/NAI, Rotterdam, 2003: 142–51, at 150–51.

50   'They hunted till darkness came on, but they found / Not a button, or feather, or mark, / By which they could tell that they stood on the ground Where the Baker had met with the Snark. // In the midst of the word he was trying to say / In the midst of his laughter and glee, / He had softly and suddenly vanished away – / For the Snark *was* a Boojum, you see.' *Snark*, 252. Deleuze notes with respect to the experiential dimension and the departure from the sense-making of 'denotation, manifestation, and signification', which as discussed characterizes the processes and narrative of intellectual property and other forms of pronouncement, that the hunting of the Snark might indeed illustrate the 'fourth dimension' (that is, in the present discussion, the field of familiar production): 'In truth, the attempt to make this fourth dimension evident is a little like Carroll's Snark hunt. Perhaps the dimension is the hunt itself, and sense is the Snark. It is difficult to respond to those who wish to be satisfied with words, things, images, and ideas. For we may not even say that sense exists either in things or in the mind; it has neither physical nor mental existence. Shall we at least say that it is useful, and that it is necessary to admit it for its utility? Not even this, since it is endowed with an inefficacious, impassive, and sterile splendour. This is why we said that *in fact* we can only infer it indirectly, on the basis of the circle where the ordinary dimensions of the proposition lead us.' Deleuze, *Logic of Sense*, 20.

51   *Wonderland*, 81.

Chapter 10

Reason

The Porpoise of this Tale

"Will you walk a little faster?" said a whiting to a snail,
"There's a porpoise close behind us, and he's treading on my tail."[1]

Purpose is meaning, and 'no wise fish would go anywhere without a porpoise'.[2] In its conventional sense, purpose *reduces* to effect, as it were. It aspires to an end, an objective, as distinct from the contingency of leaving things to chance. But meaning must come about in the telling, the use, it cannot simply be announced. And indeed, meaning has its purpose.

The Mock Turtle's song is striking in its rendition of the same kind of paradoxically 'purposeless' innovation considered in this discussion with respect to familiar production.[3] That is, it is innovation for the sake of innovation, the purpose (and product) will be reasoned later, after the fact, as it were, in the appropriation or assimilation of familiar production within conventional property models. A purpose is ordinarily understood as an object one has in view, not one taken in hindsight, and yet this is the very paradox of the product in contemporary innovation and high technology industries and indeed in the narrative presumed by intellectual property models themselves. The purpose is always divined in hindsight. Indeed, the drive to innovation in this tale is such that the whiting wishes to go faster, as the porpoise is treading on its tail, limiting their trajectory. To set out with a limiting purpose is simply to keep treading on one's tail, rather than moving forward with emergent creativity.

Nevertheless, the purpose makes itself clear in the course of the tale. To invent is to invent a use, the problem (the purpose) presenting itself once the solution is found. Purpose itself therefore entails contingency and chance. How that contingency is reconciled within the tale is crucial to the production of meaning. This circularity of the tale, of innovation, is itself intrinsic to the 'economics' of the law in the form of the relationship of exchange and disequilibrium considered

---

1   *Wonderland*, 84.

2   *Wonderland*, 86.

3   See the introduction of the value of chance in innovation (in particular, Chapters 3 and 5) and the discussion of experimental and experiential innovation and familiar production (in particular, Chapters 8 and 9).

in this inquiry: 'Besides the values of law and home, of distribution and partition, economy implies the idea of exchange, of circulation, of return.'[4]

The Purpose of Risk

Purpose is ordinarily understood as teleological, that is, it implies a final project or objective, a final good, a *telos* to which all human action is directed. But what is it to live beyond purpose? To live beyond history? Beyond the State? Beyond intellectual property? This is precisely the ethical field of chance in contrast to the moral track of risk. That is, purpose organizes risk; and with the import of meaning, of effect, comes the relation of risk. Earlier chapters examined the way in which risk is an integral part of the creditor–debtor relation in the intellectual property system, with the debtor assuming the 'risks' of the credit of innovation through an interminable debt.[5] Further, the action of risk as a relation, and therefore essential (or incidental) to meaning and use, is applied both in respect of the individual and population-wide.[6]

The critical relation here is between the attempt to mitigate risk (literally, an opportunity cost for innovation, as it were) and the possibility of embracing chance within the system itself. Intellectual property rules are arguably very closely related to the overall economic model of credit and debt, displacing some of the risk ('investment') of mitigation onto the consumer, creating the 'debt' to the 'author', the 'producer', which exists always already in credit. Indeed, part of the professionalization of creative endeavour secures this persistent and insistent 'flow' of debt. For example, recall what was seen in the case of the professionalization and organization of the fashion industry,[7] a process which is arguably equally applicable throughout the systematization of innovation and provides much insight into the way in which objectives to regulate and organize digital and familiar production have thus far been characterized. Indeed the very application of the term 'industry' facilitates a discourse of mechanization and predictability, of planned and accountable innovation. That is, it creates a context of an assessable and measurable 'economic' striated space for an otherwise immeasurable smooth space of creativity and innovation. Industry gives purpose.

Therefore, insofar as purpose is seemingly always attended by risk and the attempt to account for that risk, then it appears clear that this precludes a model based upon taking chances. In purpose there is value, and that value inheres in risk; indeed, the value in infringement.[8] Risk, as considered throughout these

---

4   J Derrida, *Given Time: I. Counterfeit Money*, P Kamuf (trans), U of Chicago P, Chicago, 1992: 6.

5   See the discussion of debt and risk in earlier chapters, particularly Chapters 2 and 5.

6   See the further discussion of risk in Chapter 5.

7   Chapter 4.

8   See the discussions of use and purpose earlier in 'Use'.

discussions, thus takes on a specific objective quality that is used to justify intellectual property rights in duration and scope. Further, risk consolidates the debt of the consumer. The overriding model and justification for intellectual property is thus, the greater the risk, the greater the stakes, the stronger the rights, the greater the imbalance (particularly the balance between chance and risk), and potentially the greater the monopoly welfare loss. Risk is presumed a 'value' in and of itself, and therefore risk itself becomes the commodity transacted in incentive models. It becomes the very 'purpose' of innovation and all this entails.

Chance or Purpose?

It would seem then that purpose, in its 'design', appears to be antagonistic to chance and accident. However, purpose also ushers in notions of discourse and conversation. In this way, purpose also presents a social function, a kind of 'public' function in that purpose arises through discourse, through use. Indeed, it follows then that purpose shepherds the very identity and activity of the user: 'And now it looks as if the use of the same word or the same piece, had a *purpose* – if the identity is not accidental, inessential. And as if the purpose were that one should be able to recognize the piece and how to play. – Are we talking about a physical or a logical possibility here? If the latter then the identity of the piece is something to do with the game.'[9] In other words, the user must be part of the construction of meaning and value within the language-game of intellectual property. Meaning has a purpose. The user is the purpose.

This notion of purpose eschews the sense of a pre-determined judgment (a kind of moral certainty or rule) and instead articulates an ethics of becoming, of imagination, that is, of innovation.[10] In other words, an ethics of familiar production is an ethics of choice and chance, an effort of taste and tribute; indeed, chance in this sense becomes itself a rule of the game: 'But, after all, the game is supposed to be defined by the rules!'[11] This is not to restore a rationalism to the game, but rather a purposive decision-making according to the consensus of the game. It is an ethics of innovation, drawing together the discussions throughout this inquiry of indebtedness, guilt, chance and taste. These will now be addressed in greater detail.

---

9   Wittgenstein, *Philosophical Investigations*, §566.

10   Brian Massumi explains: 'Becoming-other is the madness of the imagination. It is eminently ethical.' Massumi, *User's Guide*, 108.

11   Wittgenstein, *Philosophical Investigations*, §567.

An Ethics of Innovation

An ethics of innovation is fundamentally in relation to meaning through use. This principle informs relevant and legitimate characterizations of intellectual property transactions (*conversation*) through an ethics of use and thus a linguistic community with the significance of consent, accountability and reason that attends such a community. This is arguably the difference between a law that communicates with users and a law that rules and governs users that are otherwise absent from the conversation. That is, where users are absent they are in effect ruled without consent, resulting in a law that is literally without significance, a set of instructions that without meaning becomes ultimately arbitrary, a set of rules that is deemed inessential and thus inconsequential: 'What objection might one make to this? That one does not see the point of this prescription. Perhaps as one wouldn't see the point either of a rule by which each piece had to be turned round three times before one moved it. If we found this rule in a board-game we should be surprised and should speculate about the purpose of the rule.'[12] Where the law fails to engage with use, and thus meaning, its legitimacy becomes questionable: 'If I understand the character of the game aright – I might say – then this isn't an essential part of it.'[13]

Therefore, if the law attempts to deal with a 'social' relationship (the relationship between users through knowledge resources) through an administrative generality (moral rules or a seemingly 'illegitimate' law for reasons discussed earlier[14]) as distinct from the contingency of ethical relations, the belief and consensus upon which legitimacy is based is undermined.[15] In terms of the 'meaning' of intellectual products in the digital environment, values of 'access' and 'use' are in conflict with conventional models of distribution, exclusion and reproduction. Users have begun to disregard copyright and other forms of protection in the digital as simply *an inessential part* of the game. This cannot be remedied simply by dismissing such developments as anarchic and unlawful. That is, resolution cannot come from legal imperatives alone. Whether remedy comes from reconciling use with the business models based upon the law, or whether it comes from reforming the law that governs the relationship of use in the digital, it is clear that the legitimation crisis for intellectual property cannot be won by enforcement, in every sense of the word.

---

12   Wittgenstein, *Philosophical Investigations*, §567.

13   Wittgenstein, *Philosophical Investigations*, §568.

14   See the discussion of legitimacy in earlier chapters, including in particular, Chapters 6 and 8.

15   On the juridification of communication action, see J Habermas, *Theory of Communicative Action Volume II: Lifeworld and System, A Critique of Functionalist Reason*, T McCarthy (trans), Polity P, Cambridge, [1981]/1989: 356–73.

It's a Riddle

In purpose there is also the riddle, the question. Early uses of the term 'purpose' include its reference to a game of questions and answers; that is, the game of purpose is itself a literal endeavour at cross-purposes, indeed, at 'cross-examination'.[16] The riddle may be said to encompass the dialogue of innovation, of problem and solution, of purpose and tale, of heads and tails. Indeed, innovation comprises in and of itself the very character of the interlocutory, the philosophical investigation, the riddle. The dialectical nature of innovation is arguably also the fundamental principle in reforming the legal and social framework for its promotion and protection, particularly in the context of familiar production: 'We are in dialectics, and we are never in the *episteme* ... dialectics allows the judge to judge case by case.'[17]

A riddle also ordinarily contains a certain mischief and intentionality on the part of the teller, in that the wording is purposefully obfuscatory, indeed, puzzling. Paradoxically, a riddle is precisely where the purpose must always be speculated: 'For in riddles one has no exact way of working out a solution. One can only say, "I shall know a good solution if I see it."'[18] But I forget the question.[19]

Consider the tail.

Telling Tails

"Thank you, it's a very interesting dance to watch," said Alice, feeling very glad that it was over at last: "and I do so like that curious song about the whiting!"

"Oh, as to the whiting," said the Mock Turtle, "they – you've seen them, of course?"

"Yes," said Alice, "I've often seen them at dinn –" she checked herself hastily.

"I don't know where Dinn may be," said the Mock Turtle, "but if you've seen them so often, of course you know what they're like."

"I believe so," Alice replied thoughtfully. "They have their tails in their mouths – and they're all over crumbs."

"You're wrong about the crumbs," said the Mock Turtle: "crumbs would all wash off in the sea. But they *have* their tails in their mouths; and the reason is –" here the Mock Turtle yawned and shut his eyes. "Tell her about the reason and all that," he said to the Gryphon.

---

16   See further the discussion of 'cross-examination' in Chapter 11.

17   Lyotard & Thébaud, *Just Gaming*, 27.

18   Wittgenstein, *Lectures on the Foundations of Mathematics*, 84.

19   Bataille writes: 'The feeling of a decisive fight from which nothing would divert me now. I'm afraid, being certain I will no longer avoid the fight. Wouldn't the answer be: "that I forget the question?"' G Bataille, *The Impossible*, R Hurley (trans), City Lights Books, San Francisco, [1962]/1991: 139.

"The reason is," said the Gryphon, "that they *would* go with the lobsters to the dance. So they got thrown out to sea. So they had to fall a long way. So they got their tails fast in their mouths. So they couldn't get them out again. That's all."[20]

How then is one to invent a story of property without objects, that is, the story of familiar production without actual goods? This is indeed a very difficult sort to invent: 'the story had to grow out of the incidents, not the incidents out of the story'.[21] Nevertheless, this is ostensibly the issue faced by intellectual property in the digital. It is not so much that there are no products, but rather, that the products (and the product–consumer relation) have been transformed. In earlier discussion, the physiognomy of meaning showed the importance of giving a face to a name, whether time, the digital and so on.[22] That is, the 'naming' of the object (the patent, the trade mark, the copyright work, the design, intellectual property itself) authorizes those objects as reference points for the issues in question. In this way, patent, trade mark, design, copyright all operate as untranslatable proper names, as originary models.[23] However, the creative subject and the 'product' are both transformed in the digital, and the fantasy of the original is overwhelmed in a sea of simulacra. The generation of meaning is thus a memory, a repetition, a 'mock' turtle. In this way, meaning (and therefore use) is in and of itself an archive. Meaning is experience: *Meaning is use, on the face of things*.[24] It is, literally, what is sensible. Don't think, but look![25]

---

20   *Wonderland*, 85.

21   Preface *to Sylvie and Bruno*, 103. This point is also illustrated artfully in the following interlocutory from *Sylvie and Bruno Concluded*, Chapter 23 'The Pig-Tale', 257.

'Why should you always have live things in stories?' said the Professor. 'Why don't you have events, or circumstances?'

'Oh, please invent a story like that!' cried Bruno.

The Professor began fluently enough. 'Once a coincidence was taking a walk with a little accident, and they met an explanation – a very old explanation – so old that it was quite doubled up, and looked more like a conundrum –' he broke off suddenly.

'Please go on!' both children exclaimed.

The Professor made a candid confession. 'It's a very difficult sort to invent, I find.'

22   In particular, see the discussion of the process of naming in 'Use'.

23   See the discussion of the model–copy relationship in Chapter 9.

24   This interpretative space of meaning is noted by Brouwer and Mulder: 'With the recent introduction of digital databases we seem to be witnessing a shift. What used to be material archive-systems have become immaterial information-banks.' Brouwer & Mulder, 'Information is Alive', 4.

25   Wittgenstein, *Philosophical Investigations*, §66.

Sense and Sensibility

"I can tell you more than that, if you like," said the Gryphon. "Do you know why it's called a whiting?"

"I never thought about it," said Alice. "Why?"

"*It does the boots and shoes,*" the Gryphon replied very solemnly.

Alice was thoroughly puzzled. "Does the boots and shoes!" she repeated in a wondering tone.

"Why, what are *your* shoes done with?" said the Gryphon. "I mean, what makes them so shiny?"

Alice looked down at them and considered a little before she gave her answer. "They're done with blacking, I believe."

"Boots and shoes under the sea," the Gryphon went on in a deep voice, "are done with whiting. Now you know."[26]

At first Alice offers an answer based on belief (and expectations of class) – 'They're done with blacking, I believe.' But at once this is corrected with knowledge: 'Now you know.' The transformation of knowledge from representational belief to the curiosity of learning is coincident with the processes of familiar production: 'it is knowledge that is nothing more than an empirical figure, a simple result which continually falls back into experience; whereas learning is the true transcendental structure which unites difference to difference, dissimilarity to dissimilarity, without mediating between them.'[27] Alice is becoming an active user, she is moving from the prefigured solutions and products of knowledge (*I believe*) to the problems (*now you know*). At once she begins to engage, to investigate the ideas, to immerse herself 'under the sea' with the whiting:

"If I'd been the whiting," said Alice, whose thoughts were still running on the song, "I'd have said to the porpoise, 'Keep back, please: we don't want *you* with us!'"

"They were obliged to have him with them," the Mock Turtle said: "no wise fish would go anywhere without a porpoise."

"Wouldn't it really?" said Alice in a tone of great surprise.

"Of course not," said the Mock Turtle: "why, if a fish came to *me*, and told me he was going on a journey, I should say, 'With what porpoise?'"

"Don't you mean 'purpose'?" said Alice.

"I mean what I say," the Mock Turtle replied in an offended tone.[28]

If 'no wise fish would go anywhere without a porpoise', then is the Mock Turtle's story challenging the nature of the ethical journey without end, the becoming?

---

26   *Wonderland*, 86.

27   Deleuze, *Difference and Repetition*, 166–67.

28   *Wonderland*, 86.

Perhaps here for a moment Alice's 'meaning blindness'[29] defeats her understanding. But this lack of understanding is immediately rejected by the Mock Turtle: 'I mean what I say.' No wise fish would go anywhere without purpose, no user would indeed exist without purpose, as purpose always already attends (and is intended and effected by) use.

The porpoise 'attends' the journey, gives meaning to the journey. That is, purpose is not necessarily leading innovation; rather, it is meaning itself (its 'purpose') that is generated through use and application – a use not necessarily anticipated by the innovative technology but later 'devised' by it in communication with users. This is the difficulty of the story to be invented for familiar production; that is, that purpose arises through use, that use is the scene of genuine innovation. The accountability or purpose of the innovation as text is in its reading. This is the account of modern technology. Innovation has thus assumed an overwhelmingly literary demeanour, a text which is given proliferative and multiplicative life in its reading – the qualitative, directionless, infinite smooth space of familiar production.

This episode illustrates the sensibility necessary to the experience of meaning and the whole of the language-game; that is, the difference between conveying information and the wider expressive qualities and capacities of language. Without this sensibility, one is 'blind' to the perspicuity of the fuller meaning, to the sense:[30] 'If I compare the coming of the *meaning* into one's mind to a dream, then our talk is ordinarily dreamless. The "meaning-blind" man would then be the one who would always talk dreamlessly.'[31] This shared sensibility and inter-connectedness are integral to the language-game. In other words, Alice fails to appreciate the paradox and this episode brings into acute focus the difference between knowing the word and learning and understanding the language.

Sensibility to meaning is intrinsically and inextricably linked to the capacity to maximize the ethical benefits in the language-game, that is, the mutual benefits of inter-connectedness. An ethics of sensibility is therefore mutually constitutive and cooperative with an ethical drive to good health and a good life,[32] a happy

---

29   See further Wittgenstein, *Philosophical Investigations*, 213e on 'aspect blindness' and the later L Wittgenstein, *Remarks on a Philosophy of Psychology: Volume 1*, GEM Anscombe et al. (eds), Blackwell, Oxford, 1991: §198.

30   Wittgenstein, *Philosophical Investigations*, 213e.

31   Wittgenstein, *Remarks on a Philosophy of Psychology: Volume I*, §232.

32   'It's the styles of life involved in everything that make us this or that.' See Deleuze, *Negotiations*, 100. See further DW Smith, '"A Life of Pure Immanence": Deleuze's "Critique et Clinique" Project', Introduction to G Deleuze, *Essays Critical and Clinical*, DW Smith and MA Greco (trans), xi–liii: lii. See also L Wittgenstein, *Notebooks 1914–1916*, GH von Wright & GEM Anscombe (eds), GEM Anscombe (trans), 2nd ed, Blackwell, Malden MA, [1961]/1998: 81e: 'The good conscience is the happiness that the life of knowledge preserves.'

life.[33] This is a mutually life-affirming ethics, a life of connections: 'The happy life seems to be in some sense more *harmonious* than the unhappy.'[34]

> The World and Life are one.
> Physiological life is of course not 'Life'. And neither is psychological life. Life is the world.
> Ethics does not treat of the world. Ethics must be a condition of the world, like logic.
> Ethics and aesthetics are one.[35]

Thus, the aesthetics of familiar production facilitate a proliferation of story, of perspective, of experience. Familiar production thrives upon an aesthetic of experimentation and experience,[36] where 'the essential characteristic of the modern work of art' is indeed experimentation: 'This is, without doubt, the essential characteristic of the modern work of art. It is not at all a question of different points of view on one story supposedly the same; for points of view would still be submitted to a rule of convergence. It is rather a question of different and divergent stories, as if an absolutely distinct landscape corresponded to each point of view.'[37] The smooth space of familiar production is not one of resemblance but an innovative production and multiplication of identities, the power and 'the aggression of simulacra', the power of use: 'These are the characteristics of the simulacrum, when it breaks its chains and rises to the surface: it then affirms its phantasmatic power, that is, its repressed power ... Thus the conditions of real experience and the structures of the work of art are reunited.'[38]

On Perspective

> "I could tell you my adventures – beginning from this morning," said Alice a little timidly: "but it's no use going back to yesterday, because I was a different person then."

---

33 Wittgenstein writes: 'I keep on coming back to this! simply the happy life is good, the unhappy bad.' Wittgenstein, *Notebooks 1914–1916*, 78e.

34 Wittgenstein, *Notebooks 1914–1916*, 78e.

35 Wittgenstein, *Notebooks 1914–1951*, 77. See further, 'Ethics so far as it springs from the desire to say something about the ultimate meaning of life, the absolute good, the absolute valuable, can be no science. What it says does not add to our knowledge in any sense. But it is a document of a tendency in the human mind which I personally cannot help respecting deeply and I would not for my life ridicule it.' Wittgenstein, *Philosophical Occasions*, 44.

36 Recall the discussion in Chapter 8.

37 Deleuze, *Logic of Sense*, 260.

38 Deleuze, *Logic of Sense*, 261.

"Explain all that," said the Mock Turtle.

"No, no! The adventures first," said the Gryphon in an impatient tone: "explanations take such a dreadful time."[39]

Explain all that.

This exchange asserts the central concern of the continuous innovation of familiar production, as distinct from the incarnation, fixation and striation of the space of innovation through the more dominant models of property. On the one hand, the explanations or lessons of conventional representational knowledge simply prolong the inevitable and reproduce the imperative (the directive, the moral certainty, the belief): 'Ordinary repetition is prolongation, continuation or that length of time which is stretched into duration.'[40] This recalls the multiplication tables on which Alice attempted to rely in order to make sense of her unanticipated and unusual surroundings. On the other hand,, is the repetition of difference in the 'learning' the innovation of familiar production? That is, meaning is at once generated in the production of difference and repetition in the familiar, where it is possible to 'Make something new of repetition itself ... make it the supreme object of the will and of freedom.'[41]

> "Stand up and repeat ' *'Tis the voice of the sluggard,*'" said the Gryphon.
>
> "How the creatures order one about, and make one repeat lessons!" thought Alice. "I might as well be at school at once." However, she got up, and began to repeat it, but her head was so full of the Lobster Quadrille, that she hardly knew what she was saying, and the words came very queer indeed.[42]

Very queer indeed.

> "That's different from what *I* used to say when I was a child," said the Gryphon.
>
> "Well, I never heard it before," said the Mock Turtle; "but it sounds uncommon nonsense."
>
> Alice said nothing; she had sat down with her face in her hands, wondering if anything would *ever* happen in a natural way again.[43]

And indeed, nothing ever will, because the 'nature' of the original has been troubled irreversibly by the difference and repetition of simulacra.

---

39   *Wonderland*, 86–87.

40   Deleuze, *Difference and Repetition*, 201.

41   Deleuze, *Difference and Repeitition*, 6.

42   *Wonderland*, 87.

43   *Wonderland*, 87–88.

The Use of Repetition

"What *is* the use of repeating all that stuff," the Mock Turtle interrupted, "if you don't explain it as you go on? It's by far the most confusing thing *I* ever heard!"[44]

Confusion, however, is itself a character of knowledge: 'the clear is confused by itself, in so far as it is clear'.[45] Importantly, this notion of confusion is not with respect to an originary idea or model, and does not qualify any 'original' as such: 'Clear-confused does not qualify the Idea, but the thinker who thinks or expresses it.'[46] In other words, the 'confusion' of learning belies the supposed satisfaction rendered by representational knowledge which is confused itself, *confused in so far as it is clear*: 'the weakness of the theory of representation, from the point of view of the logic of knowledge, was to have established a direct proportion between the clear and the distinct, at the expense of the inverse proportion which relates these two logical values: the entire image of thought was compromised as a result.'[47]

Just like the child who 'hums to summon the strength for the schoolwork she has to hand in',[48] the refrain, the repetition, is a mechanism for organizing Alice's participation within the space of the Mock Turtle and the Gryphon, to calm chaos. However, 'A mistake in speed, rhythm, or harmony would be catastrophic because it would bring back the forces of chaos, destroying both creator and creation.'[49]

"Yes, I think you'd better leave off," said the Gryphon: and Alice was only too glad to do so.

"Shall we try another figure of the Lobster Quadrille?" the Gryphon went on. "Or would you like the Mock Turtle to sing you a song?"

"Oh, a song, please, if the Mock Turtle would be so kind," Alice replied, so eagerly that the Gryphon said in a rather offended tone, "Hm! No accounting for tastes! Sing her 'Turtle Soup,' will you, old fellow?"

The Mock Turtle sighed deeply, and began, in a voice sometimes choked with sobs, to sing this: …[50]

Punctuated by sobs, the music stammers.[51] Nevertheless, the rich repetition of the refrain is crucial to the establishment of the 'abode' of familiar production,

---

44 *Wonderland*, 88.
45 Deleuze, *Difference and Repetition*, 254.
46 Deleuze, *Difference and Repetition*, 253.
47 Deleuze, *Difference and Repetition*, 253.
48 Deleuze & Guattari, *Thousand Plateaus*, 311.
49 Deleuze & Guattari, *Thousand Plateaus*, 311.
50 *Wonderland*, 87–88.
51 On the literature of Sacher Masoch, Deleuze notes that the author makes language 'stammer' (puts into suspense) as distinct from stutters (repeats): 'whenever a language is

the musicality of smooth space: 'The song is like a rough sketch of a calming and stabilizing, calm and stable, center in the heart of chaos.'[52] In the digital, the refrain is a fundamental accompaniment, the expression and creation of the territory of familiar production, the time of innovation, the materiality of the user, the actualization of the digital product: 'The refrain may assume other functions, amorous, professional or social, liturgical or cosmic: it always carries earth with it.'[53]

The Mock Turtle's sobs literally provide a counterpoint, one which is other than the usual music. This counterpoint is both a recombination of the refrain and at the same time a process of familiarity and proximity. This familiarity offers a kind of materiality to the song, an intimacy beyond the product, an experiential innovation, a genuine affective labour. Intriguingly, the Mock Turtle's performance grapples with the relationship between the quest for materiality in the digital and the way in which this cannot be resolved through fixed goods, but rather, must be addressed through the affect, through experience.[54] And this is perhaps the crucial 'dissonance' or 'counterpoint' of familiar production: 'Is that what friendship is, a harmony embracing even dissonance?'[55] Is that what ethics is? Indeed, even while writing this, an enthusiastic bird is outside the window punctuating the thoughts with song, inviting an intimacy between the isolation of writing and nature outside. The refrain territorializes, just as for the bird, marking territory, bringing others within its domain, its abode: 'it is *ethos*, but the ethos is also the Abode.'[56] So at once the refrain creates the territory and becomes the theme that sustains the territory.

An Ethics of Innovation

Insofar as an ethics of innovation must account for the harmony and counterpoint of interactions, the potential for maximizing such interactions arguably subsists in the potential of chance: 'What is retained and preserved, therefore created, what consists, is only *that which increases the number of connections* at each

---

submitted to such creative treatments, it is language in its entirety that is pushed to its limit, to music or silence.' Deleuze, *Essays Critical and Clinical*, 55.

52   Deleuze & Guattari, *Thousand Plateaus*, 311.

53   Deleuze & Guattari, *Thousand Plateaus*, 312.

54   The Lobster Quadrille might be understood similarly in that it engages the interaction of different lines of movement between animals, across species and in concert with the environment. In this way there is a becoming through the intimacy between the music and dance and the animals themselves. See further the discussion in Deleuze & Guattari, *Thousand Plateaus*, 299–309.

55   Deleuze, *Negotiations*, 163.

56   Deleuze & Guattari, *Thousand Plateaus*, 312.

level of division or composition.'[57] Interactions within contemporary frameworks and applications of intellectual property rules have increasingly interfered with these connections. The challenge for an ethics of innovation is to maximize the opportunity for use and tribute, while at the same time sustaining the commercial and economic realities of intellectual property industries. The challenge is to embrace the chance and uncertainty of the digital in ways that will proliferate the economic and commercial opportunities upon which the industries of intellectual property are articulated. An ethics of innovation is all at once a commitment to maximizing these interactions or connections that are proliferating within the digital environment and associated technologies, the becoming of innovation.

Immediately the distinction between morality and ethics is again clear. As distinct from judging the roles within intellectual property and the 'moral' conventions that have become cemented within, appropriate ethical reform of those interactions can multiply the creative potential of enterprise within that system. A 'moral' framework interferes with the processes and behaviours of experimentation and chance that are fracturing the conventional debtor–creditor relationship, instead maintaining the 'debt' to the 'author' within the rationale of the so-called social contract or bargain of intellectual property.[58] On the other hand, ethical reform unsettles the 'transfer' of debt that underwrites the usual incarnation of consumer–producer relationships in intellectual property discourse. The ethics of familiar production promises a more pragmatic approach to the challenges of the digital environment, and to the implementation of new business models that are not only ethically appropriate but also opportunities for greater economic and commercial potential for industries otherwise vulnerable to the digital.

With respect to the ethical reform of the relationship between production and consumption, and the possible legal and policy reform of the intellectual property system itself, the individual 'self' must be maintained: 'No one can ever so utterly transfer to another his power and, consequently, his rights, as to cease to be a man; nor can there ever be a power so sovereign that it can carry out every possible wish. It will always be vain to order a subject to hate what he believes brings him advantage, or to love what brings him loss, or not to be offended at insults, or not to wish to be free from fear.'[59] Fundamentally, an ethical framework ensures the integrity of self outside the modes of production and property. As distinct from a pre-existing and unchallengeable moral certainty of conventional models of production and protection (including intellectual property), the ethics of familiar production indicates the necessity of addressing the iterative and deferred nature of justice: 'In this way, Ethics, which is to say, a typology of immanent modes of existence, replaces Morality, which always refers existence

---

57   Deleuze & Guattari, *Thousand Plateaus*, 508.

58   See the discussion of 'social contract' in 'Use' as well as in Chapters 2 and 9.

59   B de Spinoza, *A Theological-Political Treatise, and a Political Treatise*, Cosimo Books, NY, [1883]/2007: 214.

to transcendent values.'[60] Ethics is not only crucial to any programme of reform in intellectual property, it affirms innovation, it is itself the life of innovation, free from prescriptive moral imperatives and infinite guilty debt: 'Life is poisoned by the categories of Good and Evil, of blame and merit, of sin and redemption.'[61] If information really wants to be free, an ethics of innovation sets intellectual life free: free as in freedom.

> "Chorus again!" cried the Gryphon, and the Mock Turtle had just begun to repeat it, when a cry of "The trial's beginning!" was heard in the distance.
> "Come on!" cried the Gryphon, and, taking Alice by the hand, it hurried off, without waiting for the end of the song.
> "What trial is it?" Alice panted as she ran; but the Gryphon only answered, "Come on!" and ran the faster, while more and more faintly came, carried on the breeze that followed them, the melancholy words:
> "Soo–oop of the e–e–evening,
> Beautiful, beautiful Soup!"[62]

The chorus is interrupted, the refrain is deterritorialized, imitation has destroyed itself: 'since the imitator unknowingly enters into a becoming that conjugates with the unknowing becoming of that which he or she imitates. One imitates only if one fails, when one fails.'[63]

Come on! Come on! Come to know. The directive yet again drives Alice on, towards the future as yet unknown, a becoming, *'democracy to come'*,[64] ethics to come, justice to come. *Come on! Come on!* Justice is about to be served, for the trial that cannot be named: 'What's important in "democracy to come" is not "democracy," but "to come." That is, a thinking of the event, of what comes … "To come" means "future".'[65] Justice is a future present: 'Justice remains, is yet, to come … Perhaps it is for this reason that justice, insofar as it is not only a juridical or political concept, opens up for *l'avenir* the transformation, the recasting or refounding law and politics.'[66] In this way, the ethics of chance inform the circumstances for a just future.

Come on!

---

60   Deleuze, *Spinoza: Practical Philosophy*, 23.
61   Deleuze, *Spinoza: Practical Philosophy*, 26.
62   *Wonderland*, 89.
63   Deleuze & Guattari, *Thousand Plateaus*, 305.
64   Derrida, *Negotiations*, 179.
65   Derrida, *Negotiations*, 182.
66   Derrida, 'Force of Law', 27.

Chapter 11

Account

A Court of Appeal

> "I wish they'd get the trial done," she thought, "and hand round the refreshments!"
> But there seemed to be no chance of this, so she began looking about her, to pass
> away the time.[1]

Alice wishes for a speedy trial, a revolution of becoming. All at once, justice is approximating as it can never be representative; interminable but will never wait, immediate and yet always becoming: 'But justice, however unpresentable it may be, doesn't wait. It is that which must not wait. To be direct, simple and brief, let us say this: a just decision is always required *immediately*, "right away." It cannot furnish itself with infinite information and the unlimited knowledge of conditions, rules or hypothetical imperatives that could justify it.'[2]

Immediately, access to justice is presented as a matter of spectacle, of entertainment. It is a 'show' followed by refreshments. In this way, the court room and the entertainment therein is an assembly of all the creatures encountered by Alice throughout her journey. The court room is thus a site of consensus and consent to participate, an 'in common' spectacle. Kafka teaches us: 'Within the law all is accusation, advocacy, and verdict; any interference by an individual here would be a crime. It is different, however, in the case of the verdict itself; this is based on inquiries being made here and there, from relatives and strangers, from friends and enemies, in the family and public life, in town and village – in short, everywhere.'[3] The verdict is, literally, familiar.

In some respects this resonates with earlier discussions of the notion of taste and aesthetics, the spectacle of the museum, the entertainment of fashion, as elements of an ethical dialogue.[4] This is the crucially important dialectical process of justice: 'dialectics allows the judge to judge case by case. But if he can, and indeed must (he has no choice), judge case by case, it is precisely because each situation is singular, something that Aristotle is very sensitive to. This singularity comes from the fact that we are in matters of opinion and not in matters of truth.'[5]

---

1   *Wonderland*, 90.

2   Derrida, 'Force of Law', 26.

3   F Kafka, 'Advocates', T Stern & J Stern (trans), F Kafka, *The Complete Short Stories*, NN Glatzer (ed), Minerva, London, 1992: 449, at 450.

4   See the discussion of fashion, aesthetics and ethics in Chapter 4.

5   Lyotard & Thébaud, *Just Gaming*, 27.

That is, there is no moral certainty, only the dialogue of ethics: 'In every instance, one must evaluate relations: of force, of values, of quantities, and of qualities; but to evaluate them there are no criteria, nothing but opinions.'[6] This is the court of appeal, as it were; that is, it is the game of law played according to the rules of attraction.

Room for Judging

> Alice had never been in a court of justice before, but she had read about them in books, and she was quite pleased to find that she knew the name of nearly everything there. "That's the judge," she said to herself, "because of his great wig."[7]

The court provides a literal, physical space for the community and consensus of justice, the resolution of disagreement and the construction of common ground. The solving of the 'riddle' is thus in the dialogue of the collective. This recalls the notion of the therapeutic value of the philosophical interlocutory accredited at the start of this journey: 'The philosopher's treatment of a question is like the treatment of an illness.'[8] The court room may therefore be regarded as not only a therapeutic space, but also one in which the interlocutory, the 'cross' examination, the riddle, might itself traverse the discourse of transgression.

Wittgenstein sees ethics and aesthetics as intrinsically cooperative: 'I am going to use the term Ethics in a slightly wider sense, in a sense in fact which includes what I believe to be the most essential part of what is generally called Aesthetics.'[9] Further, he links aesthetics with judicial language, likening aesthetic discussions to those in a court of law in that they extend the circumstances of the object of discussion to further description, comparison and reason.[10] This gathering of

---

6   Lyotard & Thébaud, *Just Gaming*, 27.

7   *Wonderland*, 90.

8   Wittgenstein, *Philosophical Investigations*, §255.

9   Wittgenstein, *Philosophical Occasions*, 38.

10   GE Moore records the following from Wittgenstein's lecture on Ethics: 'all that Aesthetics does is "to draw your attention to a thing", to "place things side by side". He said that if, by giving "reasons" of this sort, you make another person "see what you see" but it still "doesn't appeal to him", that is "an end" of the discussion; and that what he, Wittgenstein had, "at the back of his mind" was "the idea that aesthetic discussions were like discussions in a court of law", where you try to "clear up the circumstances" of the action which is being tried, hoping that in the end what you say will "appeal to the judge". And he said that the same sort of "reasons" were given, not only in Ethics, but also in Philosophy.' GE Moore, 'Wittgenstein's Lectures in 1930–33', in L Wittgenstein, *Philosophical Occasions 1912–1951*, J Klagge & A Nordmann (eds), Hackett, Indianapolis, 1993: 46–114, at 106.

additional circumstances is deployed to increase the probability of a decision (a judgment on aesthetics, on evidence, on culpability): 'The procedure in a court of law rests on the fact that circumstances give statements a certain probability.'[11] This lessens the risk of an erroneous decision, but heightens the chance of further narrative, fuller meaning. Nevertheless, justice attempts to make sense ostensibly in relation to 'the just': 'existence is cut into lots, the affects are distributed into lots, and then related to higher forms.'[12] Within the justice of the intellectual property framework, to the credit of the 'producers' the 'consumers' of justice are rendered in infinite debt: 'The doctrine of judgment has reversed and replaced the system of affects.'[13]

Justice is thus a communication, a spectacle, sheer entertainment with refreshments later. The important thing, in spite of this, is to appeal to the court, in every sense of the word. Indeed, this is the personality of the common law system, the complex discourse of justice as articulated through cross-examination, disclosure and oral argument. This is the justice of the interlocutory, the fundamental character of ethical speech and decision, and through which the proximity and familiarity of relations of user and use may be articulated. It is a process of evaluation relying on 'nothing but opinions'.[14] In other words, this evaluative process 'cannot be resolved in terms of models',[15] but is instead to be understood in terms of decision. That is, ethical evaluation is not a judgment of morality and a moral imperative. It is an iterative justice through dialogue. Justice is a verb. That is, talking is doing, speaking is use. Argument is a key word in the ethics of justice, in justice *to come*: 'Becoming is a verb.'[16]

This sort of spoken justice, a talking cure, is not necessarily ubiquitous in jurisdictions throughout the world. Indeed, this opportunity for discursive action in the court room, as it were, is potentially an invaluable part of the intellectual property system in the United Kingdom and a crucial aspect of any mechanisms of international harmonization (whether within Europe or beyond). Indeed, judicial innovations in England and Wales, like the achievements of the Intellectual Property Enterprise Court (formerly the Patents County Court or PCC),[17] provide such an opportunity for a spoken solution together with a 'speedy trial' in time for refreshments. The Intellectual Property Enterprise Court (IPEC) deals specifically

---

11   Wittgenstein, *On Certainty*, §335.

12   Deleuze, *Essays Critical and Clinical*, 129.

13   Deleuze, *Essays Critical and Clinical*, 129.

14   Lyotard & Thébaud, *Just Gaming*, 27.

15   Lyotard & Thébaud, *Just Gaming*, 25.

16   Deleuze & Guattari, *Thousand Plateaus*, 239.

17   The Patents County Court was established in 1990 under the Copyright, Designs and Patents Act 1988, s 287(1) and was abolished (to be replaced by the IPEC) on 1 October 2013.

with intellectual property issues[18] (as an alternative to bringing such cases under the full and conventional High Court procedure with its disclosure and costs regime[19]) and provides for a streamlined system of litigation as well as a small claims track. If litigants choose to take a dispute to the IPEC, a limit on the value of claims will apply.[20] Supported by a Users' Committee, the use of the IPEC has arguably proven the need for such discursive remedy within the system, particularly to the benefit of 'indebted' users such as individuals and small- and medium-sized enterprises.[21]

With respect to an emphasis on discourse in legal and judicial reform, this resonates with the philosophical method of the interlocutory. In this context, the discursive action of the court room, the 'use' of talk, recollects earlier discussions of the presence of 'decision' in an ethics of innovation.[22] The process of judicial interrogation (or questioning) is itself a language-game, conducted according to internal conventions and process. The iterative process of the common law arrives at meaning precisely through participation in the language-game, that is, through use. In other words, justice is not fixed and referenced to some immutable certainty. Justice is changeable. The 'becoming'[23] of justice (justice *to come*) is in fact the ethical dimension of the system. The rules of justice and of judgment are thus contractual, subject to agreement (subject to use): 'their rules do not carry within themselves their own legitimation, but are the object of a contract, explicit or not, between players (which is not to say that the players invent the rules).'[24] Clearly this does not mean that justice is uncertain, impotent, indecisive; quite the contrary. Rather, the ethics of justice comes through this collective, this contractual

---

18   Before its move to become part of the High Court as the IPEC, the Patents County Court dealt only with matters within its special jurisdiction (which technically applied to patents, designs and trade marks and ancillary matters: see Patents County Court (Designation and Jurisdiction) Order 1994 (SI 1994/1609 as amended by SI 2005/587)). It also heard certain other claims related to these within its special jurisdiction. It could hear other types of claim (for example, copyright infringement) as an ordinary county court (from 1 October 2013, 'the' county court, Crime and Courts Act 2013, s 17 and Sch 9).

19   However, intellectual property cases can be transferred to the"full" general Chancery Division or Patents Court under Civil Procedure Rules (CPR) Pt 30,

20   The limit is such that only claims valued at no more than £500,000 can be heard before the IPEC. (see CPR 63,17), although with the consent of the parties higher value claims can be heard.

21   There is a concern with the cost of appeals to the Court of Appeal, if the unintended effect of this would be to override the benefits of scaled costs in the IPEC. Informally, some practitioners have suggested this could result in larger litigants simply waiting to appeal, and thus overcoming the limits on claims. However, there is as yet no evidence to support that this is happening and the restrictions on the scope of appeal might make such concerns unfounded.

22   See Chapter 8.

23   Recall the discussion of 'becoming' in Chapter 6.

24   Lyotard, *Postmodern Condition*, 10.

relationship.[25] Indeed, 'if there are no rules, there is no game'.[26] Therefore, in the language-game of judicial process, as in any other language-game, 'every utterance should be thought of as a "move" in a game'.[27]

Court Cards

> The twelve jurors were all writing very busily on slates. "What are they all doing?" Alice whispered to the Gryphon. "They can't have anything to put down yet, before the trial's begun."
>
> "They're putting down their names," the Gryphon whispered in reply, "for fear they should forget them before the end of the trial."[28]

This active memory ('no mere passive inability to rid oneself of an impression') is a crucial aspect of the '*memory of the will*'.[29] The jury writes things down, without thinking, without reflection, before anything has happened. The jury is, in a sense, *remembering the new*. In this way, the jury too becomes indebted, even in the writing down of one's own name, one's own image: 'Man himself must first of all have become *calculable, regular, necessary*, even in his own image of himself, if he is to be able to stand security for *his own future*, which is what one who promises does!'[30] And so the court procedure makes the jurors responsible: 'This precisely is the long story of how *responsibility* originated. The task of breeding an animal with the right to make promises evidently embraces and presupposes as a preparatory task that one first *makes* men to a certain degree necessary, uniform, like among like, regular, and consequently calculable … with the aid of the morality of mores and the social straitjacket, man was actually *made* calculable.'[31]

Thus, in recording their overwhelming responsibility to the archive, the jurors become 'witnesses in spite of themselves'.[32] The status and contribution of history, of testimony, is where everything is recorded, everything is monitored, everyone is a historian, is left to chance, accidental, involuntary. Everyone is an owner:

---

25   This notion of the collective in justice is also discussed further in Chapter 8. See also naming as a collective process in Negri, *Negri on Negri*, 120.

26   Lyotard, *Postmodern Condition*, 10.

27   Lyotard, *Postmodern Condition*, 10.

28   *Wonderland*, 90–92.

29   Nietzsche, *On the Genealogy of Morals*, 58.

30   Nietzsche, *On the Genealogy of Morals*, 58.

31   Nietzsche, *On the Genealogy of Morals*, 58–59.

32   Historian Marc Bloch explains that narrative evidence has assumed increased significance in historical research and that such research has come to emphasize more 'the evidence of witnesses in spite of themselves'. See M Bloch, *The Historian's Craft*, Manchester UP, Manchester, 1992: 51.

'Rather than calling into question the epistemological status of documents, it enlarges their field.'[33]

In many respects, this is similar to the process by which the social becomes overcoded by an economic structure of production, where the seemingly obligatory participation in social media is at once captured by the platforms themselves as surplus value.[34] The user becomes a producer in spite of itself, and at the same time, the social incurs the risk and infinite and inexhaustible debt of production: 'The author is only the witness or guarantor of his own absence in the work in which he is put into play, and the reader can only provide this testimony once again, making himself in turn the guarantor of the inexhaustible game in which he plays at missing himself.'[35] The almost obligatory participation in social media as a form of life, is thus attended by a much more efficient and telescopic system of surveillance, where users become 'witnesses in spite of themselves'. Social media writes everything down, as it were, in case it forgets anything and so that it forgets everything.

More startling examples of the way in which social media asserts itself as adjudicator and jury are the recent events of social media vigilance (and vigilantism) at the margins of the law – always awake, always looking. In the case of the bombs at the Boston Marathon, 15 April 2013, which killed three and wounded 264,[36] the enormous community of 'witnesses' is credited with a swift apprehension of the two suspects through so-called crowd-sourced justice. However, the response on social media also emerged in some areas as a 'farce'[37] and a 'witchhunt'[38] which, in the case of the global Reddit campaign, saw Reddit users (or Redditors) and the Reddit Bureau of Investigation (RBI)[39] 'starting the trial by public aspect'.[40]

---

33   P Ricoeur, *Time and Narrative: Volume III*, K Blamey & D Pellauer (trans), U of Chicago P, Chicago, [1985]/1988: 117.

34   Introduced in the earlier discussion in 'Use' and reviewed throughout.

35   Agamben, *Profanations*, 71.

36   D Kotz, 'Injury toll from Marathon bombs reduced to 264', *Boston Globe*, 24 April 2013.

37   P Walker, 'Boston bombing identification attempts on social media end in farce', *The Guardian*, 19 April 2013.

38   I Steadman, 'Reddit users are hosting a witch-hunt for the Boston Marathon bomber', *Wired*, 17 April 2013.

39   The Reddit Bureau of Investigation (RBI) states: 'The goal of the RBI (Reddit Bureau of Investigation) is to use the power of Reddit to solve crimes/mysteries and catch criminals.' www.reddit.co/r/RBI. The RBI establishes 'sub-reddits' on particular issues, in this case establishing one named /r/findbostonbombers, which was subsequently taken down in the aftermath of identifying incorrect suspects, putting them through significant turmoil and placing them and their families at potential risk. See further Walker, 'Boston bombing identification attempts on social media end in farce'; R Adams, 'Boston "witchhunt" on social media sites – and a bad week for the old guard', *The Guardian*, 22 April 2013.

40   Steadman, 'Reddit users are hosting a witch-hunt for the Boston Marathon bomber'; and 'Social media vigilantes cloud Boston bombing investigation', *NPR*

Social media vigilance campaigns express both the potential and the violence at the margin of the law.[41] Nevertheless, it is clear that it is crucial to direct attention to the borders: 'everything happens at the border'.[42] And notably, even the case of social media and the Boston bombing has been subsequently incorporated and striated by management discourse.[43]

An intriguing counterpoint of 'crowd-sourced' justice (and shaming) is the apparent reassurance of the crowd and a type of 'crowd-sourced' immunity through collective action. This is illustrated particularly in the case of riots, but nevertheless provides insight into the architecture of sharing or 'piracy' as well. In the case of the London riots in the summer of 2011, as well as the literal application of social media in coordinating and communicating activity, more interesting is the fact that those acting in the swell appeared to believe their actions were immune from retribution simply because of the wisdom and protection of the crowd.[44] Indeed, many looters did not even attempt to mask their identity, such was their confidence in the crowd.[45] However, social media has also emerged as the very tool by which to adjudicate on a 'crowd' scale.[46] Social media has become the space of the modern community court; everyone has privity to the event, it turns everyone into witnesses.[47] A sense of property in the social is materialized through familiarity, a virtual proximity and an intangible probability. Familiarity breeds consent.

---

broadcast, 22 April 2013.

41 A further aspect of this is the penetration of the sanctity of the court itself by the prejudice of information: 'Juror admits contempt of court over Facebook contact', *BBC News*, 14 June 2011; J Halliday, 'Facebook juror and defendant guilty of contempt', *The Guardian*, 14 June 2011; 'Juror denies contempt of court over Facebook paedophile post', *The Guardian*, 23 July 2013; 'Jurors jailed for contempt of court over internet use', *BBC News*, 29 July 2013; K Hall, 'Jurors found guilty of Facebook and Google contempts', *Law Society Gazette*, 29 July 2013.

42 Deleuze, *Logic of Sense*, 9.

43 GC Kane, 'What can managers learn about social media from the Boston Marathon Bombing?' *MIT Sloan Management Review*, 25 April 2013.

44 J Thompson states: 'Morality is inversely proportional to the number of observers. When you have a large group that's relatively anonymous, you can essentially do anything you like.' Cited in T De Castella & C McClatchey, 'UK riots: what turns people into looters?' *BBC News*, 9 August 2011. See further EP Thompson, 'The Moral Economy of the English Crowd in the Eighteenth Century', 50 *Past and Present* February 1971: 76–136; EJ Hobsbawm, *Primitive Rebels: Studies in Archaic Forms of Social Movement in the Nineteenth and Twentieth Centuries*, Manchester UP, Manchester, 1959.

45 De Castella & McClatchey, 'UK riots: what turns people into looters?'

46 For instance, social media has become instrumental in accounting for responsibility in riots and other forms of mass action. See D Finch, S McIntyre & K Sundberg, 'The Vancouver Riot hangover: crowdsourced justice and public shaming', *Centennial Reader*, June 2012.

47 For instance, Andrew Kurjata notes in a blog post: 'social media brought this to me in real time ... Radio and TV were on it soon after, but I didn't feel like I could

Revisiting earlier discussions, the smooth space of the sea of familiar production is a most spectacular crowd symbol: 'Corn and forest, rain, wind, sand, fire and the sea are such units … Although they do not consist of men, each of them recalls the crowd and stands as symbol for it in myth, dream, speech and song.'[48] The smooth space of the crowd expresses the emergent creativity and becoming of innovation and familiar production. The crowd promises both the enormous potential of innovation, the swell of the sea of familiar production, and at the same time the anti-social risk associated with anonymity and supposed immunity, the violence at the margins: 'There are also the individual drops of water. It is true that they only become drops in isolation, when they are separated from each other. Their smallness and singleness then makes them seem powerless; they are almost nothing and arouse a feeling of pity in the spectator … They only begin to count again when they can no longer be counted, when they have again become part of a whole.'[49] Indeed, they 'count' when they are part of the uncountable, immeasurable, qualitative smooth space of familiar production.

It is necessary therefore to put the 'authors' and users into genuine play: 'what binds the infamous lives to the fleshless writings that record them is not a relationship of representation or refiguration, but something different and more essential: they are "played out" or "put into play" in these sentences; their freedom and their disgrace are risked and decided.'[50] This play and decision is fundamental to the ethics of justice, 'room for judging', as it were. So the user must be perspicuous and promiscuous; the user must play the field in order to discharge the debt, that is, to enter the realm of chance and decision: 'Chance is forever at the mercy of itself. It's always at the mercy of play, always *in* play.'[51] Reiterated here is a dialogue of meaning through use as fundamental to the potential for an 'ethics' of intellectual property. Recalling the discursive and dialogical value of justice, speech is both combative and collaborative, a duty and a benefit, and is indeed just the process by which the proximity and materiality of justice is achieved: 'Negation is the heart of testimony. We do not show the sense, we show something.'[52] It is hear and say.

---

get a full picture without hearing from those in the thick of it.' A Kurjata, 'Social media, crowd-sourced justice, and the Vancouver Riots', 18 June 2011, www.andrewkurjata.ca/blog/2011/06/18/crowd-sourced-justice/.

48   Canetti, *Crowds and Power*, 75. Canetti explains: 'Crowd symbols is the name I give to collective units which do not consist of men but which are felt to be crowds. Corn and forest, rain, wind, sand, fire and the sea are such units. Every one of these phenomena comprehends some of the essential attributes of the crowd.'

49   Canetti, *Crowds and Power*, 80.

50   Agamben, *Profanations*, 67.

51   Bataille, *Guilty*, 77.

52   J-F Lyotard, *The Differend: Phrases in Dispute*, G Van Den Abbeele (trans), U of Minnesota P, Minneapolis, [1983]/1988: 54.

A Time for Justice

> "Consider your verdict," the King said to the jury.
> "Not yet, not yet!" the Rabbit hastily interrupted. "There's a great deal to come before that!"[53]

Earlier discussions have noted the relationship between the time of innovation and risk, and the consequences for linking time to markets.[54] The value of the product is thus not only in terms of its market, but also in terms of time and duration itself (particularly explicit in the context of term of protection). Therefore, conventional mechanisms of value in the product itself are articulated around time of the market as well as time to the market. This is in contrast to the smooth space-time of familiar production which is not articulated in respect of the product, but in the resilience of continuous innovation. In this way, innovation is an indefinite and inconstant event.

In some respects, in conventional constructions of value the value of the product is determined only provisionally, with its final pronouncement if and when it is infringed. In this respect, justice is deferred, but at the same time, value now becomes a construction of relation. It is itself realized through the 'promise' or debt of the consumer, ultimately calibrated within circumstances of infringement and values of claims. Further, this relation of value severs any link between the size of the enterprise (and subsequently also the 'site' of innovation) and the intellectual property. For example, in relation to patent litigation and patent infringement, where a 'business' may be the reflection of a single patent, the patent's value is in its doubling (that is, a perverse doubling of the patent insofar as the business, based on a single patent, is valued through a single patent). Value is a future event. Value is a memory of the future, limited by risk rather than realized through chance. In other words, paradoxically the execution of the law comes to be indistinguishable from its transgression.[55] Rights are determined with respect to their enforceability, and the value of a patent with respect to the value of its infringement. In this way, the transgression of the law is its very mechanism of sustainability.

Reform has therefore focused to an extent on value through rights (on value through the object) as distinct from connections (and the principles of familiar production).[56] In a sense, this is the logic of abbreviating reform from systemic change to that of a single proposition, the 'lobby' (and this is not a 'value' judgment but is used in this sense to include interest specific propositions from all different perspectives, that is, to indicate the 'method' of reform as distinct from

---

53 *Wonderland*, 92.

54 See the earlier discussion in 'Use' and Chapter 3. See further the discussion of time in innovation in Chapter 7.

55 Agamben, *Homo Sacer*, 57.

56 Policy-making is considered in more detail later in this chapter and in Chapter 12.

the content). The purpose of reform keeps treading on the tail of innovation. It is reduced to going 'hat in hand', after the verdict:

> "Take off your hat," the King said to the Hatter.
> "It isn't mine," said the Hatter.
> "*Stolen!*" the King exclaimed turning to the jury, who instantly made a memorandum of the fact.
> "I keep them to sell," the Hatter added as an explanation: "I've none of my own. I'm a hatter."[57]

Court of Justice

> "Give your evidence," said the King; "and don't be nervous, or I'll have you executed on the spot."[58]

Fundamentally these final scenes of Alice's journey in Wonderland entertain the concept of justice and its articulation with the law. Indeed, in this court room, the law itself appears to be absent. When the creatures start delivering their testimonies, the event plays out in a way that is intriguingly iterative with seemingly no endpoint. Unlike the presumed certainty and finality of a moral judgment through the law, this Court of Justice agitates with speculative and provisional responses to seemingly illogical and irrelevant testimony.

This provides a startling account of the relationship between the pragmatism and dogmatic finality of the law, and the ethical drive towards justice as an ideal *beyond the law*: 'Law is the element of calculation, and it is just that there be law, but justice is incalculable, it requires us to calculate with the incalculable; and aporetic experiences are the experiences, as improbable as they are necessary, of justice, that is to say of moments in which the decision between just and unjust is never insured by a rule.'[59]

Justice is therefore also a dissonant refrain, a stutter: 'Justice, as law, is never exercised without a decision that *cuts*, that divides.'[60] Importantly, recalling and reviewing earlier discussions, justice is learning and investigating, rather than rehearsing an imperative: 'It begins, it ought to begin, by right or in principle, with the initiative of learning, reading, understanding, interpreting the rule, and even in calculating. For if calculation is calculation, the decision to calculate is not of the order of the calculable, and must not be.'[61] Justice is crucially and fundamentally a process of decision, indeed, a logic of chance as distinct from a certainty of

---

57   *Wonderland*, 93.
58   *Wonderland*, 93.
59   Derrida, 'Force of Law', 16.
60   Derrida, 'Force of Law', 24.
61   Derrida, 'Force of Law', 24.

risk. Even in the calculation of probabilities, a decision must be approximated: 'A judge might even say "That is the truth – so far as a human being can know it". But what would this rider achieve? ("beyond all reasonable doubt").'[62] Returning to the iterative and linguistic function of justice (language is justice), the process of decision is fundamentally a dialogue, a combat, an ordeal: 'A decision that didn't go through the ordeal of the undecidable would not be a free decision, it would only be the programmable application or unfolding of a calculable process. It might be legal; it would not be just.'[63]

For justice then, Alice will always be waiting.[64]

Negative Reinforcement

> "Give your evidence," the King repeated angrily, "or I'll have you executed, whether you're nervous or not."[65]

At once the process of evidence is not only collective in the Wonderland Court of Justice, but also a phenomenal encounter between all the creatures of Wonderland. In this way, evidence is both a duty (the debt to the sovereign, the King and Queen of Hearts) and an experience, and all are always already guilty before the law, before the sovereign.

Evidence-based policy-making presents as a reasonable and desirable approach to legislative and policy development and it would seem wholly counter-intuitive to suggest otherwise. Indeed, as a matter of sense, the incalculability of justice must nevertheless engage with the calculable systems against which it operates. This is the important distinction between the incalculability of justice and the calculation of law (and judgment): 'Law (*droit*) is not justice. Law is the element

---

62   Wittgenstein, *On Certainty*, §607.

63   Derrida, 'Force of Law', 24. Derrida explains further: 'And once the ordeal of the undecidable is past (if that is possible), the decision has again followed a rule or given itself a rule, invented it or reinvented, reaffirmed it, it is no longer *presently* just, fully just. There is apparently no moment in which a decision can be called presently and fully just: either it has not yet been made according to a rule – whether received, confirmed, conserved or reinvented – which in its turn is not absolutely guaranteed by anything; and, moreover, if it were guaranteed, the decision would be reduced to calculation and we wouldn't call it just.' Derrida, 'Force of Law', 24.

64   Derrida explains: 'This "idea of justice," infinite because it is irreducible, irreducible because owed to the other, owed to the other, before any contract, because it has come, the other's coming as the singularity that is always other. This "idea of justice" seems to be irreducible in its affirmative character, in its demand of gift without exchange, without circulation, without recognition or gratitude, without economic circularity, without calculation and without rules, without reason and without rationality.' Derrida, 'Force of Law', 27.

65   *Wonderland*, 94.

of calculation, and it is just that there be law, but justice is incalculable, it requires us to calculate with the incalculable.'[66] In other words, despite the inestimable and incalculable nature of ethics and justice, it is inevitable and fundamental that the law itself intersects with a variety of traversing fields and disciplines for which an account is possible: 'all the fields from which we cannot separate it, which intervene in it and are no longer simply fields: ethics, politics, economics, psycho-sociology, philosophy, literature, etc.'.[67] Therefore, the incalculability of justice does not defeat the role of evidence, but at the same time evidence-based policy cannot proceed as a prescriptive or imperative from the calculable alone. Policy-making must 'negotiate the relation between the calculable and the incalculable', and 'we *must* take it as far as possible, beyond the place we find ourselves and beyond the already identifiable zones of morality or politics or law, beyond the distinction between national and international, public and private, and so on'.[68] Contingency of law is therefore necessarily part of the circumstantial evidence relevant to any shadow of reform: 'But do you know of such a shadow? And by a shadow do I not mean some picture of the movement – for such a picture would not have to be a picture of just *this* movement. But the possibility of this movement must be the possibility of just this movement.'[69]

> "I'm a poor man," the Hatter went on, "and most things twinkled after that – only the March Hare said –"
> "I didn't!" the March Hare interrupted in a great hurry.
> "You did!" said the Hatter.
> "I deny it!" said the March Hare.
> "He denies it," said the King: "leave out that part."[70]

Paradoxically, however, in engaging with evidence the very structure of the question (and the very evidence thus admissible and admitted) is itself engaged in the construction of a narrative of doubt. The crucial issue in the analysis of policy processes is to answer what it might be that the process is trying to exclude through evidence: 'There is no evidence, only a reprieve granted to scepticism. Not *It is certain that* ... but, *It is not excluded that* ... By naming and by showing, one eliminates. Proof is negative, in the sense of being refutable. It is adduced in debate, which is agonistic or dialogical if there is a consensus over the procedures for its being adduced.'[71] Evidence is therefore necessarily a process of negation or opposition. If pursuing a narrative of rights (and of a specious rhetoric of

---

66   Derrida, 'Force of Law', 16.
67   Derrida, 'Force of Law', 28.
68   Derrida, 'Force of Law', 28.
69   Wittgenstein, *Philosophical Investigations*, §194.
70   *Wonderland*, 94.
71   Lyotard, *Differend*, 54.

'balance' based on rights), then evidence-based policy will always be refutable by the opposed.

Evidence is therefore structured in relation to an already existing claim or pronouncement, proof with respect to a fact in dispute. In the context of policy-making the use of evidence is retrospection on a presumed fact, an authoritative proposition of value. In other words, the evidence relied upon, for the purposes of evidence-based policy-making, is always in relation to a purpose, a proposition. The question must be presumed. While any preparatory process of consultation might be said to construct an informed question, more accurately the process of consultation is still in response to a question (a presumed problem, for example) and towards identifying more refined propositions of policy for which evidence will be provided. So while a consultation may raise more questions (and the need for more evidence), it is still garnering evidence in respect of a prior proposition. This is precisely the structure of evidence through presumption and proof, whether that evidence is direct testimony (the materiality of a witness or the object itself) or indirect evidence where the proposition in issue might be inferred from the proof (or negation) of other propositions and facts. This kind of indirect evidence is perhaps the most likely and available kind in respect of policy-making. What is at stake is whether particular evidence is significant and telling, or whether it is a hazardous, presumptive wager of circumstantial probability. What can it possibly mean to bear witness to future policy, to future allocation and attribution of values? This resonates with the paradox of value in intellectual property, and the imperative of the law (the proper names of intellectual property) and its manifestation at the time of enforcement: 'Not causing us to know anything, the law teaches us what it is only by marking our flesh, by already applying punishment to us, and thus the fantastic paradox: we do not know what the law intended before receiving punishment, hence we can obey the law only by being guilty.'[72]

> "But what did the Dormouse say?" one of the jury asked.
> "That I can't remember," said the Hatter.
> "You *must* remember," remarked the King, "or I'll have you executed."
> The miserable Hatter dropped his teacup and bread-and-butter, and went down on one knee. "I'm a poor man, your Majesty," he began.
> "You're a *very* poor *speaker*," said the King.[73]

In many respects, therefore, the evidence underpinning policy deliberations will always be an evidence of opinion, that is, indirect presumptive or circumstantial evidence. This is precisely the challenge for contemporary policy reform in intellectual property, 'positioning several ways of turning reality out' in a 'plurality

---

72   G Deleuze, *Proust and Signs: The Complete Text*, R Howard (trans), Athlone P, London, [1964]/2000: 132.

73   *Wonderland*, 95.

of games without any of them being able to claim that it can say all the others'.[74] This is the challenge for evidence-based policy in that it is always policy based on evidence based on a particular game at the time. This is not a recommendation for 'wait and see', it is a consideration that any policy reform is necessarily reform in language use: 'We want to establish an order in our knowledge of the use of language: an order with a particular end in view; one out of the many possible orders; not *the* order.'[75] Policy development is perhaps the opportunity for making genuine use of language; however, where policy development simply becomes a refrain for the dominant discourse (disciplinary or otherwise), it is language when it is 'like an engine idling, not when it is doing work'.[76]

Recall the earlier discussions of the nature of economic modelling of the law, including its necessarily circumstantial nature, together with the circumstances created through the very process of such modelling.[77] That is, the evidence of such modelling is seemingly irrefutable, in that it is intangible and, necessarily in many respects, disengaged from the process of justice as well as the experience of the particular industry or creative encounter that will come to be classified within those propositions. It becomes therefore the perfect platform for a refrain of opinion as distinct from work in the use of language. Policy education strategy, deployed to accompany the message of intellectual property reform, risks engaging in similar rhetorical flourishes and strategies as various other 'advices' on intellectual property, regardless of its relationship, antagonism, belief, or otherwise towards intellectual property. It is idle language. It is not about the subject matter, but about the evidentiary process of proposition and refutation. Policy is blind.

In some respects, poor practices in evidence-based policy-making risk robbing the indebted Peter to pay Paul, as it were.[78] Reforming intellectual property based on any one particularly dominant discourse, such as economics, risks over-governing the social and overcoming use, to the detriment of the sustainability and legitimacy of reform: 'The infinite creditor and the infinite credit have replaced the blocks of mobile and finite debts. There is always a monotheism on the horizon of despotism: the debt becomes a *debt of existence*, a debt of the existence of the subjects themselves.'[79]

> "This debt will simply swallow all,
> And make my life a life of woe!"
> "Nay, nay, my Peter!" answered Paul.
> "You must not rail on Fortune so!"
> ...

---

74   Lyotard & Thébaud, *Just Gaming*, 58.
75   Wittgenstein, *Philosophical Investigations*, §132.
76   Wittgenstein, *Philosophical Investigations*, §132.
77   Chapter 1.
78   *Sylvie and Bruno*, 135–38.
79   Deleuze & Guattari, *Anti-Oedipus*, 197.

Weeks grew to months, and months to years:
Peter was worn to skin and bone:
And once he even said, with tears,
"Remember, Paul, that promised Loan!"
Said Paul, "I'll lend you, when I can,
All the spare money I have got –
Ah, Peter, you're a happy man!
Yours is an enviable lot!"

…

A man may surely claim his dues:
But, when there's money to be lent,
A man must be allowed to choose
Such times as are convenient.[80]

Cross Examination

"Give your evidence," said the King.

"Shan't," said the cook

The King looked anxiously at the White Rabbit, who said in a low voice, "Your Majesty must cross-examine *this* witness."

"Well, If I must, I must," the King said with a melancholy air, and, after folding his arms and frowning at the cook till his eyes were nearly out of sight, he said in a deep voice, "What are tarts made of?"[81]

Cross-examination, as noted earlier, is both one of the controversial aspects[82] as well as one of the more important and significant 'literary' qualities of the common law and an adversarial and iterative system of justice. The Wonderland Court of Justice demonstrates all the characteristics of a common law system of justice – cross-examination (by the Queen), disclosure (of the letter)[83] and oral argument (of all the inhabitants, including Alice, making them witnesses in spite of themselves[84]).

Cross-examination and cross purposes return the discussion full circle to the purpose of the riddle, the question: *I shall know a good solution if I see it.*[85] And indeed this is the riddle of representation, of the attempt to account for the entirety of the picture: 'if language is to formulate it, this can take place only in successive phases worked out in the dimension of time. We can never hope

---

80   *Sylvie and Bruno*, 136–37.

81   *Wonderland*, 96.

82   OG Wellbon, 'Demeanor', 76 *Cornell Law Review* 1990–1991: 1075.

83   See Chapter 12.

84   Bloch, *Historian's Craft*, 51. See the further discussion in Chapter 12.

85   Wittgenstein, *Lectures on the Foundations of Mathematics*, 84.

to attain a global view in one single supreme instant; language chops it into its component parts and connects them up into a coherent explanation.'[86] One needs the warranty of the record, the archive, the return ticket,[87] as it were, in order to guarantee the promise in relation to the debt of representation: 'The analytic presentation makes it impossible for the successive stages to coalesce.'[88] Thus, the totality of representation is an apparition, an illusion, a trick of the eyes: 'Our attention remains fixed on this whole but we can never see it in the full light of day. A succession of propositions flickering off and on merely hides it from our gaze, and we are powerless to alter this.'[89] And this indeed is the riddle, 'the riddle of existence' with no question and no answer: 'It is not necessary to answer the riddle of existence; it is not even necessary to ask it.'[90]

As introduced in earlier discussions, the memoranda of intellectual property (the patent specification, the trade mark register, the design drawing and so on) become in and of themselves necessarily objects for invalidation, disconfirmation, improvement and indeed obsolescence within the system. In other words, the 'archives' of intellectual property are themselves a kind of tribunal within the logic of the intellectual property system. However, in conjunction with the archives or registers of the intellectual property system (patents, trade marks, designs), there are the industries of drafting and narration of the invention, the trade mark, the design, all depicted in such a way that is not a guarantee of any supposed originary truth as an object, but rather to confirm that object as widely as possible in a sea of simulacra. Rather than a process of disconfirmation, the register is an archive of probability, a gathering of additional circumstances to increase the probability of a favourable decision. The register is thus both an utterly strategic as well as an ethical mechanism within the intellectual property process.

This returns the discussion to the language-game of intellectual property, the game of justice: 'Here the term "language-*game*" is meant to bring into prominence the fact that the *speaking* of language is part of an activity, or of a form of life.'[91] When it comes to documents and records of intellectual property, their roles in the game can change from orders (to users), to reports (to the court), to imperatives (to policy-makers). They may appear as stories and solutions, or riddles and jokes.[92] In every respect, such documents and records are utterances with different roles and different probabilities in different circumstances. Intellectual property rules may rely on strict concepts or delimitations in order (grammar), however it is in their use (by all participants in the system, particularly in the way in which that *relationship* or *dialogue* of use is articulated through limitations and exceptions)

---

86   Bataille, *Erotism*, 274.

87   See further the later discussion in '*Re* Use'.

88   Bataille, *Erotism*, 274.

89   Bataille, *Erotism*, 274.

90   Bataille, *Erotism*, 274.

91   Wittgenstein, *Philosophical Investigations*, §23.

92   Wittgenstein, *Philosophical Investigations*, §23.

that the meaning of the system emerges: 'And is there not also the case where we play and – make up the rules as we go along? And there is even one where we alter them – as we go along.'[93]

> "Never mind!" said the King, with an air of great relief. "Call the next witness." And, he added in an undertone to the Queen, "Really, my dear, *you* must cross-examine the next witness. It quite makes my forehead ache."
>
> Alice watched the White Rabbit as he fumbled over the list, feeling very curious to see what the next witness would be like, "– for they haven't got much evidence *yet*," she said to herself. Imagine her surprise, when the White Rabbit read out, at the top of his shrill little voice, the name "Alice!"[94]

Alice!

---

93 Wittgenstein, *Philosophical Investigations*, §83.

94 *Wonderland*, 96.

Chapter 12

Witness

Here!

> "Here!" cried Alice, quite forgetting in the flurry of the moment how large she
> had grown in the last few minutes, and she jumped up in such a hurry that she
> tipped over the jury-box with the edge of her skirt, upsetting all the jurymen on
> to the heads of the crowd below, and there they lay sprawling about, reminding
> her very much of a globe of gold fish she had accidentally upset the week before.[1]

Alice is immediately and successfully interpellated by the court. That is, she
recognizes herself in the language-game of the court room, she occupies space –
the space of judgment, of indebtedness, of culpability. Alice has been summoned
and her identity questioned throughout her journey, and at last she is rendered
accountable: *Here!* If Alice felt as though she were a mere visitor when she first
fell to Wonderland, she nevertheless now cannot but have an effect on the system
itself, and so is called as a witness. The user cannot help but change the product
through use. She cannot maintain a distance outside that which she observes,
cannot stand perfectly outside the world to describe it, when she herself must
be part of that world. She cannot stand apart as witness to the crowd; instead
she invariably changes the game through that very same process. She cannot
describe the world without also taking account of her process of description, her
perspective, her situation and so on in an infinite regress of representation. The
user cannot make use without also making.

At the same time as attempting to represent her knowledge of the judicial
process, it is wholly inevitable that Alice herself must participate in the oral
argument of the court. Her entire journey (and indeed the shadowed journey of
innovation) has proceeded by way of interlocutory, of riddle, of philosophical
investigation. Alice's journey has proceeded (as indeed has this journey) by way
of dialogue and debate. In such a dialogue, a certain responsibility (a debt) is
thus inscribed on the inhabitants, which Alice is duty-bound to reciprocate. In this
way, Alice has entered into an ethical relationship with the other inhabitants of
Wonderland. She is duty-bound by the language-game.

Alice is experienced with the ordinary but is sensible to the extraordinary.
Continuously marching ahead in this tale is the impossibility of representational
knowledge, the illogicality of representation. Alice cannot finish her journey as
it is literally an impossible object. It is this unrepresentable other, the endless

---

1   *Wonderland*, 97.

ethical journey, the remainder of any attempt to account for the world (legal, social, political, economic) that fills Wonderland. This is the space of the truly extraordinary that guarantees the continuing reinvention of the ordinary. Innovation is in itself an ethical process of becoming, guaranteed by the fact that it can never finally finish. Innovation is in and of itself just. That is, it is not finally presentable and can never actually appear. It is named (it is in the name of justice that law is applied) but it cannot be experienced as such as it is always unfinished business: 'Justice is an experience of the impossible.'[2]

Alice's Evidence could be taken to refer both to her observational account and to her own personal clarity, perspicuity and becoming. She accounts her own personal 'evidence' or illumination in this journey with no end: 'A perspicuous representation produces just that understanding which consists in "seeing connexions" ... The concept of a perspicuous representation is of fundamental significance for us. It earmarks the form of account we give, the way we look at things.'[3] But at the same time, 'The difficulty is to realize the groundlessness of our believing.'[4]

These 'perspicuous connexions' provide the account of ethics in a language-game, such as that of intellectual property, and the conditions of experience itself in familiar production. These are the interactions among participants in ways that are mutually constitutive and beneficial. While morality prescribes a pre-established standard (the manners and etiquette with which Alice struggles throughout her journey), an ethics of intellectual property must appear through maximizing such connections. Ethics is therefore intrinsic to becoming; it is life-affirming, an endless journey in which *I don't know my way about*. That is, ethics does not prescribe a destination, an intention, a recipient of the message. Morality, on the other hand, dictates a way. The ethics of intellectual property in a digital environment will be anchored in meaning through use.

The Ethics of Evidence

"What do you know about this business?" the King said to Alice.

"Nothing," said Alice.

"Nothing *whatever*?" persisted the King.

"Nothing whatever," said Alice.

"That's very important," the King said, turning to the jury. They were just beginning to write this down on their slates, when the White Rabbit interrupted: "*Un*important, your Majesty means, of course," he said in a very respectful tone, but frowning and making faces at him as he spoke.

---

2 Derrida, 'Force of Law', 16.
3 Wittgenstein, *Philosophical Investigations*, §122.
4 Wittgenstein, *On Certainty*, §166.

"*Un*important, of course, I meant," the King hastily said, and went on to himself in an undertone, "important – unimportant – unimportant – important –" as if he were trying which word sounded best.

Some of the jury wrote it down "important," and some "unimportant." Alice could see this, as she was near enough to look at their slates; "but it doesn't matter a bit," she thought to herself.[5]

Returning to the question of evidence-based policy raised in the previous chapter, there is not only the issue of the quality of circumstantial evidence, but also the risk in legislating upon that evidence. Further, there is the troubling risk of a tendency towards policy based upon single propositions, single issues, and thus gathering evidence that is limited to the refutation of any objection to that proposition.[6] This is the propositional momentum of the lobby. Lobbying is by definition in respect of a special interest as distinct from systemic and systematic fields of change. In the complex commercial, cultural and economic fields of intellectual property, there is a genuine danger that the process of evidence is simply inadequate – important, unimportant, it does not matter a bit, the hats are not our own. When it comes to extrapolating across the enormous diversity of activity that comes within intellectual property, or within patents, copyright, trade marks, designs and so on, the problem of single propositions being applied to specific business models becomes compounded and chaotic. Furthermore, evidence is open to interpretation – important, unimportant. Does it matter? Is it ultimately just a question of how it is intended to be used; that is, making the evidence fit the proposition? That is, is it a case of simply rationalizing and submitting different points of view to convergence and agreement?[7]

This process of evidence may also be extrapolated to the character of intellectual property harmonization in the context of international trade. Here, the proposition is convergence on the principles of trade and commerce to the benefit of all. However, in many respects this 'benefit' is inconsistent in terms of vastly different states of development, resources, industries and wealth.[8]

---

5   *Wonderland*, 98.

6   There has been a long tradition in the United Kingdom, in particular in the area of intellectual property, whereby a committee would be appointed to investigate what changes were needed to an area of law. This would involve an expert committee hearing evidence and making specific discrete recommendations. Thus, the Copyright, Designs and Patents Act 1988 was preceded a decade earlier by the report of the Whitford Committee on Copyright and Designs (1977) (Cmnd. 6732) and the Patents Act 1977 by a report of the Banks Committee on the Patent System and Patent Law (1970) (Cmnd. 4407).

7   Lyotard notes that 'For Kant, the idea of justice is associated with that of finality. But "finality" means a kind of convergence, of organization, of a general congruence, on the part of a given multiplicity moving toward its unity.' Lyotard & Thébaud, *Just Gaming*, 94.

8   Indeed, the implementation period for TRIPS for the least developed countries has just been extended again until July 2021: see Decision of Council of 11th June 2013 on

[O]ne may still find inspiration in the Marxist 'spirit' to criticize the presumed autonomy of the juridical and to denounce endlessly the *de facto* take-over of international authorities by powerful Nation-States, by concentrations of techno-scientific capital, symbolic capital, and financial capital, of State capital and private capital. A 'new international' is being sought through these crises of international law; it already denounces the limits of a discourse on human rights that will remain inadequate, sometimes hypocritical, and in any case formalistic and inconsistent with itself as long as the law of the market, the 'foreign debt,' the inequality of techno-scientific, military, and economic development maintain an effective inequality as monstrous as that which prevails today, to a greater extent than ever in the history of humanity.[9]

This repentence, this forgiveness (and so forgetting) is instrumental in the homogeneity of the international 'state' subject to international law. This is the end of language, the 'end of history',[10] the end of the archive, as it were; but more importantly, the end of law with significance (culturally, politically, economically).[11] Without consent and consensus on the law, the law is without legitimacy for users. The paradox is that the end of borders, the end of history in a transnational world will bring with it the risk of the end of change.

Nevertheless, principles of international treaty negotiation comprise the potential for genuine justice in a 'Great Work' of the otherwise cacophonous landscape of international law: 'It is rather a question of different and divergent stories, as if an absolutely distinct landscape corresponded to each point of view. There is indeed a unity of divergent series insofar as they are divergent, but it is always a chaos perpetually thrown off center which becomes one only in the Great

---

the Extension of the Transition period for Least Developed Countries under the TRIPS Agreement (Document IP/C/64).

9   J Derrida, *Spectres of Marx: The State of the Debt, The Work of Mourning, and the New International*, P Kamuf (trans), Routledge, New York, [1993]/1994: 85.

10   Alexandre Kojève states: 'If, then, Man's complete satisfaction is the goal and the natural end of history, it can be said that history completes itself by Man's perfect understanding of his death … Therefore, if this Science, which is Wisdom, could appear only at the end of History, only through it is History perfected and definitively completed. For it is only by understanding himself in this Science as mortal – that is, as a historical free individual – that Man attains fullness of consciousness of a self that no longer has any reason to negate itself and become other.' A Kojève, *Introduction to the Reading of Hegel: Lectures on the Phenomenology of Spirit*, JH Nichols Jr (trans), Cornell UP, Ithaca, [1947]/1980: 258.

11   Agamben notes 'Alexandre Kojève's idea of the end of history and the subsequent institution of a new homogeneous state presents many analogies with the epochal situation we have described as law's being in force without significance … What, after all, is a State that survives history, a State sovereignty that maintains itself beyond the accomplishment of its telos, if not a law that is in force without signifying?' Agamben, *Homo Sacer*, 60.

Work. This unformed chaos, the great letter of *Finnegans Wake*, is not just any chaos: it is the power of affirmation.'[12]

Within the World Trade Organization (WTO), disputes regarding a vast range of subject matter, including intellectual property, are adjudicated by the Dispute Settlement Body (DSB)[13] according to an appeal to 'trade principles' as the ultimate value, the breach of which is 'unforgiveable' in a globalized world of international trade. In so doing, the DSB undertakes a 'holistic' approach to treaty interpretation[14] according to the jurisprudence of the International Court of Justice.[15] To an extent, principles of international treaty interpretation are thus entirely in rhythm with the understanding of the work of language, of work through 'use'.[16]

Similarly, in the European context, in clarifying the interpretation of an EU instrument or treaty, reference to the travaux préparatoires (the official record of negotiation)[17] and acquis communautaire[18] may be made. In a very real sense, the justice of the international system is the open-ended nature of the evidence,

---

12   Deleuze, *Logic of Sense*, 260.

13   Established by the Dispute Settlement Understanding, Annex 2 of the WTO Agreement.

14   See the extensive discussion in S Frankel & D Gervais, 'Plain Packaging and the Interpretation of the TRIPs Agreement', *Vanderbilt Journal of Transnational Law*, forthcoming.

15   Anthony Aust explains: 'The International Law Commission rejected the view that in interpreting a treaty one must give greater weight to one particular factor, such as the text ("textual" or "literal" approach), or the supposed intentions of the parties, or the object and purpose of the treaty ("effective" or "teleological" approach). Reliance on one to the detriment of the others was contrary to the jurisprudence of the International Court of Justice.' A Aust, *Modern Treaty Law and Practice*, Cambridge UP, Cambridge, 2000: 185.

16   Indeed, Alan Boyle and Christine Chinkin note the greater legitimacy of the WTO: 'The present WTO represents a major evolution in the law-making machinery for international trade. Compared to the arrangements before 1994, its membership is more inclusive, although still not universal, its negotiations are more open and transparent, even if there is room for further improvement, and the dominance of the US and EU has been replaced by the need for a broader consensus involving developing states. Moreover the agreements it adopts now have a genuinely legal character, and trade relations among the parties are largely governed by rules ... Dispute settlement is also no longer determined by the secretariat but proceeds within a recognisably judicial and rules-based structure. As an international regulatory body WTO can now be compared to UN specialised agencies such as IMO or ICA.' A Boyle & C Chinkin, *The Making of International Law*, Oxford UP, Oxford, 2007: 140.

17   Under the Vienna Convention of the Law of Treaties, reference will frequently be made to the travaux préparatoires in order to clarify the intentions of a European instrument (pursuant to Art 32, Supplementary means of interpretation), albeit as a secondary concern (see Art 31, General rule of interpretation).

18   The acquis communautaire (also called the Community acquis, EU acquis or simply acquis) is the accumulated body of legislation, decisions and opinions comprising European Union law. It is thus a sense of agreement of consent to meaning 'of the community'.

the sense of justice *to come*, striving towards an idea of justice through the work of language and through the affirmation of the community and communities of negotiation (including that inside and outside the negotiated text): 'A word or term may have more than one meaning or shade of meaning but the identification of such meaning in isolation only commences the process of interpretation, it does not conclude it.'[19] Indeed, the 'proper names' of interpretation, the 'proper names' of intellectual property (patent, trade mark, design, copyright) are merely starting points, commencing the process of interpretation, of meaning.

Despite this precedent of 'communities' of negotiation, there is perhaps the informal view that the process of European harmonization and law-making (as well as increasing levels of international harmonization) means the complexity of issues is such that examination of the terrain must proceed by isolated propositions, in ways similar to the attenuation of policy debate at the domestic level.[20] However, there is no reason international legislative process cannot continue to be similarly systemic in its approach to innovation and change, affirming various different, divergent and distinct stories through communication, 'in common'. An example of this process is the establishment within the World Intellectual Property Organization (WIPO) of the Intergovernmental Committee on Intellectual Property and Genetic Resources, Traditional Knowledge and Folklore (IGC),[21] which proceeded not by way of propositions but by way of dialogue. Indeed, to an extent this did prolong justice, with the problems themselves not being fully articulated and affirmed for approximately seven years. However, considering the enormous complexity of the questions and the hugely divergent landscape, the current progress towards a text-based solution in the form of a treaty is a considerable achievement.

The Letter of the Law

> The King turned pale, and shut his notebook hastily. "Consider your verdict," he said to the jury, in a low trembling voice.

19　Appellate Body Report, *United States: Continued Existence and Application of Zeroing Methodology* ¶269, WT/DS350/AB/R (Feb 4, 2009). See further, for an insightful discussion of this approach in the context of plain packaging, Frankel & Gervais, 'Plain Packaging and the Interpretation of the TRIPs Agreement'.

20　See the discussion in Chapter 11.

21　The WIPO IGC was established in the 26th (12th Extraordinary Session) of the WIPO General Assembly, Geneva, 25 September to 3 October 2000 to consider and advise on appropriate actions concerning the economic and cultural significance of tradition-based creations, and the issues of conservation, management, sustainable use and sharing of the benefits from the use of genetic resources and traditional knowledge, as well as the enforcement of rights to traditional knowledge and folklore. See also the earlier discussion in Chapter 2.

"There's more evidence to come yet, please your Majesty," said the White Rabbit, jumping up in a great hurry: "this paper has just been picked up."

"What's in it?" said the Queen.

"I haven't opened it yet," said the White Rabbit, "but it seems to be a letter, written by the prisoner to – to somebody."

"It must have been that," said the King, "unless it was written to nobody, which isn't usual, you know."

"Who is it directed to?" said one of the jurymen.

"It isn't directed at all," said the White Rabbit; "in fact, there's nothing written on the *outside*." He unfolded the paper as he spoke, and added, "It isn't a letter, after all: it's a set of verses."

"Are they in the prisoner's handwriting?" asked another of the jurymen.

"No, they're not," said the White Rabbit, "and that's the queerest thing about it." (The jury all looked puzzled.)

"He must have imitated somebody else's hand," said the King. (The jury all brightened up again.)

"Please your Majesty," said the Knave, "I didn't write it, and they can't prove I did: there's no name signed at the end."

"If you didn't sign it," said the King, "that only makes the matter worse. You *must* have meant some mischief, or else you'd have signed your name like an honest man."

There was a general clapping of hands at this: it was the first really clever thing the King had said that day.

"That *proves* his guilt," said the Queen.

"It proves nothing of the sort!" said Alice. "Why, you don't even know what they're about!"[22]

The documentary evidence of the letter is potentially hearsay (hear and say). Not only is its intended destination and meaning unclear, but also its author. Without an addressee it is a dead letter, as it were, literally without meaning, nonsense. And yet it is recuperated within the negotiation of the court room. The letter is opened and read in the Wonderland Court of Justice (as distinct from remaining a mere marker of an undeciphered message, merely the derivative, unoriginal and reported invitation to the Duchess seen earlier[23]). In this respect, the content of the letter is finally privileged over the object, the affect over the physical, the service over the goods; take care of the sense and the sounds will take care of themselves. Nevertheless, this is not to constrain the potential proliferation of meaning through use, as the meaning of the opened letter similarly multiplies and confuses throughout the court. Like the words of a treaty considered in isolation, the letter is a starting-point for meaning, open-ended evidence on the road to meaning. Meaning comes about through hear and say.

22 *Wonderland*, 98–99.

23 See Chapter 6.

Additionally, in contrast to the letter of invitation to the croquet game, this letter has no sender and apparently no intended recipient. Indeed, this is its source of creativity and rhizomatic potential: 'The instability of criteria, even in fashions, comes from this experimental situation. It also makes for the fact that today the majority of people who write interesting things, write without knowing to whom they are speaking. That is part of the workings of this society, and it is very good. There is no need to cry about it.'[24] The letter is also unsigned: it is seemingly authorless, 'dishonest'. The consumer, the user, is similarly without signature, rendered subject to the author, indebted to the author's expression, always already guilty before the law. It is up to the readers in the court room to adjudicate on the meaning, the culpability of the verse, the debt to society. They are indeed involuntary witnesses, 'witnesses in spite of themselves',[25] tasked with inventing a story out of the incidents, not the incident out of the story.[26]

As a matter of evidence and as a repository of value, the letter's treatment in the court is insightful. With neither sender nor receiver identified by the letter, there is a multiplication of intended recipients and arguably a dislocation (but at the same time, a proliferation) of meaning. Nevertheless, the letter is attributed some intentionality of meaning through a kind of moral certainty – through judgment (pronounced by the King) as distinct from justice (urged by Alice). That is, the evidence is treated in such a way by virtue of a certainty of foreseeing a particular kind of behaviour and meaning on behalf of the letter and presumed sender. The evidence is made to fit the proposition (if it is not in the prisoner's handwriting, then he must have imitated someone else's). In this way, the evidence is treated with the same certainty as any other 'moral', prefiguring roles and meaning. Broadening the concept of evidence (in the ordinary sense), the meaning of the letter is illuminated by the way evidence may be treated according to experience, that is, the rules and grammar of the particular language-game: 'The question of evidence for what is experienced has to be connected with the certainty or uncertainty of foreseeing someone else's behaviour.'[27]

The dispute, then, is the language-game itself. If the juridical setting proceeds by 'moral certainty', as in the verses presented before the Court of Justice in Wonderland, then the outcome is both unsurprising and ridiculous. This is the dilemma of arbitrarily infusing morality as distinct from embracing an ethics of chance and the unforeseeable in the intellectual property system: 'I think unforeseeability must be *an* essential property of the mental. Just like the endless

---

24   Lyotard & Thébaud, *Just Gaming*, 9.

25   Bloch, *Historian's Craft*, see also the discussion in the previous chapter.

26   Preface to *Sylvie and Bruno*, 103.

27   Wittgenstein, *Last Writings on the Philosophy of Psychology: The Inner and Outer, Volume 2*, 65e.

multiplicity of expression.'[28] It is the inevitable arbitrariness of an unreflective dogma, as distinct from the infinite 'in common' of ethics.

The turn of evidence returns the consideration to the balance of risk and chance. The 'effect' of innovation within this balance becomes clearer in the context of the judgment; that is, the judgment appeals to the oversight of 'moral certainty' or the foreseeable as a counter to any presumptive risk of evidence. Recalling Virilio's admonition: 'To invent something is to invent an accident ... And to invent the electronic superhighway or the Internet is to invent a major risk which is not easily spotted because it does not produce fatalities like a shipwreck or a mid-air explosion. The information accident is, sadly, not very visible. It is immaterial like the waves that carry information.'[29] However, the risk is 'mitigated' by the State in the form of the debt to a moral certainty of progress.

It is all too easy to dismiss Virilio's position as pessimism, without recognizing its significant import and insight with respect to the balance between risk and chance, and the import of evidence in what are necessarily predictive policy futures. That is, the perspective of accident is not simply a form of technophobia, but rather a perspective upon the necessary corollary of any communicative event – that is, any invention, creation, design and so on. Any communicative, innovative event must be put to use, must acquire meaning. In the dislocation and decommissioning of intent in the digital policy environment, the 'product' is revealed not through naming, but through use. This may lead to accidents, but in many ways this will be the only way for legislators to learn.

So, to invent something is to invent an accident, an unforeseen event elsewhere, encompassing both risk and, literally, further circumstances for change. As in earlier discussion, to innovate is not to invent, to create, to author a product as such, but rather, it is to create the circumstances for use to emerge with respect to a new innovation, a new technology, a new experience. It is use itself which ultimately is the innovation from which a new product emerges. The distraction in reading Virilio is to concentrate on the discourse of accident as part of a technological phobia. In stark contrast to this, what is emerging is the nature of innovation and not risk in relation to the product, but risk in relation to the social, in relation to policy, in relation to use:

> If large corporations such as Time Warner, Microsoft, Disney, etc., are in the process of becoming giants, it is because they must be competitive on the worldwide level. The multinationals did not aspire to worldwide status. But, today, a multinational corporation is necessarily faced with becoming worldwide. Hence, a considerable increase in publicity investment and an inevitable propaganda effect. The second aspect of this propaganda: the origin of technologies such as the Internet. They derive from deterrence ... This mixture

---

28   Wittgenstein, *Last Writings on the Philosophy of Psychology: The Inner and Outer, Volume 2*, 65e.

29   Virilio, 'Virilio: Cyberesistance Fighter'.

is not to be trusted: on one side an investment in publicity; on the other a silence concerning the control of information by the military powers.[30]

To invent a language is thus to invent a game.[31]

Hits and Missives

"Begin at the beginning," the King said gravely, "and go on till you come to the end; then stop."
    These were the verses the White Rabbit read:

"They told me you had been to her,
And mentioned me to him:
She gave me a good character,
But said I could not swim.

He sent them word I had not gone,
(We know it to be true):
If she should push the matter on,
What would become of you?

I gave her one, they gave him two,
You gave us three or more;
They all returned from him to you,
Though they were mine before.

If I or she should chance to be
Involved in this affair,
He trusts to you to set them free.
Exactly as we were.

My notion was that you had been
(Before she had this fit)
An obstacle that came between
Him, and ourselves, and it.

---

30   Virilio, 'Virilio: Cyberesistance Fighter'.

31   Wittgenstein, *Philosophical Investigations*, §492. To invent a language could mean to invent an instrument for a particular purpose on the basis of the laws of nature (or consistently with them); but it also has the other sense, analogous to that in which we speak of the invention of a game. Here I am stating something about the grammar of the word 'language', by connecting it with the grammar of the word 'invent'.

Don't let him know she liked them best,
For this must ever be
A secret, kept from all the rest,
Between yourself and me."[32]

The verses are an astonishing 'message in a bottle', neither sender nor addressee be. In many respects this is the 'digital' of contemporary social and political life, the intoxication of virtual indifference: 'Even at that time the hope of leaving behind messages in bottles on the flood of barbarism bursting on Europe was an amiable illusion: the desperate letters stuck in the mud of the spring of rejuvenescence and were worked up by a band of Noble Human-Beings and other riff-raff into highly artistic but inexpensive wall-adornments.'[33]

This recalls the 'message in a bottle' of the works of Banksy and the controversy surrounding the removal and sale of *Slave Labour* from the wall of a North London shop, and the attempt to sell it at auction in Miami, Florida.[34] The dislocation of the work deprives it of meaning and integrity, precisely because it deprives it of use. It becomes, literally, a dead letter, undelivered, meaningless, and without legitimacy.

At the same time, however, the very process of removal and the attempt to sell the 'community' work in a rich art auction generated a whole new discourse of use around the work. Arguably this is the very 'process' of deterioration and restoration that inheres in Banksy works as well as other forms of familiar production. Indeed, the verses read out by the White Rabbit contain this very same agitation of familiar production, the impossibility of tracing the source, the originary moment of deception and honest dissemination:

I gave her one, they gave him two,
You gave us three or more;
They all returned from him to you,
Though they were mine before.

After all that, chance will set them free.

If I or she should chance to be
Involved in this affair,
He trusts to you to set them free.
Exactly as we were.

---

32   *Wonderland*, 99–100.

33   Adorno, *Minima Moralia*, 209. As Deleuze notes, 'The only form of communication one can envisage as perfectly adapted to the modern world is Adorno's model of a message in a bottle, or the Nietzschean model of an arrow shot by one thinker and picked up by another.' Deleuze, *Negotiations*, 154.

34   See the more detailed discussion of this in 'Use'.

I Mean What I Use

> "That's the most important piece of evidence we've heard yet," said the King, rubbing his hands; "so now let the jury –"
>
> "If any one of them can explain it," said Alice (she had grown so large in the last few minutes that she wasn't a bit afraid of interrupting him), "I'll give him sixpence. *I* don't believe there's an atom of meaning in it."
>
> The jury all wrote down on their slates, "*She* doesn't believe there's an atom of meaning in it," but none of them attempted to explain the paper.[35]

Throughout this inquiry and in making sense and materiality of familiar production, the meaning of the language-game of intellectual property – its rules, its grammar, its practice – is revealed by its use: 'the use of a word in the language is its meaning'.[36] Meaning in use is the fundamental ethical character of familiar production, an ethics of connectedness, proximity and chance. This relationship between meaning and use imports relations of ethics into the intellectual property system. In terms of ethics, policy and reform, the overriding question in the context of 'digital' innovation is the possibility and probability of connecting all users (whatever their role within the game) more meaningfully, more usefully.

> It's jurisprudence, ultimately, that creates law, and we mustn't go on leaving this to judges. Writers ought to read law reports rather than the Civil Code. People are already thinking about establishing a system of law for modern biology; but everything in modern biology and the new situations it creates, the new courses of events it makes possible, is a matter for jurisprudence. We don't need an ethical committee of supposedly well-qualified wise men, but user-groups. This is where we move from law into politics.[37]

While policy reform might agitate business models and prescribe imperatives for the law, ultimately its reform is to be achieved through the use of the system, the process of justice, the iterative dialogue of the law *between* adjudicators and users. That is the just reform of the system. And for that we need 'user-groups' in the widest sense of the term.

Some Meaning After All

> "If there's no meaning in it," said the King, "that saves a world of trouble, you know, as we needn't try to find any. And yet I don't know," he went on, spreading

---

35   *Wonderland*, 100.

36   Wittgenstein, *Philosophical Grammar*, 60. See also Wittgenstein, *Philosophical Investigations*, §43.

37   Deleuze, *Negotiations* 169–70.

out the verses on his knee, and looking at them with one eye; "I seem to see some meaning in them, after all."[38]

As noted throughout, the language of reform is critical – the notion of the digital, the priority of the creative, the question of policy. In order to see meaning, however, policy engagement must spread out the verses on its knees. This will play out not only in legislative reform but also, and perhaps more strategically and effectively, in the reform of business practices and behaviours. Rather than fixating upon the form and the technology (for instance, the preoccupation with digital technology as opposed to the digital in use, the preoccupation with the counterfeit as opposed to the creative[39]), it is essential to engage with the narrative of innovation and creativity. All industries are creative industries if creativity, and not technology, is still at the heart of creative. The term is arguably now so inattentive to its subject matter that it is no longer curious, but utterly distracted, entirely negligent. Fundamentally, also, all industries are digital, in the same way that all industries use language. Digital technology is merely a mechanism, a medium, for recording and communicating language. It cannot become the entire embodiment of creativity in and of itself without obliterating the user and meaning completely.

Spread the verses out on your knee, and see some meaning in them, after all.

"Then the words don't *fit* you," said the King, looking round the court with a smile. There was a dead silence.[40]

And at once, the King relaxes his quest to make the evidence fit the proposition.

"Let the jury consider their verdict," the King said, for about the twentieth time that day.

"No, no!" said the Queen. "Sentence first – verdict afterwards."[41]

The fundamental exchange and equilibrium, the 'economics', that is presupposed at the centre of the law are incarnated in this moment. Recalling the relationship between the proposition and the evidence, the imperative and its enforcement, the guilt of the law is always already upon the prisoner:[42] 'The law is indistinguishable from its sentence, and the sentence is indistinguishable from its implementation or execution. If the law is primary, it no longer has any way of distinguishing between the "accusation," the "defence,", and the "verdict."'[43]

---

38   *Wonderland*, 100.
39   Recall the example of the exhibition of counterfeits in Bucharest in honour of World Intellectual Property Day, discussed in Chapter 9.
40   *Wonderland*, 101.
41   *Wonderland*, 101.
42   Deleuze, *Proust and Signs*, 132.
43   Deleuze, *Essays Critical and Clinical*, 32.

> "Stuff and nonsense!" said Alice loudly. "The idea of having the sentence first!"
>
> "Hold your tongue!" said the Queen, turning purple.
>
> "I won't!" said Alice.
>
> "Off with her head!" the Queen shouted at the top of her voice. Nobody moved.
>
> "Who cares for you?" said Alice (she had grown to her full size by this time.) "You're nothing but a pack of cards!"[44]

Again, Alice is experienced with the ordinary but is sensible to the extraordinary, to innovation, to experience. The problem of representation has always been just out of reach. Alice's journey is impossible and yet replete with chance. It is this unrepresentable other, the endless ethical journey, the remainder of any attempt to account for the world (legal, social, political, economic) that fills Wonderland. This is the space of the truly extraordinary that guarantees the continuing reinvention of the ordinary, the infinite play of familiar production. Innovation is in itself an ethical process of becoming, guaranteed by the fact that it can never finally finish, frustrated by the notion that it can. Innovation is named in the interests of the law, but only ever experienced in the unfinished business of justice.

Life Objects

> So she sat with closed eyes, and half believed herself in Wonderland, though she knew she had but to open them again, and all would change to dull reality.[45]

The challenge ahead would seem to be with respect to the very referents of justice themselves, that is, the fixity of the object of protection in the context of the innovation environment, the 'digital' of social and cultural life. Recuperating the commercial within this dialogue and chance of innovation is not a question of assimilating the digital to the materiality of the law, nor is it a notion of disabling the law in the interests of technology. As discussed at the outset of this inquiry, not all uses of intellectual property can necessarily be justified (or assumed to be based) upon incentives to innovate. Even if it could be assumed, despite the lack of clear causal evidence, that commercial innovation enterprises are motivated by the opportunities presented by intellectual property, there are potentially innumerable cases of creativity and innovation not motivated by the promise of intellectual property. And indeed, in the digital environment such examples are proliferative.

Rather, intellectual property laws should be understood in terms of incentives to transact (that is, to achieve appropriate and effective relationships of access and use). In this respect, the transaction need not necessarily be in terms of a fixed object or good, but could perhaps more effectively be achieved through

---

44   *Wonderland*, 102.

45   *Wonderland*, 103.

the multiplication of value through access, experience and tribute. Rather than emphasizing the integrity of the object in a world without barriers, without originals, or exaggerating the centrality of technology over the culture of innovation, it is now more than ever critical to engage with the process, the social, the familiar.

Familiar production is not merely a mechanism of communication, it is not a privileging of technology over the *techne*. Familiar production is a form of life. Reform and policy in the creative should not be coerced by the rhetoric of the technology, confused by the disguise of the mechanism, or seduced by the tool over the language itself. The issue is not technology, it is creative and innovative process. Creativity and innovation must not be reduced to mere technicalities, and the familiar abandoned to the technical. It is therefore necessary to find some way to understand and benefit from the value in repetition.

If we just sit with closed eyes, I seem to see some meaning after all.

RE USE

Re Use

Suppose it were asked: "*When* do you know how to play chess? All the time? or just while you are making a move? – How queer that knowing how to play chess should take such a short time, and a game so much longer.[1]

Preface

> Child of the pure unclouded brow
> And dreaming eyes of wonder!
> Though time be fleet, and I and thou
> Are half a life asunder,
> Thy loving smile will surely hail
> The love-gift of a fairy-tale.[2]

On reflection

I wonder.

Having journeyed through this inquiry into the logic and identity of innovation, and the nature of familiar production, there is the opportunity for some 'self'-reflection and the imaginings of other worlds and new reforms. Looking-glass House exists as reflection; that is, it presents as it cannot be and yet as it is. Similarly, to reflect on debate and discussion in intellectual property, it becomes clear that the opportunity and circumstances for reform subsist within the system itself: 'It is as though we were in Lewis Carroll's mirror where everything is contrary and inverted on the surface, but "different" in depth.'[3]

In deciphering the character of the language-game of intellectual property, predominantly the academic and professional literature is nevertheless a comment on the spectacle provided by counterparts in the game – that is, the accounts of

---

1   Wittgenstein, *Philosophical Investigations*, §59.

2   *Through the Looking-Glass*, 108. With respect to the progression in *Through the Looking-Glass*, Deleuze notes, 'Here events, differing radically from things, are no longer sought in the depths, but at the surface, in the faint incorporeal mist which escapes from bodies, a film without volume which envelops them, a mirror which reflects them, a chessboard on which they are organized according to plan. Alice is no longer able to make her way through to the depths. Instead, she releases her incorporeal double. It is by following the border, by skirting the surface, that one passes from bodies to the incorporeal.' Deleuze, *Logic of Sense*, 9–10.

3   Deleuze, *Difference and Repetition*, 51.

moves of others through perspectives provided by the various players in the game (industry, consumer groups, politicians, gamers, pirates, civil society, libertarians and so on). Stepping through the Looking-glass is no longer merely to reflect; it is to step into the game, into the world of familiar production, *inverted on the surface, but different in depth*. It is to undertake an analysis from the perspective of the piece within the game itself, the user of the system; it is to take an adventure in the performance of use. Stepping through the Looking-glass is to participate in the game played on the other side of conventional production, discharging the debt of use through a meaningful interaction with familiar production. It is to explore familiar production through the rules of the game of intellectual property and the potential circumstances for reform. Just as the chess problem through the Looking-glass may have puzzled some of Carroll's readers, this journey too is nevertheless 'strictly in accordance with the laws of the game'.[4] Here, through the Looking-glass, the user becomes queen.

Looking-glass House

One thing was certain

> One thing was certain, that the *white* kitten had had nothing to do with it: – it was the black kitten's fault entirely.[5]

One thing is certain, the game ensures accountability to meaning in every respect: '*Denying* responsibility means, not *holding* anyone responsible.'[6] It may seem that any inquiry that proceeds from the question of the inquiry itself might have been 'all knots and tangles', like a 'kitten running after its own tail in the middle'.[7] Similarly in this journey into the possible potentialities and pitfalls for innovation within the language-game of the intellectual property system, there seem to be uncovered simply more questions.

At this juncture in the history of intellectual property development and in the digital transformation of labour and products, it is therefore almost inevitable to consider alternatives to the current worldviews presented by the individual perspectives on intellectual property and use, all presumed as total and complete by each proponent, but inevitably always already incomplete. Through the Looking-glass is the universe beyond mere reflection of what we know; it is the ethical dimension of ought, an ethics to come: 'You *ought*, Dinah, you know you ought!'[8] This may appear to be like a moral certainty for reform: 'When a general

---

4   *Through the Looking-Glass*, 107.
5   *Through the Looking-Glass*, 109.
6   Wittgenstein, *Culture and Value*, 73e.
7   *Through the Looking-Glass*, 109.
8   *Through the Looking-Glass*, 109.

ethical law of the form "Thou shalt ..." is set up, the first thought is: Suppose I do not do it?'[9] However, ethics is not prescriptive and transcendent, but rather is always becoming: 'it is clear that ethics has nothing to do with punishment and reward. So this question about the consequences of an action must be unimportant. At least these consequences cannot be events. For there must be something right about that question after all. There must be a *kind* of ethical reward and of ethical punishment but these must be involved in the action itself.'[10] Ethical inquiry thus demands performance and experiment, experience and chance, a journey through the Looking-glass.

Indeed, it is to begin to untangle not merely the answers and solutions to the intellectual property debate and its questions, but rather, it is the examination of the very cultural and political landscape in which those questions are raised and indeed in which they acquire any meaning at all. It is to ask questions about questions. That is, in order to produce an analysis 'upon reflection' it becomes prudent now to explore the 'imaginary scenarios' of use: *can you play*?

Can you play chess?

"Kitty, can you play chess? Now, don't smile, my dear, I'm asking it seriously."[11]

Alice's game of looking into the mirror and imagining the other world is a critical philosophical and analytical game to play in relation to the questions and interests at stake in the present discussion. This is the game here – looking into the mirror of intellectual property and imagining that other world. This game at once copies the world of intellectual property and innovation, the system through which this journey has been made, and at the same time, in repeating the world it is rendered anew.[12]

---

9   Wittgenstein, *Notebooks 1914–1916*, 78e.

10   Wittgenstein, *Notebooks 1914–1916*, 78e.

11   *Through the Looking-Glass*, 111.

12   In *Philosophical Investigations*, §197, Wittgenstein describes this tension between seeing everything and nothing at once: "'It's as if we could grasp the whole use of a word in a flash." – And that is just what we say we do. That is to say: we sometimes describe what we do in these words. But there is nothing astonishing, nothing queer, about what happens. It becomes queer when we are led to think that the future development must in some way already be present in the act of grasping the use and yet isn't present. – For we say that there isn't any doubt that we understand the word, and on the other hand its meaning lies in its use. There is no doubt that I now want to play chess, but chess is the game it is in virtue of all its rules (and so on). Don't I know, then, which game I want to play until I have played it? or are all the rules contained in my act of intending? Is it experience that tells me that this sort of game is the usual consequence of such an act of intending? so is it impossible for me to be certain what I am intending to do? And if that is nonsense – what kind of super-strong connexion exists between the act of intending and the thing intended? – Where is the connexion effected between the sense of the expression "Let's play a game of chess" and

Arguably (and quite intentionally) the nature of this inquiry has been experiment, or excuses, for the terms themselves – it is to *regard* use as well as to speak in regard of use. The relationship or conflict between familiar production and conventional modes of production and protection is not necessarily resolved or operable through assimilation of the cultures of familiar production to the technology of intellectual property protection. Rather than characterizing the relationship between familiar production and intellectual property as one of antagonism and conflict, it is useful to explore the articulation of use as an exchange in a language-game: 'Language is a labyrinth of paths. You approach from *one* side and know your way about; you approach the same place from another side and no longer know your way about.'[13] To navigate the labyrinth of language, it is necessary to immerse oneself in the game, so *let's pretend*.

Let's pretend

> "Kitty, dear, let's pretend –" And here I wish I could tell you half the things Alice used to say, beginning with her favourite phrase, "Let's pretend." She had had quite a long argument with her sister only the day before – all because Alice had begun with, "Let's pretend we're kings and queens"; and her sister, who like being very exact, had argued that they couldn't, because there were only two of them, and Alice had been reduced at last to say, "Well, *you* can be one of them then, and *I'll* be all the rest."[14]

The Looking-glass presents again the question of representation itself, that with which this discussion has been preoccupied inside and out. That is, this inquiry into the question of innovation is implicated in the very subject of innovation itself: 'In a work of art I rather like to find transposed on the scale of the characters, the very subject of that work. Nothing throws a clearer light upon it or more surely establishes the proportions of the whole.'[15] The intrigue through the Looking-glass is the way in which the very subject of the system (that of use, of familiar production) might be transposed upon the users, the producers, the characters of intellectual property itself.

> "Now, if you'll only attend, Kitty, and not talk so much, I'll tell you all my ideas about Looking-glass House. First, there's the room you can see through the glass – that's just the same as our drawing-room, only the things go the other way. I can see all of it when I get upon a chair – all but the bit just behind the

---

all the rules of the game? – Well, in the list of rules of the game, in the teaching of it, in the day-to-day practice of playing.'

13   Wittgenstein, *Philosophical Investigations*, §203.

14   *Through the Looking-Glass*, 111–12.

15   A Gide, *The Journals of André Gide*, J O'Brien (trans), Alfred A Knopf, New York, 1955: 29.

fire-place. Oh! I do so wish I could see *that* bit! I want so much to know whether they've a fire in the winter: you never *can* tell, you know, unless our fire smokes, and then smoke comes up in that room too – but that may be only pretence, just to make it look as if they had a fire. Well then, the books are something like our books, only the words go the wrong way; I know that, because I've held up one of our books to the glass, and then they hold up one in the other room."[16]

Thus, the Looking-glass House reflects all that is known: "'It's as if we could grasp the whole use of a word in a flash.'"[17] But at the same time there is the desire and curiosity to attempt to see outside the conventional referencing, outside the frame. This is the transformative value of travelling through the Looking-glass, of the imagination of new scenarios, of pretending: 'A child has much to learn before it can pretend.'[18] In other words, rather than the representation of the world presented in the Looking-glass being accepted as complete and natural, Alice attempts to see beyond the frame.[19] Similarly, in order to interrogate the questions raised in intellectual property discourse and debate, it is necessary to go beyond reflection, to interrogate the frame of that debate, to transgress that 'worldview' of innovation itself. That is, instead of accepting as natural and commonsensical the ongoing characterization of the innovation narrative that is presented in conventional modes of production and intellectual property frameworks, what emerges is that this very representation of originals and copies, innovation and production, is itself artificial and open to interrogation.

The Looking-glass thus offers the doubling and opening up of telescopic difference through the interlocutory. This is the very nature of the aesthetic dimension of ethics: 'like the surrounds of the work of art, or at most its outskirts: frame, title, signature, museum, archive, reproduction, discourse, market, in short: everywhere where one legislates on *the right to painting* by marking the limit'.[20] This present inquiry has agitated at the limits of the disciplinary and discursive framing by intellectual property and its institutions – the author, the inventor, the register, the museum, the archive, the copy, the market, the public and the private, the idea and the fixation, the memory of all that has gone before. But through the Looking-glass is an opportunity to penetrate the frame of intellectual property, towards a theory of familiar production, towards an understanding of the field of innovation as distinct from the objects: *you know it may be quite different on beyond.*

16   *Through the Looking-Glass*, 112.

17   Wittgenstein, *Philosophical Investigations*, §197.

18   Wittgenstein, *Philosophical Investigations*, 229e.

19   Indeed, Alice's first clues came in Wonderland, when 'she had sat down with her face in her hands, wondering if anything would ever happen in a natural way again'. *Wonderland*, 88.

20   Derrida, *The Truth in Painting*, 11.

You know it may be quite different on beyond

> "How would you like to live in Looking-glass House, Kitty? I wonder if they'd
> give you milk, there? Perhaps Looking-glass milk isn't good to drink – But
> oh, Kitty! Now we come to the passage. You can just see a little *peep* of the
> passage in Looking-glass House, if you leave the door of our drawing-room
> wide open: and it's very like our passage as far as you can see, only you know it
> may be quite different on beyond."[21]

This incomplete reflection is the iconography of innovation, as it were, the always
already copy, the very problem of representation. That is, the innovation of the
intellectual property system (or the landscape of which intellectual property is a
part) is in fact this reversion itself, this remembering, this iconography of the past
in order to recognize the new, to harmonize the different, to innovate upon the
invention, to venture beyond the frame. Through the Looking-glass it is possible
to reflect upon the artificial nature of the frame, the necessary forgetting in order
to presume the total picture: 'What my eyes saw was simultaneous: what I shall
transcribe is successive, because language is successive. Nevertheless, I shall cull
something of it all.'[22]

Enticingly, the Looking-glass motivates a necessary self-consciousness, a
reflection and reflexion upon the way in which the classification and hierarchization
of knowledge is an inevitably selective exercise, a 'culling' that betrays the
incomplete and entirely flexible nature of the system and of representation itself:[23]
'Every classification is similar: they are flexible, their criteria vary according to the
cases presented, they have a retroactive effect, and they can be infinitely refined
or reorganized.'[24] Rather than 'forget' familiar and continuous production, perhaps
it is possible to consider a reconfiguring of the frame through which it might be
understood in a commercial context alongside conventional modes of production.
Indeed, this reconfiguring, this blurring, is the very nature of the game: 'One might
say that the concept "game" is a concept with blurred edges. – "But is a blurred
concept a concept at all?" – Is an indistinct photograph a picture of a person at all?
Is it even always an advantage to replace an indistinct picture by a sharp one? Isn't
the indistinct one often exactly what we need?'[25]

Indeed, the unfinished and indistinct is exactly what is needed.

---

21   *Through the Looking-Glass*, 112–13.
22   Borges, 'The Aleph', 150.
23   See further Gibson, *Creating Selves*, chapter 1, particularly 31–34.
24   Deleuze, *Two Regimes of Madness*, 285.
25   Wittgenstein, *Philosophical Investigations*, §71.

It'll be easy enough to get through

> "Oh Kitty! how nice it would be if we could only get through into Looking-glass House! I'm sure it's got, oh! such beautiful things in it! Let's pretend there's a way of getting through into it somehow, Kitty. Let's pretend the glass has got soft like gauze, so that we can get through. Why, it's turning into a sort of mist now, I declare! It'll be easy enough to get through –"[26]

Not only is the validity of the objects successively disconfirmed and replaced by new objects (most explicitly seen in technological progress, the patent and prior art[27]), but also the reflection of the system itself is displaced in passing through the Looking-glass. Whereas conventionally the objects of capital (the patent, copyright, trade mark, design) have been variously accepted as a kind of tribunal or authentication of value within the logic of the intellectual property system, this reflection of a creativity and innovation is displaced through the Looking-glass. As the present interrogation of innovation has progressed, it has emerged that the indices of innovation point not to a truth of innovation (of the invention, of originality, of novelty and the various criteria in the taxonomy of innovation), but instead to a necessary 'forgetting' of aspects of innovative and creative activity in order to confirm the transactable products as widely as possible; that is, a necessary blurring, an indistinction.

For example, the very practices of skilled drafting in patent law necessarily and fundamentally, in accordance with the logic of the system, attempt not to refine the 'truth' of the invention, but rather to confirm it as broadly as the language will allow in order to ensure more successfully its claims to inventiveness and novelty in a range of as yet unpredicted or unpredictable 'aesthetics' (that is, comparison). Despite rhetoric or mythology to the contrary, it is the question of the probability of value (validity and enforcement) as distinct from truth that is crucial to the logic of the system. The 'chance' of innovation is almost in direct conflict with the 'risk' of replacement: 'The genius of a philosophy must first be measured by the new distribution which it imposes on beings and concepts', that is, its genius is that it will 'displace all reflection'.[28]

Into the Looking-glass room

> Then she began looking about, and noticed that what could be seen from the old room was quite common and uninteresting, but that all the rest was as different as possible.[29]

---

26 *Through the Looking-Glass*, 113.
27 See the earlier discussion in 'Use'.
28 Deleuze, *Logic of Sense*, 6.
29 *Through the Looking-Glass*, 113.

Now that innovation has been explored, the concepts and assumptions have been mainstreamed and have become *quite common and uninteresting*. In a flash the necessary presumptions of the system's logic appear clear: *so that's what it was*. However, all the rest outside the frame – that which lies ahead in terms of reform, whether reform of the law, business practice, social practice or all – *all the rest is as different as possible*.

> "Here are the Red King and the Red Queen," Alice said (in a whisper, for fear of frightening them), "and there are the White King and the White Queen sitting on the edge of the shovel – and here are two Castles walking arm in arm – I don't think they can hear me," she went on, as she put her head closer down, "and I'm nearly sure they can't see me. I feel as if I were invisible –."[30]

The irony is that the silence and invisibility of her venture through the Looking-glass mark Alice's entry as author, her eligibility and her 'illegibility',[31] her ethical and unfinished becoming: 'The author marks the point at which a life is offered up and played out in the work. Offered up and played out, not expressed or fulfilled.'[32] Finally, we are in play in the language-game of innovation: 'The secret is to become invisible and to make a rhizome without putting down roots.'[33] The user becomes accountable for meaning, an authority for production, a protagonist for value.

Memorandum of feelings

> Alice looked on with great interest as the King took an enormous memorandum-book out of his pocket, and began writing. A sudden thought struck her, and she took hold of the end of the pencil, which came some way over his shoulder, and began writing for him.
>
> The poor King looked puzzled and unhappy, and struggled with the pencil for some time without saying anything; but Alice was too strong for him, and at last he panted out, "My dear! I really *must* get a thinner pencil. I can't manage this one a bit; it writes all manner of things that I don't intend –"
>
> "What manner of things?" said the Queen, looking over the book (in which Alice had put, "The White Knight is sliding down the poker. He balances very badly"). "That's not a memorandum of *your* feelings!"[34]

While the King feels burdened by the overwhelming responsibility to the archive, the Queen immediately points out the fallacious and partial nature of that record:

---

30   *Through the Looking-Glass*, 113–14.
31   Agamben, *Profanations*, 69–70.
32   Agamben, *Profanations*, 69.
33   Deleuze, *Two Regimes of Madness*, 66.
34   *Through the Looking-Glass*, 116.

'the *memorandum* … is at the same time afflicted with an essential forgetting.'[35] It is not possible to record the continuous nature of familiar production in a memorandum-book; indeed, just as it is also becoming less relevant to record the use of intellectual and creative output in objects, that is, to exact and transact value in goods: 'Affects aren't feelings, they're becomings that spill over beyond whoever lives through them (thereby becoming someone else).'[36] The memorandum-book, on the other hand, presents itself as an autonomous record of indebtedness, 'the bookish doctrine of judgment' in which '[w]e are dispossessed, expelled from our territory, inasmuch as the book has already collected the dead signs of a Proprietorship that claims to be eternal'.[37] That is, the book is a memorandum of the infinite debt to the property of the producer, as distinct from the affect of familiar production. The proprietary debt, recorded in the memorandum-book, is thus in contrast to the 'justice' of familiar production. It is not a memorandum of the user's feelings.

Conventional intellectual property models operate as for the proprietorship of the memorandum-book, where the 'bookish' register is complicit in a kind of taxonomy of judgment: 'This is the essential effect of judgment: existence is cut into lots, the affects are distributed into lots, and then related to higher forms.'[38] That is, innovation, as it were, is classified into objects, where the object itself is testimony to the existence of the debt of users to the producer. On the other hand, the continuous innovation of familiar production upsets the logic of objects and proprietary existence, as conceived within the doctrine of judgment.

Judgment is articulated on the language of moral certainty, appealing to a higher value in order to present certain distinctions or exclusions as valid. Indeed, this has at times been the very language of the intellectual property debate, a 'war' (of judgment) as distinct from a 'combat' of the just.[39] In contrast to any reference to a pre-existing order, the just is a combat of decisions,[40] the process of which is crucial to justice: 'A decision is not a judgment, nor is it the organic consequence of a judgment: it springs vitally from a whirlwind of forces that leads us into combat. It resolves the combat without suppressing or ending it.'[41]

Immediately, therefore, this passage restores the question of the continuous 'ethics' of the archive in the context of familiar production.[42] Rather than a repository within the selective nature of the frame, of the memorandum-book, the

---

35   Deleuze, *Difference and Repetition*, 140.

36   Deleuze, *Negotiations*, 137.

37   Deleuze, *Essays Critical and Clinical*, 128.

38   Deleuze, *Essays Critical and Clinical*, 129.

39   Deleuze explains the difference between judgment as a war and justice as combat, that is, a more dialogical, iterative and 'athletic' process. See Deleuze, *Essays Critical and Clinical*, 132–34.

40   See the discussion in Chapter 11.

41   Deleuze, *Essays Critical and Clinical*, 134.

42   Recall the discussion in Chapter 7.

affect of familiar production is a becoming, a production of chance and ambiguity. The fundamental question is whether it is possible, or indeed desirable, to construct a way in which to conduct the business of familiar production, using the tools of intellectual property, and indeed to reconcile the culture and subjectivity of familiar production alongside existing business models.

And so the concept of 'taste' returns as a crucial element in the universe of familiar production. On the one hand, there is the continuous and transitive 'affective' relationships through familiar production, and on the other hand the assumption and posturing of certainty in the 'aesthetics' of intellectual property models of taste: 'how does feeling overcome its inconstancy and become an aesthetic judgment? ... taste is a feeling of the imagination, not of the heart. It is a rule, and what grounds a rule in general is the distinction between power and the exercise of power.'[43] Intrinsic to this power relation is the 'kinship' of meanings through the aesthetics and comparison of production.[44] In the repetition of difference through familiar production there is the blurring of the edges[45] of aesthetic comparison. The aesthetics of familiar production is the comparative performance between products, between tastes, as distinct from discretion, as it were. That is, tastes are 'the product of an encounter (a pre-established harmony) between goods and a taste'.[46]

The conflict between 'feelings' and 'memoranda' presents (indeed, right at the beginning of the interrogation through the Looking-glass) the shift between the seemingly objective record of innovation (as curated through goods comprising intellectual property rights[47]) and the labour of the affect (as experienced through the immateriality of the field of enterprise, expertise and 'services'). The 'incentive' of intellectual property is an incentive to record, to curate, to provide the debt of heritage for innovation. The repetition of the system 'remembers' the new, and the obligation of heritage consolidates the relationship between 'own self' and property that is instrumental to the discourse of intellectual property.

If not the memorandum-book, then how might the King's feelings be reconciled? Inevitably, this would appear to be achieved by setting the King into play, by putting the user into the language-game of intellectual property. Only by

---

43   G Deleuze, *Empiricism and Subjectivity: An Essay on Hume's Theory on Human Nature*, CV Boundas (trans), Columbia UP, New York, [1953]/1991: 57.

44   See Chapter 4.

45   Wittgenstein, *Philosophical Investigations*, §71.

46   Bourdieu, *Sociology in Question*, 108. Bourdieu notes further: 'Tastes are the product of this encounter between two histories, one existing in the objectified state, the other in the incorporated state, which are objectively attuned to one another' and describes the producer as the 'absent third party' with respect to 'the encounter between a work of art and the consumer' and who transforms taste into an object. Bourdieu, *Sociology in Question*, 109.

47   Considered throughout, but see in particular the discussion in 'Use' as well as Chapters 4 and 7.

ensuring realistic, effective and legitimate accountability for the user within the game will it be possible for familiar and continuous production to be played out in any meaningful way (for all users, producers, consumers) into a commercial, economic model. The user becomes Queen.

So what of the product through the Looking-glass? What of the book?

A Looking-glass book

> "Why it's a Looking-glass book, of course! And if I hold it up to a glass, the words will all go the right way again."[48]

The Looking-glass poem, 'Jabberwocky' introduces Alice to portmanteau words, later explained to her more meaningfully by Humpty Dumpty.

> Beware the Jabberwock, my son![49]

The poem's monster, the Jabberwock, is itself a portmanteau word. To jabber is to gabble and chatter rapidly in indistinct and unintelligible gibberish. And woc is an obsolete form of woke, reminding again of the questions of reflection, awakening, memory and dreams that punctuate the journey.

> So rested he by the Tumtum tree,
> And stood awhile in thought.[50]

The onomatopoeic Tumtum is variously understood as an imitation, an impersonation, as it were, of a stringed instrument played (strummed) monotonously and repetitively: *tumtum, tumtum*. Tumtum suggests the imperfection and inadequacy of imitation and repetition (of sound and instrument), but here that repetition is creating something new and indeed organic (not only the tree but also new language, and new *affect*). And in a tale of ingestion and oral fascination, it is of course all too curious that the Tumtum also refers to the stomach!

The music, the refrain of the Tumtum tree is perhaps also the habitat, the territory, of the Jubjub bird in the poem: 'Beware the Jubjub bird'.[51] Recalling the earlier discussion of the ethics of the refrain,[52] the musical refrain at once both marks and sustains the territory, the becoming, the abode: 'Now we are at home.'[53] The refrain is the music of familiar production: 'A musical "nome" is a little tune, a melodic formula that seeks recognition and remains the bedrock or

---

48 *Through the Looking-Glass*, 116.
49 *Through the Looking-Glass*, 117.
50 *Through the Looking-Glass*, 117.
51 *Through the Looking-Glass*, 117.
52 Chapter 10.
53 Deleuze & Guattari, *Thousand Plateaus*, 311.

ground of polyphony ... it is *ethos*, but the ethos is also the Abode.'[54] The Jubjub is the counterpoint, the refrain that at once secures the territory and kinship of familiar production.

Opposed to the territory of familiar production is the 'land' of capital; that is, the way in which intellectual property, both conceptually and in concert with the apparatus of capital, is indeed intrinsically linked with the geography of land and land ownership. In other words, the immateriality of intellectual property does not in itself defy the construction of value through what are essentially similar processes of commodification: the rent from land or the taxation of commodities. This process of capturing value equally assimilates the value of production from users: 'In a word, money – the circulation of money – *is the means for rendering the debt infinite.*'[55] Taxation therefore is the great success of capitalism: 'money is fundamentally inseparable, not from commerce, but from taxes as the maintenance of the apparatus of the State. Even where dominant classes set themselves apart from this apparatus and make use of it for the benefit of private property, the despotic tie between money and taxes remains visible.'[56]

While there are regular arguments raised regarding the cost of production (for example, of a CD, of medicines), or the alignment of value with benefits (for example, to the patient), these constructions of value have not gained currency within the system. Problematically, the process of commodification and the construction of value and price of commodities is 'quite distinct from their palpable and real bodily form',[57] whether that is the form of land or the materiality of intellectual property. However, this dematerialized construction of value within intellectual property industries is in fact wholly consistent with familiar production and the quality of experience and experiment. It is an opportunity for reform in the way in which use is accountable and responsible within the system, rather than an obstacle to transformation. Arguments focusing on the costs of producing the product (as distinct from the innovation) in many ways fail to engage with the actual mechanism of capture of which intellectual property is a tool. Therefore, it is necessary to investigate ways in which the object of exchange might be reconfigured in the context of the digital and familiar production, in the realm of the affect, because regardless of whether value within the market is ascribed through comparison or through monopolistic scarcity, it has nevertheless little to do with the physical carrier of value (the land, the book, the CD, the medicine and so on) and everything to do with access to the means of comparison or exchange. *Money literally buys nothing.*

---

54   Deleuze & Guattari, *Thousand Plateaus*, 312.

55   Deleuze & Guattari, *Anti-Oedipus*, 197.

56   Deleuze & Guattari, *Anti-Oedipus*, 197.

57   Marx, *Capital*, 189. See further the development of Marx's theory of money and commodities in Deleuze & Guattari, *Thousand Plateaus*, 437–48. Deleuze and Guattari offer a 'trinity formula' whereby the value (through comparison or monopolistic appropriation) is captured in land (as rent), in work and labour (as profit), and in objects (as taxation).

Fill my head with ideas

> "It seems very pretty," she said when she had finished it, "but it's *rather* hard to understand!" (You see she didn't like to confess even to herself, that she couldn't make it out at all.)[58]

So what of the Jabberwock and the capture of sense? In attempting to make sense of the poem, the relationship between innovation and memory is clear. Memory both organizes innovation (socially and conceptually) as well as maintains the credit to the producer, guaranteeing, as it were, the infinite debt of the user. It follows that at the same time, this social organization through memory is also the paradox of innovation. That is, recalling the discussion throughout, the curious nature of innovation (invention and creativity) is that its very process is to remember, to recall, to situate within the narrative and position with respect to the 'prior art', with respect to the archive. Indeed, to invent is to remember.

However, as this adventure has revealed, to use is to acquire meaning. On the Looking-glass book, Alice resolves:

> "Somehow it seems to fill my head with ideas – only I don't exactly know what they are! However, *somebody* killed *something*: that's clear, at any rate –"[59]

To use is to 'fill my head with ideas'. It is not to direct new products ('I don't exactly know what they are!') but to motivate new 'technologies'. The crucial character of familiar production is the provocation of new ideas, notwithstanding any relationship to intellectual property or the product. Familiar production is a fundamental and ubiquitous structure of creativity in a contemporary digital society. Therefore, if the intellectual property system is to be confronted with changes in consumer expectations of access and use, it is also being confronted with transformations in the relationship of the individual to creative processes. The creative self is digital: digital as in freedom, not digital as in technology.

The affect produces in the current language of innovation 'a kind of foot stomping, a stammering, an obsessional tom-tom, like a repetition that never ceases to create something new. Under the impulse of the affect, our language is set whirling, and in whirling it forms the language of the future, as if it were a foreign language, an eternal reiteration, but one that leaps and jumps. We stomp within the turning question, but this turning is the bud of the new language.'[60] It is the monotonous strumming of the Tumtum tree, the ferocity of the Jubjub bird, the fabulousness of the Jabberwock, the nonsense of familiar production. Whatever it is, *it seems very pretty.*

---

58   *Through the Looking-Glass*, 117.
59   *Through the Looking-Glass*, 117.
60   Deleuze, *Essays Critical and Clinical*, 98.

The Garden of Live Flowers

It's no use talking about it

> "But how curiously it twists! It's more like a corkscrew than a path! Well, this turn goes to the hill, I suppose – no it doesn't! This goes straight back to the house! Well then, I'll try it the other way."
>
> And so she did: wandering up and down, and trying turn after turn, but always coming back to the house, do what she would. Indeed, once, when she turned a corner rather more quickly than usual, she ran against it before she could stop herself.
>
> "It's no use talking about it," Alice said, looking up at the house and pretending it was arguing with her. "I'm *not* going in again yet. I know I should have to get through the Looking-glass again – back into the old room – and there'd be an end of all my adventures!"[61]

But there is no use talking about it. Naming is in fact no preparation for talk,[62] it alone does not provide the way about in the language-game. *Don't think, but look!* It is not possible simply to point to the names and thus enter language: 'One cannot guess how a word functions. One has to *look at* its use and learn from that.'[63] The discourse on intellectual property has been plagued by the branding of slogans, buzzwords, passwords and product names ('digital', 'creative industries' and so on[64]). The functionality of use and innovation has been overrun by the discursive products of 'digital' and 'technology'. The capture and diminishing of innovation and creativity to the audacity of technology alone is not only damaging to producers, but also entirely dismissive of all users. *Don't think, but look!*

> "Oh, it's too bad!" she cried. "I never saw such a house for getting in the way! Never!"[65]

Despite the lack of preparation, as it were, Alice is nevertheless progressing far more quickly into the garden through the Looking-glass than she ever did during her adventure in Wonderland. Despite the cataclysmic 'revolution' in a rethinking of immaterial labour and property and the becoming of familiar production, it is all nevertheless starting to make sense. This is the refrain of territory, of becoming. Ethics has an abode: 'Now we are at home.'[66]

---

61   *Through the Looking-Glass*, 118.

62   Wittgenstein, *Philosophical Investigations*, §25 and §27.

63   Wittgenstein, *Philosophical Investigations*, §340.

64   See the further discussion in 'Use'.

65   *Through the Looking-Glass*, 119.

66   Deleuze & Guattari, *Thousand Plateaus*, 311.

I should advise you to walk the other way

> "O Tiger-lily," said Alice, addressing herself to one that was waving gracefully about in the wind, "I *wish* you could talk!"
>
> "We *can* talk," said the Tiger-lily: "when there's anybody worth talking to."
>
> Alice was so astonished that she couldn't speak for a minute: it quite seemed to take her breath away. At length, as the Tiger-lily only went on waving about, she spoke again, in a timid voice – almost in a whisper. "And can *all* the flowers talk?"
>
> "As well as *you* can," said the Tiger-lily. "And a great deal louder."
>
> "It isn't manners for us to begin, you know," said the Rose, "and I really was wondering when you'd speak!"[67]

Alice's wish betrays her presumption about the nature of thinking and of speech. It is not that the flowers do not have language, but that they simply do not talk until someone participates in their language-game: 'Commanding, questioning, recounting, chatting, are as much a part of our natural history as walking, eating, drinking, playing.'[68] As soon as Alice participates in the game, and enters into dialogue with the flowers, her progress is assured. The pawn enters the language-game and her 'way' is revealed. Even if at first Alice defies the advice on the way, she learns very quickly that sometimes acting counter-intuitively will succeed beautifully:

> "You can't possibly do that," said the Rose: "*I* should advise you to walk the other way."
>
> This sounded nonsense to Alice, so she said nothing, but set off at once towards the Red Queen. To her surprise, she lost sight of her in a moment, and found herself walking in at the front-door again.
>
> A little provoked, she drew back and, after looking everywhere for the Queen (whom she spied out at last, a long way off), she thought she would try the plan, this time, of walking in the opposite direction.
>
> It succeeded beautifully.[69]

Look up, speak nicely

> "Where do you come from?" said the Red Queen. "And where are you going? Look up, speak nicely, and don't twiddle your fingers all the time."
>
> Alice attended to all these directions and explained, as well as she could, that she had lost her way.

---

67   *Through the Looking-Glass*, 119.

68   Wittgenstein, *Philosophical Investigations*, §25.

69   *Through the Looking-Glass*, 121–22.

> "I don't know what you mean by *your* way," said the Queen: "all the ways
> about here belong to *me* – but why did you come here at all?" she added in a
> kinder tone.[70]

Alice is again confronted with the many ways of the living rhizome, the many
ways and linguistic connections of familiar production: 'Language is a labyrinth
of paths.'[71] For the pawn, all the ways belong to the producer, all the credit is with
the Queen. However, to unlock the rhizomatic potential of familiar production the
user must enter the language-game. The user must become Queen.

Don't think, but look! Only then is it possible to participate in the language-
game. *Look up, speak nicely.*

> "When you say 'hill,'" the Queen interrupted, "*I* could show you hills, in
> comparison with which you'd call that a valley."
>
> "No, I shouldn't," said Alice, surprised into contradicting her at last: "a hill
> *can't* be a valley, you know. That would be nonsense –"
>
> The Red Queen shook her head. "You may call it 'nonsense' if you like," she
> said, "but *I've* heard nonsense, compared with which that would be as sensible
> as a dictionary!"[72]

This illustrates the artificial nature of language, and the way in which naming
is indeed no preparation for language. To point to a name, to give it a proper
name (patent, copyright, trade mark, design) might indicate different meanings
in use and for different users. Intellectual property, patent, copyright, trade mark,
design – these have all become proper names and thus untranslatable. This is the
preliminary presumption of any discourse or debate on the value of intellectual
property and, as such, intellectual property, as a term, has ensured its import is
fundamentally outside the language-game of innovation. It resists translation, it is
beyond challenge. However, 'the proper name is no way the indicator of a subject'.[73]
      This is the crucial articulation of the interaction of intellectual property and
familiar production. In translating the untranslatable, it is essential not to disregard
the meaning in use in a preference for the object bearing the name: 'It is important
to note that the word "meaning" is being used illicitly if it is used to signify the
thing that "corresponds" to the word. That is to confound the meaning of a name
with the *bearer* of the name. When Mr N. N. dies one says that the bearer of the
name dies, not that the meaning dies.'[74] In other words, the conflict between use,
access and production with respect to intellectual property and on the part of the
range of users (consumers, producers) will achieve nothing but nonsense and dead

70   *Through the Looking-Glass*, 122.
71   Wittgenstein, *Philosophical Investigations*, §203.
72   *Through the Looking-Glass*, 122–23.
73   Deleuze & Guattari, *Thousand Plateaus*, 263–64.
74   Wittgenstein, *Philosophical Investigations*, §40.

letters unless there is genuine participation in the game over and apart from the assumptions and perspectives of any one particular player in the game: 'Should it be said that I am using a word whose meaning I don't know, and so am talking nonsense? – Say what you choose, so long as it does not prevent you from seeing the facts. (And when you see them there is a good deal that you will not say.)'[75] Distracted by objects, arguably the intellectual property system is missing the user.

A great game of chess

> "It's a great game of chess that's being played – all over the world – if this *is* the world at all, you know. Oh, what fun it is! How I *wish* I was one of them! I wouldn't mind being a Pawn, if only I might join – though of course I should *like* to be a Queen, best."
>
> She glanced rather shyly at the real Queen as she said this, but her companion only smiled pleasantly, and said, "That's easily managed. You can be the White Queen's Pawn, if you like, as Lily's too young to play; and you're in the Second Square to begin with: when you get into the Eighth Square you'll be a Queen –"[76]

Alice is all at once inside the game, yet at the same time alluding to the effable inadequacy of representation: 'if this *is* the world at all, you know'. Alice is both observing the encyclopaedic knowledge of the language-game of intellectual property, that *great game of chess*, and a participant in the game. The user enters the game, but at the same time the position of the subject of knowledge, the perspective on the game, is at issue. In other words, Alice, the user, must include herself in the game in order to address the game. This is the crucial difficulty with the discourse of the intellectual property debate, and the institutional processes of consultation, development and policy; that is, examining the debate within the terms and from the positions of all users, as distinct from proceeding from what must always be accepted as provisional assumptions at best, not incomparable schema.

Now! Now! Faster! Faster!

> Alice never could quite make out, in thinking it over afterwards, how it was that they began: all she remembers is, that they were running hand in hand, and the Queen went so fast that it was all she could do to keep up with her: and still the Queen kept crying, 'Faster!' but Alice felt she *could not* go faster, though she had no breath to say so.
>
> The most curious part of the thing was, that the trees and the other things round them never changed their places at all: however fast they went, they never seemed to pass anything.[77]

---

75   Wittgenstein, *Philosophical Investigations*, §79.

76   *Through the Looking-Glass*, 123.

77   *Through the Looking-Glass*, 123–24.

Through the Looking-glass Alice achieves the speed of becoming. Through the Looking-glass the usual momentum from product to use is reflected in disarray. This is the speed of familiar production, the here and now, and now, and now: "'and then you will see it now *this* way, now *this*" – *What* way? There *is* no further qualification.'[78] Familiar production is thus a production without dominion: it is territory without land; relations without products; products without authors. This is not, however, a dismantling of authorship, of attribution and of the integrity of the work in a practical sense. Rather, it is coming to know familiar production as a wider range of phenomena and use that will be crucial to achieving a greater legitimacy of new intellectual property models across a range of production in the digital. Familiar production, as introduced at the outset, is a kinship through use, an inconstant, iterative and proliferative 'smooth space'[79] of innovation, of becoming.

So for Alice, the trees and other things never changed their places, never gave an inch[80] (itself an interesting perspective upon her own movement with respect to the trees). This is the 'movement' of the nomad: 'The nomad distributes himself in a smooth space; he occupies, inhabits, holds that space; that is his territorial principle. It is therefore false to define the nomad by movement.'[81] Indeed, the nomad establishes territory not by representation but by appearances: 'nomads have no points, paths, or land, even though they do by all appearances.'[82] Familiar production has territory but that territory is dismembered by the conventions in place in order to capture value. In understanding the provisional nature of those conventions, it becomes possible to reject them as incomparable and inexhaustible schema for innovation, and begin to account for an affinity with familiar production by addressing the relations to users and to use within intellectual property models: 'The land ceases to be land, tending to become simply ground (*sol*) or support.'[83]

Now, here

   "Faster! Don't try to talk!"
   Not that Alice had any idea of doing *that*. She felt as if she would never be able to talk again, she was getting so out of breath: and still the Queen cried,

---

78   Wittgenstein, *Philosophical Investigations*, 200e.

79   The concept of 'smooth space' comes from Deleuze and Guattari and is introduced as the aggregating, associational space of the social, as distinct from more stable space of institutional norms and processes. The concept of smooth space is addressed in more detail in Chapter 2.

80   On the forest as a crowd symbol (considered in more detail later in this chapter), Elias Canetti notes the 'multiple immovability' of the forest, 'Its resistance is absolute; it does not give an inch.' Canetti, *Crowds and Power*, 84.

81   Deleuze & Guattari, *Thousand Plateaus*, 381.

82   Deleuze & Guattari, *Thousand Plateaus*, 381.

83   Deleuze & Guattari, *Thousand Plateaus*, 381.

"Faster! Faster!" and dragged her along. "Are we nearly there?" Alice managed to pant out at last.

"Nearly there!" the Queen repeated. "Why, we passed it ten minutes ago! Faster!" And they ran on for a time in silence, with the wind whistling in Alice's ears, and almost blowing her hair off her head, she fancied.

"Now! Now!" cried the Queen. "Faster! Faster!" And they went so fast that at last they seemed to skim through the air, hardly touching the ground with their feet, till suddenly, just as Alice was getting quite exhausted, they stopped, and she found herself sitting on the ground, breathless and giddy.

The Queen propped her against a tree, and said kindly, "You may rest a little now."

Alice looked round her in great surprise. "Why, I do believe we've been under this tree all the time! Everything's just as it was!"

"Of course it is," said the Queen: "what would you have it?"

"Well, in *our* country," said Alice, still panting a little, "you'd generally get to somewhere else – if you ran very fast for a long time, as we've been doing."

"A slow sort of country!" said the Queen. "Now, *here*, you see, it takes all the running *you* can do, to keep in the same place. If you want to get somewhere else, you must run at least twice as fast as that!"[84]

'Breathless and giddy', through the Looking-glass Alice is thrilled with the *giddy seductiveness of chance*.[85] 'Hardly touching the ground', she is at once in the 'vertical motion' of the smooth space of becoming.[86] To be *now, here*, the instant of innovation, the always already present, takes all of Alice's effort; it takes all the running one can do to stay in one place. The time of innovation is the time of the instantaneous, the time of becoming, and yet it is always yet *to come*.[87]

---

84   *Through the Looking-Glass*, 124.

85   Bataille, *Guilty*, 72.

86   Deleuze and Guattari explain: 'Laminar movement that striates space, that goes from one point to another, is weighty; but rapidity, celerity, applies only to movement that deviates to the minimum extent and therefore assumes a vertical motion, occupying a smooth space, actually drawing smooth space itself.' Deleuze & Guattari, *Thousand Plateaus*, 371.

87   On the promise of democracy to come, the always inaccessible, out of reach, higher shelf, Derrida notes: 'the event of that promise takes place here, now in the singularity of a here-now that, as paradoxical as it might seem, I believe I must dissociate from the value of presence … It presents itself only in losing or undoubling itsef in iterability, thus in the mark and the generality or ideality that, moreover (threat or luck), will allow later for a calculated negotiation between the presentable and the nonpresentable, the subject and a-subjective singularity, rights and a justice beyond rights and ethics, and perhaps even beyond politics … The here-now indicates that this is not simply a question of utopia. There is constant and concrete renewal of the democratic promise as there is of the relation to the other as such, of the relation to infinite distance, incalculable heterogeneity, etc.' Derrida, *Negotiations*, 180.

Familiar production, most triumphantly in the digital, is the world of faster and faster innovation but no movement, as it were; the smooth space of the *now, here*. All time becomes productive labour time. The challenge is to resist the reterritorialization of familiar production such that it is simply a monopolistic appropriation of all labour, all life, where the individual is simply a commodity for exchange.

Recalling the time of ethics, the smooth space-time of the ethical dimension preserves instead the 'to come', that is, the anticipation of future choice, decision-making and agency. This is what may provide the ethical character to intellectual property; the aesthetics or 'taste' of intellectual property, not only in its application but also in its interpretation and analysis, including in respect of familiar production and the digital. The ethical 'to come' cannot be reconciled within an infinite economic sphere, nor can it be reduced to the calculation and assessment of models of economic rationality. Policy and legislative reform must not be abbreviated and attenuated by quantitative, numerical assessments alone (where 'one counts in order to occupy'), but must negotiate the smooth space-time of divergent and emergent creativity, the time 'without counting'.[88] And so, all through Wonderland, Alice was urged to come on, to pursue the future. Through the Looking-glass, she has the opportunity not to live without time, as such, but to explore the game, the *now, here*. The future is not frozen, or neutralized, rather it is something for which Alice is *now, here* enabled to make decisions.

Move as fast as you can, to stand still: 'It is thus necessary to make a distinction between *speed* and *movement*: a movement may be very fast, but that does not give it speed; a speed may be very slow, or even immobile, yet it is still speed. Movement is extensive; speed is intensive.'[89] The Queen and Alice run faster and faster and yet stay in the same place, their speed is qualitative, rather than quantitative: 'What qualifies a deterritorialization is not its speed (some are very slow) but its nature.'[90] In other words, '*Slow and rapid are not quantitative degrees of movement but rather two types of qualified movement*',[91] not occupying space through counting but occupying through use. Alice cannot perceive the absolute speed of becoming because 'Movements, becomings, in other words, pure relations of speed and slowness, pure affects, are below and above the threshold of perception ... So that movement in itself *continues* to occur elsewhere.'[92]

This is the speed of familiar production, of continuous innovation. The increments or changes are understood (by users and producers alike) as qualitative, not quantitative. Familiar production is not a linear progress as such; it is the smooth space of emergent creativity, of absolute becoming: *now, here*. Thus,

---

88   Deleuze & Guattari, *Thousand Plateaus*, 477.

89   Deleuze & Guattari, *Thousand Plateaus*, 381. As Brian Massumi explains, 'Becoming concerns speed, but speed is relative.' Massumi, *User's Guide*, 104.

90   Deleuze & Guattari, *Thousand Plateaus*, 56.

91   Deleuze & Guattari, *Thousand Plateaus*, 371.

92   Deleuze & Guattari, *Thousand Plateaus*, 281.

it is impossible to speak about change in terms of adjustments of the products of familiar production to demand, in that familiar production is not in respect of objects but in respect of use, the affect. Supply not only precedes demand, it exceeds and reveals it.

Looking-glass Insects

A grand survey

> Of course the first thing to do was to make a grand survey of the country she was going to travel through.[93]

It would appear that the first thing to do is to undertake the conventional cartography and assimilation of the environment, the universe of innovation. But in doing so, the very localities of the issues are displaced. Indeed, Alice eliminates all character from the place: no rivers, no other mountains, no towns. Is Alice inside or outside the map? Is Alice inside or outside the story? Is Alice inside or outside the game?

Alice's map very quickly transforms from a lesson in geography (a quantitative and measurable survey) into a survey of creatures, activities, appearances, and experience (a qualitative and immeasurable array). This juxtaposes immediately the survey of conventional modes of production and their registers, with the production of appearances and affect in familiar production.

Tickets, please!

> "Tickets, please!" said the Guard, putting his head in at the window. In a moment everybody was holding out a ticket: they were about the same size as the people, and quite seemed to fill the carriage.
>
> "Now then! Show your ticket, child!" the Guard went on, looking angrily at Alice. And a great many voices all said together ("like the chorus of a song," thought Alice), "Don't keep him waiting, child! Why, his time is worth a thousand pounds a minute!"
>
> "I'm afraid I haven't got one," Alice said in a frightened tone: "there wasn't a ticket-office where I came from." And again the chorus of voices went on. "There wasn't room for one where she came from. The land there is worth a thousand pounds an inch!"[94]

Alice, the user, must pursue the smooth space-time of *now then*; the instant of innovation and the 'to come' of ethics. Nevertheless, the smooth space of familiar

---

93   *Through the Looking-Glass*, 126.
94   *Through the Looking-Glass*, 127.

production will always be subjected to the striation of dominant conventions and discourse and the attempt to render all relations to commodities for exchange: the time is worth a thousand pounds a minute, the land is worth a thousand pounds an inch. But the smooth space of familiar production is time without passage, territory without land.

In this scenario, the ticket is a physical object in which one acquires the right to travel. It is literally a right of admission (to the game). A ticket is also a memorandum, an instrument of public information. Tickets, please! Alice is presented with an imperative. She is commanded, directed, ordered to present the evidence of her right to proceed, her right to participate in the game. As a pawn, a mere user, Alice does not have such a right. There are no ticket offices for users, there is no official memorandum or documentation of a right to use. There is no way for Alice to produce evidence of use other than from the perspective of the producer. Users are still without a right to travel. No admission.

At the same time, this puts the user in a curious position within the intellectual property system. The user is both excluded and at the same time in a position (imposed or otherwise) to write her own ticket, to stipulate its own conditions of use. This is the crisis of legitimacy of the intellectual property system. With no ticket office for users, what else can be expected? Writing the user into the system not only addresses the vagueness with respect to the user's admission to the system, but also provides the user with the invitation (literally). That is, the user is accommodated with the incentive and inducement to engage legitimately in the game.

Nevertheless, a ticket to travel can also be resumed by the system, indeed acting also as an acknowledgement of the user's indebtedness, as a promise, a guarantee, a memorandum of the 'debt' to innovation (received on credit). The admission of the user thus also consolidates the relationship of trust. But with no ticket, and no prospect of buying one, Alice will never acknowledge the debt. She will never become a legitimate traveller. She cannot perform the correct action, the expected (or fashionable) act.

> "Don't make excuses," said the Guard: "you should have bought one from the engine-driver." And once more the chorus of voices went on with, "The man that drives the engine. Why, the smoke alone is worth a thousand pounds a puff!"[95]

Money literally buys nothing, nothing but puff.

Language is worth a thousand pounds a word

> Alice thought to herself, "Then there's no use in speaking." The voices didn't join in this time, as she hadn't spoken, but, to her great surprise, they all *thought* in chorus (I hope you understand what *thinking in chorus* means – for *I* must

---

95   *Through the Looking-Glass*, 127.

confess that *I* don't), "Better say nothing at all. Language is worth a thousand pounds a word."[96]

Language is worth a thousand pounds a word, a picture is worth a thousand words and what is the 'use' of a book without pictures? The question might be how much one must pay in order to find use in speaking: 'It soon appears that the corresponding problem is no longer that of the unique and withdrawn voice, and that it has rather become the problem of multiple discourse: what must one pay, how much must one pay in order to be able to speak?'[97] Language is worth a thousand pounds a word. However, individual words simply point to names, it is the meaning that characterizes the language-game: 'Every word has a meaning. This meaning is correlated with the word. It is the object for which the word stands.'[98] Meaning comes not from the word itself but from use: 'Meaning is not a process which accompanies a word. For no *process* could have the consequences of meaning.'[99]

The cacophony of the train is, to Alice, nonsense, noise. It is no use speaking because she is at once at cross-purposes with the language-game on the train. She cannot afford nor can she even access the conditions for admission to the journey. If intellectual property discourse is the dominant voice in the debate, familiar production will be reduced to silence in a train filled with noise. It is essential that in order to create the circumstances for dialogue and the realization of the revolutionary opportunities in thinking, creating and speaking that arise in the digital, there is the need for reform of the language-game of intellectual property, and the granting of admission to users.

Wrong way

> All this time the Guard was looking at her, first through a telescope, then through a microscope, and then through an opera-glass. At last he said, "You're travelling the wrong way," and shut up the window and went away.[100]

In Wonderland, Alice herself folded up like a telescope, grew like unrest, shrank like a violet, and changed so regularly that it became expected. In Wonderland Alice was alienated from her usual custom and 'familiar production'. Away from her family she, literally, could not find her feet.[101] Without legitimate admission to the train, similarly, there is no use speaking. Without meaningful admission

---

96   *Through the Looking-Glass*, 128.
97   Deleuze, *Logic of Sense*, 236.
98   Wittgenstein, *Philosophical Investigations*, §1.
99   Wittgenstein, *Philosophical Investigations*, 218e.
100   *Through the Looking-Glass*, 128.
101   'We cannot find our feet with them.' Wittgenstein, *Philosophical Investigations*, 223e.

of the user to the language-game of intellectual property, to the conventions and logic of intellectual property, the user is quite firmly 'disowned'. At this point it is important to re-orientate the narrative. Rather than proceeding from presumptions of the logic of policy and reform, it is time for users to find their feet.

It is time to find our way.

As luggage

"She'll have to go back from here as luggage!"[102]

On this train journey, this striation of the territory of innovation, the user is luggage, baggage, a portmanteau. On these terms, if use is admitted to the intellectual property journey at all, that admission will be qualified through a portmanteau conjunction of meaning; for example, user-led or user-generated innovation (where indeed use reveals the product, rather than demand resolving a product for passive consumption or audience).

However, the term 'use' itself stores a wealth of meaning.[103] Fundamentally, use distresses and excites the very relationship between innovation, product and use. Recalling the trilogy of invention-innovation-diffusion introduced at the outset of this journey, it is now very clear that this houses a complexity of economic logic that far exceeds the notion of inventing, commercializing and disseminating to a passive audience of consumers, to mere spectators. Indeed, at times use is itself in contradistinction to the objective product. Even in conventional modes of production, it is important to see the way in which use is a necessary corollary of the product's existence but at the same time contrary to the product's subsistence, and further, the way in which the user can be both an audience and an operator. More remarkably perhaps, in familiar production use is always in terms of a means, rather than an end in itself. Familiar production is the sphere of the affect.

In this way, 'end-user' itself becomes a curious and quaint term, almost to the point of nonsense but always to the benefit of nostalgia. In every condition the user in familiar production is a source as well as a target for innovation. Arguably, the most important actor in familiar production is the *username*: 'She must be labelled'.[104] The user has a head on her, 'She must go by post'.[105] The user is a message,[106] not merely a letter. The user cannot be returned, re-posted, sent back to her proper place. The user is always already *now, here*. The user can never fail to arrive because the message is itself manifest at the time of 'use', at the inception of meaning. The user is indeed the very engine of the train itself.[107]

---

102   *Through the Looking-Glass*, 128.
103   See the extensive discussion in 'Use'.
104   *Through the Looking-Glass*, 128.
105   *Through the Looking-Glass*, 128.
106   *Through the Looking-Glass*, 128–29.
107   *Through the Looking-Glass*, 129.

Throughout this journey there has been the ambiguity of destination, the iterability of messages, the repetition of error. But the user cannot be returned. The user is *now, here* and *now then*, upcast in the dialectic of innovation. Thus, the nonsense of a return-ticket at every station of the journey's progress is in fact an affirmation of the difference in repetition: 'Never mind what they all say, my dear, but take a return-ticket every time the train stops.'[108]

This is the impossibility of movement in the speed of becoming: *Faster! Faster!* But the ground never moves. Notions of progress, causality, movement in innovation are impossible, just as it is impossible for Achilles to overtake the tortoise in Zeno's paradox: 'that in a race the quickest runner can never overtake the slowest, since the pursuer must first reach the point whence the pursued started, so that the slower must always hold a lead'.[109] Innovation is potentially infinite because of its 'credit' to use, that is, to its infinite 'prior art' or 'precursors'.

In the essay, 'Kafka and His Precursors' Borges explores exactly this impossibility of movement in innovation: 'I once premeditated making a study of Kafka's precursors. At first I had considered him to be as singular as the phoenix of rhetorical praise; after frequenting his pages a bit, I came to think I could recognize his voice, or his practices, in texts from diverse literatures and periods.'[110] Borges notes Zeno's paradox against movement and states 'the moving object and the arrow and Achilles are the first Kafkian characters in literature'.[111] The essay characterizes the phenomenon of meaning and innovation through 'use' itself – all the texts which go before, all stations which return the progress of the train, and,, in turn, repay the debt of innovation: 'If I am not mistaken, the heterogeneous pieces I have enumerated resemble Kafka; if I am not mistaken, not all of them resemble each other. This second fact is the more significant. In each of these texts we find Kafka's idiosyncrasy to a greater or lesser degree, but if Kafka had never written a line, we would not perceive this quality; in other words, it would not exist.'[112]

Carroll is himself alarmed at a very similar phenomenon: 'Perhaps the hardest thing in all literature – at least *I* have found it so: by no voluntary effort can I accomplish it: I have to take it as it comes – is to write anything *original*.'[113] Carroll is addressing the fallacy of the primacy of the original, and affirming the play of simulacra. Further, innovation as such cannot be planned, 'by no voluntary effort can I accomplish it', but 'perhaps the easiest is, when once an original line has

---

108   *Through the Looking-Glass*, 129.

109   Aristotle, *Physics*, Book VI, 9 in *The Complete Works of Aristotle*, J Barnes (ed), rev Oxford trans, Princeton UP, Princeton, 1984, 404.

110   JL Borges, 'Kafka and His Precursors', JE Irby (trans), in *Labyrinths: Selected Stories and Other Writings*, DA Yates & JE Irby (eds), New Directions, New York, 1964: 199–201, at 199.

111   Borges, 'Kafka and His Precursors', 199.

112   Borges, 'Kafka and His Precursors', 201.

113   Preface to *Sylvie and Bruno*, 103.

been struck out, to follow it up, and to write any amount more to the same tune'.[114]
This is precisely the sonorous, rhythmic musicality of familiar production.

You might make a joke on that

> "I don't belong to this railway journey at all – I was in a wood just now – and I
> wish I could get back there!"
>
> "You might make a joke on *that*" said the little voice close to her ear:
> "something about 'you *would* if you could,' you know."
>
> "Don't tease so," said Alice, looking about in vain to see where the voice
> came from; "if you're so anxious to have a joke made, why don't you make
> one yourself?"[115]

It is at this point that the joke erupts: 'We are now beginning to understand why
laughter must arise at this point as the articulation of the inarticulable, as the
presentation of the unrepresentable. By prescribing that no game, especially not
that of prescription, should dominate the others, one is doing exactly what it is
simultaneously claimed is being avoided: one is dominating the other games in
order to protect them from domination.'[116] *I don't belong to this railway journey
at all.*

With the user's admission into the game, there is also a necessary dissatisfaction,
the unacceptable nature of acceptance, the inadmissible evidence of admission:
'The laughter that arises is, perhaps, a witness to the uneasiness of the former
player, who has turned referee, but is not yet entirely comfortable with his new
role. For as referee, he must ask himself if, or in what way, he is still in the game.
Does the referee play? Or does he only judge? And what if judgment is also a kind
of game? If so, how should it be played?'[117]

Make the joke yourself.

What's the use of their having names?

> "I don't *rejoice* in insects at all," Alice explained, "because I'm rather afraid of
> them – at least the large kinds. But I can tell you the names of some of them."
>
> "Of course they answer to their names?" the Gnat remarked carelessly.
>
> "I never knew them do it."
>
> "What's the use of their having names," the Gnat said, "if they won't answer
> to them?"

---

114   Preface to *Sylvie and Bruno*, 103.

115   *Through the Looking-Glass*, 129.

116   S Weber, 'Literature – Just Making It', B Massumi (trans), Afterword in J-F
Lyotard & J-L Thébaud, *Just Gaming*, W Godzich (trans), U of Minnesota P, Minneapolis,
[1979]/1985: 101–20, at 105.

117   Weber, 'Literature – Just Making It', 105.

"No use to *them*," said Alice, "but it's useful to the people that name them, I suppose. If not, why do things have names at all?"[118]

Names are relevant only in the course of the language-game, only in use: 'But what does it mean to say that we cannot define (that is describe) these elements, but only name them?'[119] In other words: 'Naming is so far not a move in the language-game – any more than putting a piece in its place on the board is a move in chess. We may say: *nothing* has so far been done, when a thing has been named. It has not even *got* a name except in the language-game.'[120]

It is therefore not the patent, the trade mark, the design, the copyright itself that is of value, but rather the object to which each is applied and its use in the language-game: 'this gives the object a role in our language-game; it is now a *means* of representation.'[121] Names do not prove existence of such a thing as copyright, a patent, a trade mark, a design; names are simply something against which the real may be compared. And indeed proper names, as it is clear, are simply untranslatable: 'And to say "If it did not *exist*, it could have no name" is to say as much and as little as: if this thing did not exist, we could not use it in our language-game. – What looks as if it *had* to exist, is part of the language. It is a paradigm in our language-game; something with which comparison is made. And this may be an important observation; but it is none the less an observation concerning our language-game – our method of representation.'[122]

> After this, Alice was silent for a minute or two, pondering. The Gnat amused itself meanwhile by humming round and round her head: at last it settled again and remarked, "I suppose you don't want to lose your name?"
>
> "No, indeed," Alice said, a little anxiously.
>
> "And yet I don't know," the Gnat went on in a careless tone: "only think how convenient it would be if you could manage to go home without it! For instance, if the governess wanted to call you to your lessons, she would call out, 'Come here –,' and there she would have to leave off, because there wouldn't be any name for her to call, and of course you wouldn't have to go, you know."
>
> "That would never do, I'm sure," said Alice: "the governess would never think of excusing me lessons for that. If she couldn't remember my name, she'd call me 'Miss!' as the servants do."
>
> "Well, if she said 'Miss,' and didn't say anything more," the Gnat remarked, "of course you'd miss your lessons. That's a joke. I wish *you* had made it."[123]

---

118   *Through the Looking-Glass*, 130.
119   Wittgenstein, *Philosophical Investigations*, §49.
120   Wittgenstein, *Philosophical Investigations*, §49.
121   Wittgenstein, *Philosophical Investigations*, §50.
122   Wittgenstein, *Philosophical Investigations*, §50.
123   *Through the Looking-Glass*, 131–32.

Innovation is always missed, just out of reach; always altogether later, and yet always already past. At the outset, the narrative conventions of innovation were considered; that is, its linear, narrativized and thus monological progression. The conventional representation of innovation (utilizing the intellectual property system) censors the dialectic, 'forgets' the user, in order to pronounce an absolute value of innovation. However, this absolute value is not irrefutable; in other words, the system must account for itself, for its own catalogue, and for its own perspective which is at once inside and outside the system. The user must 'miss' the lesson.

> "This must be the wood," she said thoughtfully to herself, "where things have no names. I wonder what'll become of *my* name when I go in? I shouldn't like to lose it at all – because they'd have to give me another, and it would be almost certain to be an ugly one. But then the fun would be, trying to find the creature that had got my old name! That's just like the advertisements, you know, when people lose dogs – 'answers to the name of "Dash": had on a brass collar' – just fancy calling everything you met 'Alice,' till one of them answered! Only they wouldn't answer at all, if they were wise."[124]

Not only must the user enter the game, but also the basic permutations must be interrogated: *then the fun would be, trying to find the creature that had got my old name!* That is, while it is clear that the relationship between innovation and use belies the chronological, economic structure conventionally attributed to it, nevertheless it is necessary not merely to acknowledge this but to experience it. This poses particular challenges to the notion of the signifying product (the object of use) and the affect (the meaning through use). The product becomes little more than an index of the crucial social and familiar production of affect. *Money literally buys nothing.* However, unless the system can come to terms with the affective labour and value of experience, to the commodity it remains enslaved. *Know your place.*

The wisdom of the crowd of the wood, the forest, is not its depth, but its foliage. Even in Alice's fall at the very beginning of her journey, her quest to get to the bottom of things, it was the old dry leaves of the surface that marked the advent of meaning: 'its real density, that which makes it a forest, is its foliage ... it is the foliage which shuts out the light and throws a universal shadow.'[125] Everything is at the surface, in sensation, in use: 'The problems are solved, not by giving new information, but by arranging what we have always known.'[126] This inquiry has made its way from the pretence of depth and the original, to the interiority of surfaces, the profundity of appearances, to intellection and promise that

---

124   *Through the Looking-Glass*, 132–33.

125   Canetti, *Crowds and Power*, 84.

126   Wittgenstein, *Philosophical Investigations*, §109.

'A perspicuous representation produces just that understanding which consists in "seeing connexions".'[127]

It is in the wood, in the crowd, that Alice literally loses her name, her 'own' self and everything to play for is at the surface, in appearances, in simulacra. What if everyone responded to, answered to Alice? To Dash? Intriguingly, Dash implies not just the mundanity of a common name. It is in fact the potential of the unnamed: ' – ' insert new name here. Dash is both speed and space, marking territory, making connections, holding its own. Dash is the quintessential username – inscribe here, name through use. The user answers to the name of everything, and nothing, *if they were wise.*

Here then! Here then!

> Just then a Fawn came wandering by: it looked at Alice with its large gentle eyes, but didn't seem at all frightened. "Here then! Here then!" Alice said, as she held out her hand and tried to stroke it; but it only started back a little, and then stood looking at her again.[128]

From the *now then* to the *here then*, innovation has a time and a place, the *now* and the *here*, always to come, always *then*. But the here is territory, not place as such. While the grand survey Alice undertook transformed into an interaction with activity instead, similarly the place of innovation is discourse, language, use. It does not matter where the user is, as long as the user is here: 'One opens the circle not on the side where the old forces of chaos press against it but in another region, one created by the circle itself.'[129] All at once the user's name is not just forgotten, it is substantially other: '"Then it really *has* happened, after all! And now, who am I?"'[130] The user is translated into the smooth space of innovation in which use is both the moment and momentum of the product: '"I" is not the name of a person, nor "here" of a place, and "this" is not a name. But they are connected with names. Names are explained by means of them.'[131] Thus, Alice is determined to remember her name, but can explain it only through recognition and dialogue with the Fawn:

> "What do you call yourself?" the Fawn said at last. Such a soft sweet voice it had!
>
> "I wish I knew!" thought poor Alice. She answered, rather sadly, "Nothing, just now.""Thank again," it said: "that won't do."
>
> Alice thought, but nothing came of it. "Please, would you tell me what *you* call yourself?" she said timidly. "I think that might help a little."

---

127   Wittgenstein, *Philosophical Investigations*, §122.
128   *Through the Looking-Glass*, 133.
129   Deleuze & Guattari, *Thousand Plateaus*, 311.
130   *Through the Looking-Glass*, 133.
131   Wittgenstein, *Philosophical Investigations*, §410.

> "I'll tell you, if you'll come a little further on," the Fawn said. "I can't remember here."[132]

In the here and now of innovation, of the social, of the digital, names cannot be remembered other than through the dialectic, through dialogue. The construction of subjectivity is thus always in respect of use. At the moment of classification, when the taxonomic structure is applied, the Fawn and Alice can no longer be fellow travellers.

> "I'm a Fawn!" it cried out in a voice of delight. "And, dear me, you're a human child!" A sudden look of alarm came into its beautiful brown eyes, and in another moment it had darted away at full speed.
>
> Alice stood looking after it, almost ready to cry with vexation at having lost her dear little fellow-traveller so suddenly. "However, I know my name now," she said: "that's *some* comfort."[133]

And as sure as nonsense itself: 'feeling sure that they must be...'[134]

Tweedledum and Tweedledee

You ought to pay

> They stood so still that she quite forgot they were alive, and she was just looking round to see if the word "TWEEDLE" was written at the back of each collar, when she was startled by a voice coming from the one marked "DUM."
>
> "If you think we're wax-works," he said, "you ought to pay, you know. Wax-works weren't made to be looking at for nothing. Nohow!"
>
> "Contrariwise," added the one marked "DEE," "if you think we're alive, you ought to speak."[135]

The 'DUM' waxworks command a payment, but if alive, then there is the obligation to speak, to acquire meaning through use of language and the rules in the language-game. In other words, the 'equivalence' of the commodity (the waxwork) and the payment is upset entirely by the 'life' of the subjects. Alice ought to speak, but it is impossible to decipher a relationship based on 'balance' or 'equivalence' as these are not the rules of the language-game. With the admission of the user, the 'ethics' of innovation arise through the capacity for speech, for language, for use. If it is just about products, then pay and be off. But if this is living use, then speak.

---

132   *Through the Looking-Glass*, 133.
133   *Through the Looking-Glass*, 134.
134   *Through the Looking-Glass*, 134.
135   *Through the Looking-Glass*, 135.

That's logic

> "I know what you're thinking about," said Tweedledum: "but it isn't so, nohow."
>
> "Contrariwise," continued Tweedledee, "if it was so, it might be; and if it were so, it would be: but as it isn't, it ain't. That's logic."[136]

At this point, perhaps it was so (indicating the past, a condition that for the purposes of this discussion is fact), but it might also be (in the future). However, if it were so (and here, in the use of the subjective, there is the suggestion that the incorporation of the user, of familiar production, is a condition that is contrary to fact), nevertheless it would be (which is far more likely indeed). Tweedledee makes it very clear: the user is here to stay. This is the logic of innovation, the 'precursors' of innovation, the paradox of movement. Reasoning references what has gone before, that with which we are already familiar, in order to make sense of the new.

In this journey the discussion has considered the way in which intellectual property law does something enticingly similar, but the 'censorship' of the dialogical nevertheless occurs in the presentation of conventional modes of production. In other words, the intellectual property framework has all the flexibility to account for familiar production and the user-innovation, still documenting the new but in ways in which we can reference the past in order to ensure that innovation is at once indicated by the evidence (criteria for protection) and at the same time relevant (in practice, in use). But *as it isn't, it ain't.* Maybe, when pigs have wings. One thing is certain, the time has come.

> "'The time has come,' the Walrus said,
> 'To talk of many things:
> Of shoes – and ships – and sealing-wax –
> Of cabbages – and kings –
> And why the sea is boiling hot –
> And whether pigs have wings.'"[137]

Only one of the things in his dream

> "Are there any lions or tigers about here?" she asked timidly.
>
> "It's only the Red King snoring," said Tweedledee.
>
> "Come and look at him!" the brothers cried, and they each took one of Alice's hands, and led her up to where the King was sleeping.
>
> "Isn't he a *lovely* sight?" said Tweedledum.

---

136   *Through the Looking-Glass*, 135.
137   *Through the Looking-Glass*, 139.

Alice couldn't honestly say that he was. He had a tall red night-cap on, with a tassel, and he was lying crumpled up into a sort of untidy heap, and snoring loud – "fit to snore his head off!" as Tweedledum remarked.

"I'm afraid he'll catch cold with lying on the damp grass," said Alice, who was a very thoughtful little girl.

"He's dreaming now," said Tweedledee: "and what do you think he's dreaming about?"

Alice said, "Nobody can guess that,"

"Why, about *you*!" Tweedledee exclaimed, clapping his hands triumphantly. "And if he left off dreaming about you, where do you suppose you'd be?"

"Where I am now, of course," said Alice.

"Not you!" Tweedledee retorted contemptuously. "You'd be nowhere. Why, you're only a sort of thing in his dream!"

"If that there King was to wake," added Tweedledum, "you'd go out – bang! – just like a candle!"

"I shouldn't!" Alice exclaimed indignantly. "Besides, if *I'm* only a sort of thing in his dream, what are *you*, I should like to know?"

"Ditto," said Tweedledum.

"Ditto, ditto!" cried Tweedledee.

He shouted this so loud that Alice couldn't help saying, "Hush! You'll be waking him, I'm afraid, if you make so much noise."

"Well, it's no use *your* talking about waking him," said Tweedledum, "when you're only one of the things in his dream. You know very well you're not real."[138]

Alice is not real; that is, she has passed from the depths of reality to the incorporeal play of simulacra. She cannot penetrate the wood, only appreciate the foliage.[139] The user cannot penetrate the fallacious depths of the measured conventions of intellectual property, because it is illogical to do so; the 'real density' is at the surface, in experience, in use.

The notion of a landed property in the intellectual is precisely this logic of simulacra. Whether in terms of an exhibition of counterfeits to explain the real,[140] or the very notion of an 'original' in industries of mass reproduction and in the course of technologies of seemingly infinite dissemination. Indeed, what is being reproduced, in an intellectual product for which there is no 'original', if not the simulacrum itself? That is, what is the original of intellectual property if not a model itself? It is a picture of a picture. The 'game' of the 'reality show', the 'experience' of Google glass, the 'aesthetic' of 3-D printing – what is emerging is

---

138   *Through the Looking-Glass*, 140–41.

139   On the forest as crowd symbol, Canetti notes: 'The forest is higher than man. It may be enclosed and overgrown with all kinds of scrub; it may be hard to penetrate, and still harder to traverse, but its real density, that which makes it a forest, is its foliage.' Canetti, *Crowds and Power*, 84.

140   Chapter 9.

a universe of simulation in which there is no referent, no real, but for which the 'real' value comes not from objects but from the familiar. And as seen in the course of the digitations of social and cultural life, individual subjectivities are indeed networked through familiar production, not through relations to objects. In this sense then, everything to play for is in the affect of familiar production – whether that is in terms of publicity, privacy, emotion, sensation. It is object as meaning, not object as thing.

The dream makes representation possible, this is 'the distancing characteristic of the dream'.[141] In other words, the dream creates the necessary distance from the object in order to cast a perspective. This is again the refrain of representation that drums through and through this inquiry, the necessary forgetting or 'game of censure'[142] that allows for a seemingly complete picture of innovation. Dreams are perhaps neither a simple experience, nor an unproblematic and delayed recollection. There is thus no issue of misremembering a dream in that the dream itself is its own judgment: 'Now must I make some assumption about whether people are deceived by their memories or not; whether they really had these images while they slept, or whether it merely seems so to them on waking? And what meaning has this question? – And what interest? … Does this mean that it is nonsense ever to raise the question whether dreams really take place during sleep, or are a memory phenomenon of the awakened? It will turn on the use of the question.'[143] Indeed, 'Is all our Life, then, but a dream …?'[144]

Wittgenstein maintains, 'You can't *construct* clouds. And that is why the future you dream of never comes true.'[145] But arguably this is not a correction of the notion of a distinction between the dream and the real. Rather, it is the emergent creativity *to come*, a correction of the possibility of representation, of judgment. Even if a dream is ultimately unable to be realized, it may nevertheless be useful: 'Perhaps it spurs others to a different sort of work … This so far says *nothing* about the value of these activities. The dreamer's *may* be worthless – & so may the others.'[146] The dream is thus an affective labour, a 'sleep of insomnia',[147] as it were: 'What is intriguing about a dream, is not its *causal* connection with events in my life, etc., but rather this, that it affects us like part of a story, & indeed a very *vivid* part, the rest lying in darkness.'[148] It is not a simple dream, but an intoxicating becoming.

141   J Baudrillard, *Simulations*, P Foss et al. (trans), Semiotext(e), New York, 1983: 147.

142   Baudrillard, *Simulations*, 147.

143   Wittgenstein, *Philosophical Investigations*, 184e.

144   *Sylvie and Bruno*, 101.

145   Wittgenstein, *Culture and Value*, 48e.

146   Wittgenstein, *Culture and Value*, 62e.

147   Deleuze, *Essays Critical and Clinical*, 135.

148   Wittgenstein, *Culture and Value*, 78e.

Wool and Water

To be in the way

> She caught the shawl as she spoke, and looked about for the owner: in another
> moment the White Queen came running wildly through the wood, with both
> arms stretched out wide, as if she were flying, and Alice very civilly went to
> meet her with the shawl.
>
> "I'm very glad I happened to be in the way," Alice said, as she helped her to
> put on her shawl again.[149]

The user literally gets in the way of property, of the shawl, of the Queen's 'own'
self. The pawn is *in* the game. There is an interruption of property by the user and
at the same time a restoration (the shawl is returned to the Queen's shoulders). But
already this connection between judgment (the Queen) and the justice of the user
(Alice) in the negotiation of the proprietary existence (the shawl) has rendered the
Queen insensible. If the user is to be admitted, the doctrine of judgment must be
renegotiated, as it were. The user must manage the conversation herself.

> The White Queen only looked at her in a helpless frightened sort of way, and
> kept repeating something in a whisper to herself that sounded like, "Bread-and-
> butter, bread-and-butter," and Alice felt that if there was to be any conversation
> at all, she must manage it herself.[150]

The Queen's refrain recalls the instantaneity of innovation, first explored with the
Mad Hatter, and *now, here*: *It's always tea-time.*[151]

Addressing

> So she began rather timidly: "Am I addressing the White Queen?"
>
> "Well, yes, if you call that a-dressing," the Queen said. "It isn't *my* notion
> of the thing, at all."
>
> Alice thought it would never do to have an argument at the very beginning
> of their conversation, so she smiled and said, "If your Majesty will only tell me
> the right way to begin, I'll do it as well as I can."
>
> "But I don't want it done at all!" groaned the poor Queen. "I've been
> a-dressing myself for the last two hours."[152]

---

149   *Through the Looking-Glass*, 145.
150   *Through the Looking-Glass*, 145.
151   *Wonderland*, 61.
152   *Through the Looking-Glass*, 145.

Alice addresses the Queen. Through familiar production, the user, quite literally, addresses the producer, enters into dialogue with the producer and forces the producer to recognize the other, the user. In other words, the producer and the user enter into a kinship of familiar production: 'a kinship with the case of *knowing*'.[153] Rather than a moral representation of the innovative world, of originals and copies, of good and bad, what emerges is an ethical kinship of knowledge in the networked play of simulacra, of appearances, of a-dressing.

So at the same time, this exchange with use a-dresses and adorns production, in that through new and emergent forms of production there are at the same time new aesthetics, new fashions, new tastes, new intoxications. As seen in earlier chapters,[154] tasting and sampling are the catalysts for Alice's journey, of the user's becoming. Familiar production is thus not merely a simple progression from producer to user, but rather a recirculation of tribute and provocation, a field of creativity, an array of innovation. This combat between tastes, between producer and consumer, is the vitality generating meaning and value through use.

Living backwards

> Alice couldn't help laughing, as she said, "I don't want you to hire *me* – and I don't care for jam."
>
> "It's very good jam," said the Queen.
>
> "Well, I don't want any *to-day*, at any rate."
>
> "You couldn't have it if you *did* want it," the Queen said. "The rule is, jam to-morrow and jam yesterday – but never jam to-day."
>
> "It *must* come sometimes to 'jam to-day,'" Alice objected.
>
> "No, it can't," said the Queen. "It's jam every *other* day: to-day isn't any *other* day, you know."
>
> "I don't understand you," said Alice. "It's dreadfully confusing!"
>
> "That's the effect of living backwards," the Queen said kindly: "it always makes one a little giddy at first –"
>
> "Living backwards!" Alice repeated in great astonishment. "I never heard of such a thing!"
>
> "– but there's one great advantage in it, that one's memory works both ways."
>
> "I'm sure *mine* only works one way," Alice remarked. "I can't remember things before they happen."
>
> "It's a poor sort of memory that only works backwards," the Queen remarked.[155]

---

153   Wittgenstein, *Philosophical Investigations*, 224e.

154   See in particular Chapter 4.

155   *Through the Looking-Glass*, 146.

Living backwards always makes one a little giddy at first, giddy with the seductiveness of chance,[156] the instantaneity of becoming, the time of innovation – the *now, then*: 'the indefinite time of the event, the floating line that knows only speeds and continually divides that which transpires into an already-there that is at the same time not-yet-here, a simultaneous too-late and too-early, a something that is both going to happen and has just happened'.[157] *It's always tea-time.*

The doctrine of judgment, and its conventional representation of innovation through causal intellectual property frameworks, applies what are presumed to be pre-existing criteria that not only expect and deem that progress works in one direction (and that memory and the record necessarily 'only works backwards'), but also render the 'digital' and familiar production otherwise and nonsensical within the narrative. However, in familiar production, the justice of 'life' is inscribed on the body, on the surface, through the use and the user. As distinct from the infinite debt of judgment to the credit of the producer, familiar production is personal, physical and finite, where 'obligation, accusation, defense, and verdict all merge together',[158] as indeed they did in the Wonderland Court of Justice.

A memory that 'only works backwards' is a memory that is thus subjugated to a higher 'moral certainty', as it were – that is, to the moral value of representation as ideal. In this way, the punishment is applied before the crime, both in the form of the debt to the producer, incorporated in the object prior to any use, and perhaps more explicitly in business models which may be regarded as punitive or articulated so explicitly on the credit–debt relationship and its enforcement, that they risk disenfranchising users and thus delegitimizing the system. It is not so much that the punishment precedes the crime, but rather, that the debt is always already incurred and infinite.

> "For instance, now," she went on, sticking a large piece of plaster on her finger as she spoke, "there's the King's Messenger. He's in prison now, being punished: and the trial doesn't even begin till next Wednesday: and of course the crime comes last of all."
>
> "Suppose he never commits the crime?" said Alice.

---

156   Bataille, *Guilty*, 72.

157   Deleuze & Guattari, *Thousand Plateaus*, 262.

158   In reading Artaud, Deleuze describes the opposition to the judgment (writing of the book) by justice (a writing of book and life) as 'provoking a veritable inversion of the sign'. He notes further, with respect to Kafka: 'Is this not also the case with Kafka, when to the great book of *The Trial* he opposes the machine of "The Penal Colony" – a writing in bodies that testifies both to an ancient order and to a justice in which obligation, accusation, defense, and verdict all merge together? The system of cruelty expresses the finite relations of the existing body with the forces that affect it, whereas the doctrine of infinite debt determines the relationships of the immortal soul with judgments. The system of cruelty is everywhere opposed to the doctrine of judgment.' Deleuze, *Essays Critical and Clinical*, 128.

"That would be all the better, wouldn't it?" the Queen said, as she bound the plaster round her finger with a bit of ribbon.

Alice felt there was no denying *that*. "Of course it would be all the better," she said: "but it wouldn't be all the better his being punished."

"You're wrong *there*, at any rate," said the Queen: "were *you* ever punished?"

"Only for faults," said Alice.

"And you were all the better for it, I know!" the Queen said triumphantly.

"Yes, but then I *had* done the things I was punished for," said Alice: "that makes all the difference."

"But if you *hadn't* done them," the Queen said, "that would have been better still; better, and better, and better!"[159]

The parity presumed by the punishment and the action is thus not made out, despite the notion of an equivalence between injurious act and the punishment or retribution, that is, 'the contractual relationship between *creditor* and *debtor*'.[160] Nevertheless, for the crime that is always already underway within the world of judgment, the punishment will always fit the crime.

So, if it is to be possible to account for and to maximize familiar production, how might business models and cultural models be revolutionized in order to bring new production into existence? How will it be possible 'to bring into existence and not to judge'?[161] *Let's look all around.*

A little dark shop

"What is it you want to buy?" the Sheep said at last, looking up for a moment from her knitting.

"I don't *quite* know yet," Alice said very gently. "I should like to look all round me first, if I might."

"You may look in front of you, and on both sides, if you like," said the Sheep; "but you can't look *all* round you – unless you've got eyes at the back of your head."[162]

In the process of consumption, the user has transformed from a passive recipient or audience for new products, to an autonomous decision-maker with respect to the way in which innovation is circulated in new products. The user, as it were, is liberated from the process of consumption as habit, custom, direction, and is now free not only to exercise decisions but also to produce and innovate

---

159   *Through the Looking-Glass*, 146–47.

160   Nietzsche, *On the Genealogy of Morals*, 63.

161   Deleuze, *Essays Critical and Clinical*, 135.

162   *Through the Looking-Glass*, 149–50.

through use. The user has entered the language-game and is transforming the rules and conventions.

This relationship between consumption (and a logical economic construction of consumption) and use (as a social and linguistic performance of meaning) is encapsulated by the shopkeeper, the mediator between the product and the consumer, the producer and the user. The 'shopkeeper' is literally, as it were, the common person, the 'middle' class, the conduit of taste and commerce. As such, the shopkeeper is the merchant of the language-game: 'Now think of the following use of language: I send someone shopping. I give him a slip marked "five red apples". He takes the slip to the shopkeeper, who opens the drawer marked "apples"; then he looks up the word "red" in a table and finds a colour sample opposite it; then he says the series of cardinal numbers – I assume that he knows them by heart – up to the word "five" and for each number he takes an apple of the same colour as the sample out of the drawer. – It is in this and similar ways that one operates with words.'[163] In seeing the way in which habits confine the language-game, and the way in which language encompasses actions and words, it becomes clearer the way in which the user may move beyond those habits and assumptions in order to transform the game.

Use has thus emerged not as a simple passive demand (or audience) for new products, but rather, it is apparent as the source of new tastes, new aesthetics, new production. Digital use[164] and familiar production reappropriate use as productive capacity. And so, just as for the Looking-glass shop, emerging markets in familiar production do not create desire through lack (how can one judge a product *to come*?), but through anticipation: *What is it you want to buy?*[165]

> The shop seemed to be full of all manner of curious things – but the oddest part of it all was, that whenever she looked hard at any shelf, to make out exactly what it had on it, that particular shelf was always quite empty: though the others around it were crowded as full as they could hold.[166]

Things flow about so

> "Things flow about so here!" she said at last in a plaintive tone, after she had spent a minute or so in vainly pursuing a large bright thing, that looked sometimes like a doll and sometimes like a work-box, and was always in the shelf next above the one she was looking at. "And this one is the most provoking of all – but I'll

---

163   Wittgenstein, *Philosophical Investigations*, §1.

164   Again, 'digital' is applied here in its broadest, philosophical sense to refer not merely to the technology (simply indexical of a departure from the mechanical context of consumption), but to the cultural and social transformation of use, consumption and production. See further the discussion of 'digital' in 'Use'.

165   *Through the Looking-Glass*, 149.

166   *Through the Looking-Glass*, 150.

tell you what —" she added, as a sudden thought struck her, "I'll follow it up to the very top shelf of all. It'll puzzle it to go through the ceiling, I expect!"

But even this plan failed: the "thing" went through the ceiling as quietly as possible, as if it were quite used to it.[167]

Recalling the discussion, earlier in this chapter, of Zeno's paradox of movement and the problem of representation, innovation is in and of itself always out of reach, it is a representation of a representation, a model of a model, a picture of a picture: indeed, 'If the shopkeeper wanted to investigate each of his apples without any reason, for the sake of being certain about everything, why doesn't he have to investigate the investigation?'[168] And what would that prove?

As distinct from the conventional notions of 'use' in conjunction with purpose and profit,[169] and thus with beneficiaries and debt, through this iterative and interlocutory journey, use has emerged not as a tool of depletion (and thus exhaustion and lack in the context of economic models), but rather as a source of new production. As distinct from waste, use has emerged as creative. In a way the shopkeeper sells potential, the future, the always higher shelf. The shopkeeper sells chance: 'Intellectual curiosity puts chance beyond my reach. I seek it and it escapes, as if I just missed it.'[170] What the shopkeeper sells in terms of objects is immaterial, as it were. Meaning and value are in use, as will soon become explicit in the model of transaction in the Looking-glass shop.[171]

Beliefs and desires do indeed flow about so in the Looking-glass shop in a way that eschews direct economic models of consumption (including conventional economic structures assisted by intellectual property rules). The flows are erratic, creative, and all over the shop. Instead of an organized conjugation of goods, what Alice (and the user) experiences is the relations and kinship between flows in the Looking-glass shop, 'sometimes like a doll and sometimes like a work-box' and 'always in the shelf next above'. The great invention of intellectual property reform will be to connect the different flows of innovation and production,[172] to maintain the norm of proprietary models of intellectual property and its protection, while at the same time connecting familiar production and user innovation, as distinct from conjugating and assimilating familiar production within the dominant model

---

167   *Through the Looking-Glass*, 150.

168   Wittgenstein, *On Certainty*, §459.

169   See the discussion of conventional concepts of use and purpose in 'Use'.

170   Bataille, *Guilty*, 80.

171   See further the discussion of the smooth space of innovation in Chapter 2.

172   Deleuze and Guattari explain: 'Imitation is the propagation of a flow; opposition is binarization, the making binary of flows; invention is a conjugation or connection of different flows. What, according to [Gabriel] Tarde, is a flow? It is belief or desire (the two aspects of every assemblage; a flow is always of belief and of desire).' Deleuze & Guattari, *Thousand Plateaus*, 219.

of intellectual property commercialization and exploitation, and thus simply stopping it in its tracks.[173]

The astounding achievement of capitalism is its ability to regenerate and reform, to rewrite its own limits: 'an immanent system that's constantly overcoming its own limitations, and then coming up against them once more in a broader form, because its fundamental limit is Capital itself ... any society is defined not so much by its contradictions as by its line of flight, it flees all over the place, and it's very interesting to try and follow the lines of flight taking shape at some particular moment or other.'[174] In this respect, familiar production is a kind of 'direction' in capital, namely an invention in new sites, times and consequences of production.[175] The user has occupied production.

The digital war machine, the assemblage that effectuates familiar production, ensures innovation is always already 'in the shelf next above', propelled with 'an incredible velocity, a catapulting force',[176] right through the ceiling of the 'conquering State'.[177] Forming the smooth space of innovation,[178] the digital war machine upsets the economic conventions and equilibrium of conventional modes of production, countering the hierarchical, classificatory, organizing stratification of innovation: '[the war machine] exists in an industrial innovation as well as in a technological invention, in a commercial circuit as well as in a religious creation, in all flows and current that only secondarily allow themselves to be appropriated by the State. It is in terms not of independence, but of coexistence and competition *in a perpetual field of interaction*, that we must conceive of exteriority and interiority, war machines of metamorphosis and State apparatuses of identity, bands and

---

173   Deleuze and Guattari explain the difference between connection and conjugation of flows: '"Connection" indicates the way in which decoded and deterritorialized flows boost one another, accelerate their shared escape, and augment or stoke their quanta; the "conjugation" of these same flows, on the other hand, indicates their relative stoppage, like a point of accumulation that plugs or seals the lines of flight, performs a general reterritorialization, and brings the flows under the dominance of a single flow capable of overcoding them.' Deleuze & Guattari, *Thousand Plateaus*, 220.

174   Deleuze, *Negotiations*, 171.

175   Deleuze notes three directions in capital: (1) lines of flight; (2) minorities; and (3) 'war machines'. He explains, 'a third direction, which amounts to finding a characterization of "war machines" that's nothing to do with war but to do with a particular way of occupying, taking up, space-time, or inventing new space-times: revolutionary movements ... but artistic movements too, are war-machines in this sense.' Deleuze, *Negotiations*, 172. The war machine was introduced in Chapter 2 in the discussion of smooth space.

176   Deleuze & Guattari, *Thousand Plateaus*, 356.

177   Deleuze and Guattari ask: 'Could it be that it is at the moment the war machine ceases to exist, conquered by the State, that it displays to the utmost its irreducibility, that it scatters into thinking, loving, dying, or creating machines that have at their disposal vital or revolutionary powers capable of challenging the conquering State?' Deleuze & Guattari, *Thousand Plateaus*, 356.

178   See the discussion in Chapter 2.

kingdoms, megamachines and empires.'[179] This is precisely the interaction and familiarity that propels familiar production, that is, its coadaptation of meaning.[180]

The prettiest are always farther!

> "Oh, please! There are some scented rushes!" Alice cried in a sudden transport of delight. "There really are – and *such* beauties!"
>
> "You needn't say 'please' to *me* about 'em," the Sheep said, without looking up from her knitting: "I didn't put 'em there, and I'm not going to take 'em away."
>
> "No, but I meant – please, may we wait and pick some?" Alice pleaded. "If you don't mind stopping the boat for a minute."
>
> "How am *I* to stop it?" said the Sheep. "If you leave off rowing, it'll stop of itself."[181]

Alice is indeed adrift uponthe smooth space of innovation, the procession of the river, but a procession not so much under control, as it is in control of itself:[182] the 'rushes of breath and cries'[183] that effectuate the flows of creativity, the productions within productions. Her journey has all the cataclysmic power of effect; the dashing, the rushing, faster and faster.[184] Everything does flow about so here.

> "I only hope the boat won't tipple over!" she said to herself. "Oh, *what* a lovely one! Only I couldn't quite reach it." And it certainly *did* seem a little provoking ("almost as if it happened on purpose," she thought) that, though she managed

---

179   Deleuze & Guattari, *Thousand Plateaus*, 360–61.

180   See discussion in Chapter 4.

181   *Through the Looking-Glass*, 151–52.

182   Compare the controlled direction of the river considered by Canetti as a limited crowd symbol: 'It is the symbol of a movement which is still under control, before the eruption and the discharge; it contains the threat of these rather than their actuality.' Canetti, *Crowds and Power*, 84.

183   Deleuze & Guattari, *Anti-Oedipus*, 243.

184   Deleuze and Guattari identify a similar rhythm of the affect in the literature of Heinrich von Kleist: 'Why is it, then, that the most uncanny modernity lies with him? It is because the elements of his work are secrecy, speed, and affect. And in Kleist the secret is no longer a content held within a form of interiority; rather, it becomes a form, identified with the form of exteriority that is always external to itself.' Thus, like Alice, the attention is to the activity at the borders, at the surface, and creative transformation is in the smooth space-time of becoming: 'This element of exteriority – which dominates everything, which Kleist invents in literature, which is the first to invent – will give time a new rhythm: an endless succession of catatonic episodes or fainting spells, and flashes or rushes. Catatonia is "This affect is too strong for me," and a flash is: "The power of this affect sweeps me away."' Deleuze & Guattari, *Thousand Plateaus*, 356.

> to pick plenty of beautiful rushes as the boat glided by, here was always a more
> lovely one that she couldn't reach.[185]

Alice is more interested in the spectators (the rushes) than in the spectacle (her own journey),[186] in what is at the borders, on the banks, at the surface, rather than in the body of water itself. It may be possible to stop and pick the rushes, but the flows are in motion, the boat could tipple over, and there is always another one, more lovely, just out of reach. The sea of innovation can never be fulfilled.

The crucial task, in accounting for familiar production in relation to the organization of intellectual property, is that this does not simply replace a regime with another. The boat may tipple over: 'You didn't take enough precautions. This is the "black hole" phenomenon: a supple line rushes into a black hole from which it will not be able to extricate itself ... We have left behind the shores of rigid segmentarity, but we have entered a regime which is no less organized where each embeds himself in his own black hole and becomes dangerous in that hole, with a self-assurance about his own case, his role and his mission, which is even more disturbing than the certainties of the first line.'[187] True revolution in the way in which intellectual property models are devised and applied in the context of business and commerce must not degenerate into the tedium and indifference of polarized rhetoric and debate, which will achieve nothing but the embedding of convention and the entrenchment of inflexibility. This is not a war; it is combat, it is dialogue.

*

> "The prettiest are always farther!" she said at last, with a sigh at the obstinacy
> of the rushes in growing so far off, as, with flushed cheeks and dripping hair
> and hands, she scrambled back into her place, and began to arrange her new-
> found treasures.[188]

The prettiest things are always farther, the brightest things are always on the shelf next above. Returning to the paradox of innovation, the impossibility of movement, the presentation of the unrepresentable, Achilles may rush but the tortoise will always be ahead. All of a sudden, it is not the rush for the new product, but the steady and meaningful engagement with use.

> What mattered it to her just then that the rushes had begun to fade, and to lose all
> their scent and beauty, from the very moment that she picked them? Even real
> scented rushes, you know, last only a very little while – and these, being dream-

---

185   *Wonderland*, 152.

186   This is a striking reflexion upon the notion of the river as procession, as an opportunity to be seen. See Canetti, *Crowds and Power*, 83–84.

187   Deleuze & Parnet, *Dialogues*, 138-39. See also the discussion in G Deleuze & F Guattari, *On the Line*, J Johnston (trans), Semiotext(e), New York, 1983: 96–98.

188   *Through the Looking-Glass*, 152.

rushes, melted away almost like snow, as they lay in heaps at her feet – but Alice hardly noticed this, there were so many other curious things to think about.[189]

This is less about the obsolescence of the product, once picked, but the quality of the value in use itself, the very process of picking, the sweeping power of the affect, the very rush itself. Indeed, the prettiest is always farther, there is 'an endless succession of catatonic episodes or fainting spells, and flashes or rushes',[190] so move on.

"Crabs, and all sorts of things," said the Sheep: "plenty of choice, only make up your mind. Now, what *do* you want to buy?"[191]

Two are cheaper than one

"I should like to buy an egg, please," she said timidly. "How do you sell them?"
    "Fivepence farthing for one – twopence for two," the Sheep replied.
    "Then two are cheaper than one?" Alice said in a surprised tone, taking out her purse.
    "Only, you *must* eat them both, if you buy two," said the Sheep.
    "Then I'll have *one*, please," said Alice, as she put the money down on the counter.[192]

The doubling of the eggs presents one of the most important business models through the Looking-glass. All at once the play of appearances in innovation and the affirmation of difference in repetition is at work: 'They say we are almost as like as eggs.'[193] However, no two eggs are ever alike: 'The form of the field must be necessarily and in itself filled with individual differences … no two eggs or grains of wheat are identical.'[194]

The transaction in the Looking-glass shop involving the eggs indeed illustrates the transformation of production seen in familiar production and the repetition of difference introduced by the Mock Turtle. That is, value is derived not from increased supply, but rather, from increased use. Thus, although two eggs are cheaper in total, there can be no wastage. In this way it is not price (and access limited by price) that ensures the scarcity or rarity of the commodity (with respect to the wider notions of rarity within aesthetics[195]), but rather the affective relationship to the object; that is, the relevance of the objects, their meaning

---

189   *Through the Looking-Glass*, 152.
190   Deleuze & Guattari, *Thousand Plateaus*, 356.
191   *Through the Looking-Glass*, 153.
192   *Through the Looking-Glass*, 153.
193   Shakespeare, *Winter's Tale*, Act I, Scene ii.
194   Deleuze, *Difference and Repetition*, 252.
195   See the discussion in Chapter 4.

through use. Demand is thus not governed by the objects themselves, but rather, by the surrounding context (the affect).[196] Therefore, the meaning and affective value does not occur through purchase; instead, the significant generation of value is in use. The question for contemporary business models transacting in a digital culture is how to generate product value not in the object (and the frame, as it were) but rather in the field, in the 'service'. The cost is use.

Fundamental to this model is deciphering the way in which artificial scarcity might be delimited not by price, or even by physical access, but rather by the ability of the consumer to use the material. For certain applications of such a business model, the limiting factor will therefore be an element of prestige or expertise that must influence the ability to use the object. The user, and the act of use, thus becomes instrumental not only in identifying what is exchanged in any transaction model, but also in producing value according to that model. This shares aspects of fan-based business models in music, user-led innovation strategies in high technology products and similar. However, even in these models the user remains peripheral in terms of the production of value. The product is a vehicle or tool for value that is created elsewhere.

The importance of non-consumable objects and rights also being part of any transaction is something identified clearly in anthropological considerations of so-called 'primitive' models of commodity relations and markets. Leach notes, 'It is very important to distinguish between consumable and non-consumable materials; it is also very important to appreciate that quite intangible elements such as "rights" and "prestige" form part of the total inventory of "things" exchanged.'[197] There is thus much from traditional cultural models to inform new business models in familiar production and indeed in fulfilling the 'kinship' of familiar production. Rather than fixating, as it were, upon the notion of creating exhaustible, 'perishable wealth' out of intellectual property, it is possible to generate 'imperishable prestige through the medium of spectacular feasting'.[198] In other words, 'gorge' on two eggs and it will cost less than one, and 'The ultimate consumers are in this way the original producers.'[199]

Earlier discussions noted the relationship between taste and scarcity as intrinsic to the production of affective value in the digital environment,[200] where access is not merely an economic question, but a question of aesthetics. Within these models, access restricted on price (and its gluttonous counterpart mobilized by means) will be of limited relevance in the digital environment. Immediately access is limited by expertise (either in the product, aesthetics and taste, or in the technology). This is partly a model based on a value in service and affective

---

196   See the discussion in Chapter 4. See further Bourdieu, *Sociology in Question*, 113.
197   ER Leach, *Rethinking Anthropology*, Berg, Oxford, [1961]/2004: 100.
198   Deleuze & Guattari, *Anti-Oedipus*, 150.
199   Deleuze & Guattari, *Anti-Oedipus*, 150.
200   Chapter 4.

labour, as in models currently proposed and deployed in this way. However, this is not the only or even the primary characteristic of the use of eggs.

What does this have to do with the price of eggs? For the eggs, it is not about guaranteeing a sale at all, but rather, it is about warranting use. In this way, there is indeed no waste and at the same time no surplus value created in the product. Instead, it is a question of surplus value created in social power as distinct from labour power. Surplus value is created beyond use and in familiar production. In a sense, this is use generating production and selling back to the provider. That is, using two eggs costs less than one, where one egg has much more in common with conventional transactions. Two eggs, on the other hand, is a relationship of use.

Thus, the eggs make very clear the difference between use as consumption (as in conventional economic models) and use as production (as in familiar production). In earlier discussion,[201] it was identified the way in which use as consumption is use attached to the creation of necessity or demand (the market), whereby the scarcity (such as that created artificially by intellectual property rights)[202] or lack (through making knowledge products limited, exhaustible and rivalrous) will manipulate consumption so as to create desire or even a sense of exigency with respect to products (old and new). However, in familiar production, there is not necessarily the clear advantage that is ordinarily understood to vest in a particular person in respect of a profit to be derived from property (whether that property is intellectual property or real property, as in land).

Consumption, in the model of the eggs, literally generates use. And use generates need, not through waste (and a fallacious equilibrium[203]), but through obligation (cultural, social, affective). Therefore, demand (the market) is not driven by the product itself, but rather, by the affective relationship to the product and the 'scarcity'. One must be 'expert' enough to use all of the product. The scarcity of the product occurs after and in respect of use, creating a certain expertise or elitism with respect to that use. This is in stark contrast to the creation of scarcity *before* the consumer's access to the product, through an artificial scarcity in respect of that access through price, availability and similar barriers that have become conventional ways in which the intellectual property system is applied. In other words, neither model is in conflict with the intellectual property system. Indeed, reform and revolution is possible with respect to the business as distinct from the 'name' intellectual property. Tell me your name *and* your business.

---

201   See 'Use' in particular.

202   Recall the discussion of the principle of scarcity in 'Use'.

203   See the discussion in 'Use'.

Humpty Dumpty

Tell me your name and your business

> "Don't stand chattering to yourself like that," Humpty Dumpty said, looking at her for the first time, "but tell me your name and your business."
> "My *name* is Alice, but –"
> "It's a stupid name enough!" Humpty Dumpty interrupted impatiently. "What does it mean?"
> "*Must* a name mean something?" Alice asked doubtfully.[204]

Humpty Dumpty is 'himself the Stoic master'.[205] Alice has journeyed 'from the depth of bodies to the surface of words' and has endured 'the troubling experience of ethical ambiguity: the ethics of bodies or the morality of words'.[206] It is not enough merely to point to the bearer of a name. It is not enough any longer simply to appeal to the higher values, to the moral form of representation, or to the higher order of intellectual property. In familiar production, in accounting for the user, it is now clear that it is essential to address the meaning in order to account for value.

Proper names, as noted in previous discussions, present as untranslatable: 'My *name* is Alice.' This is intellectual property. This is copyright. However, a name must mean something. What is your business?

> "Don't you think you'd be safer down on the ground?" Alice went on, not with any idea of making another riddle, but simply in her good-natured anxiety for the queer creature. "That wall is so *very* narrow!"
> "What tremendously easy riddles you ask!" Humpty Dumpty growled out. "Of course I don't think so! Why, if ever I *did* fall off – which there's no chance of – but *if* I did –" Here he pursed up his lips, and looked so solemn and grand that Alice could hardly help laughing. "*If* I did fall," he went on, "*the King has promised me* – ah, you may turn pale, if you like! You didn't think I was going to say that, did you? *The King has promised me – with his own mouth –* to – to –"
> "To send all his horses and all his men" Alice interrupted, rather unwisely.[207]

After all the riddles with no answers, here is a riddle where Alice knows the answer almost before it is asked. The ultimate consumer might indeed be the original producer. The answer is always already here: *now, then.*

---

204   *Through the Looking-Glass*, 155.
205   Deleuze, *Logic of Sense*, 142.
206   Deleuze, *Logic of Sense*, 142. Note also in particular the discussions in Chapters 9 and 10.
207   *Through the Looking-Glass*, 155–56.

When I use a word

> "I don't know what you mean by 'glory,'" Alice said.
>
> Humpty Dumpty smiled contemptuously. "Of course you don't – till I tell you. I meant 'there's a nice knock-down argument for you!'"
>
> "But 'glory' doesn't mean 'a nice knock-down argument,'" Alice objected.
>
> "When *I* use a word," Humpty Dumpty said in rather a scornful tone, "it means just what I choose it to mean – neither more nor less."
>
> "The question is," said Alice, "whether you *can* make words mean different things."
>
> "The question is," said Humpty Dumpty, "which is to be master – that's all."[208]

Once again there is the refrain that has marked the territory of familiar production every rhizomatic root and branch of this journey: 'the meaning of a word is its use in language'.[209] As Humpty Dumpty explains, while words can be made to mean different things, it is in their use that mastery and meaning is achieved: 'To understand a sentence means to understand a language. To understand a language means to be master of a technique.'[210] It is not nonsense, far from it: 'Say what you choose, so long as it does not prevent you from seeing the facts.'[211] So say what you choose: 'To obey a rule, to make a report, to give an order, to play a game of chess, are *customs* (uses, institutions).'[212] But what then is to be made of the language-game of intellectual property? For Humpty Dumpty it is 'to property' a word, to make it one's own, to master the word. But is this in the sense of his 'own' self, or is this in the sense of the game?

> "They've a temper, some of them – particularly verbs, they're the proudest – adjectives you can do anything with, but not verbs – however, *I* can manage the whole lot! Impenetrability! That's what *I* say!"
>
> "Would you tell me, please," said Alice, "what that means?"
>
> "Now you talk like a reasonable child," said Humpty Dumpty, looking very much pleased. "I meant by 'impenetrability' that we've had enough of that subject, and it would be just as well if you'd mention what you mean to do next, as I suppose you don't intend to stop here all the rest of your life."
>
> "That's a great deal to make one word mean," Alice said in a thoughtful tone.
>
> "When I make a word do a lot of work like that," said Humpty Dumpty, "I always pay it extra."[213]

---

208   *Through the Looking-Glass*, 159.
209   Wittgenstein, *Philosophical Investigations*, §43.
210   Wittgenstein, *Philosophical Investigations*, §199.
211   Wittgenstein, *Philosophical Investigations*, §79.
212   Wittgenstein, *Philosophical Investigations*, §199.
213   *Through the Looking-Glass*, 159.

Keep your temper.[214] Becoming is a verb,[215] live it. Once again, use is foregrounded here as the primary value of exchange in language. Humpty Dumpty always pays a word extra when he makes it work harder. But arguably, the currency of this payment, just as for the eggs, is the use itself. Once again, the 'tribute' in terms of extensive value and meaning in the language-game, the mastery of language, is the value for the producer. And in this language-game, the ultimate consumer is all at once the original producer. This resonates with earlier discussions of the archive and the museum as living memory, and the notion of culture and 'to culture'.[216] That is, the museum has emerged as the site of innovation itself. Innovation is dislocated from the firm and incarnated in users.

Fundamentally, and in many respects, reform of business models or models of transactions in intellectual property, as well as any reform in the law itself, will be a question of reform of language, of the language-game. Use itself is conducted within the language-game. Whether the user is rendered peripheral, absent, or late, nevertheless that use is demarcated and influenced by the language-game. Through the Looking-glass, the user is admitted to the game, the game that makes a wager on the use, the talk, the noise of the digital. Reform is thus a question of creating speech for the user and space for the digital.

I see it now

> "You see it's like a portmanteau – there are two meanings packed up into one word."
> "I see it now," Alice remarked thoughtfully.[217]

I see it now: 'One often makes a remark and only later sees *how* true it is.'[218] Use and production are thus brought into conjunction. Notably the portmanteau was originally a case by which to carry around clothing, ushering in notions of aesthetics, bourgeois 'mobility', preference and taste with respect to the creation of 'portmanteau words'. Indeed, a word means what we *choose it to mean*. But it is more than just a contraction of meanings; rather, it is an extension of chance: 'It seems then that the portmanteau word is grounded upon a strict disjunctive synthesis. Far from being confronted with a particular case, we discover the law of the portmanteau word in general, provided that we disengage each time the disjunction which may have been hidden.'[219]

There is also an almost ubiquitous pairing of terms throughout intellectual property discourse and the surrounding debate, terms that together appear to elude

---

214   As the Caterpillar advised Alice, *Wonderland*, 39.
215   Deleuze & Guattari, *Thousand Plateaus*, 239.
216   See further the discussion in Raqs Media Collective, 'To Culture', 99–115.
217   *Through the Looking-Glass*, 160.
218   Wittgenstein, *Notebooks 1914–1916*, 10e.
219   Deleuze, *Logic of Sense*, 46.

the conjunction of meaning within their borders: creative and industries; creative and economy; digital and economy; digital and landscape; digital and environment. Rather than reading such terms as organizing disjunctions into attenuated sense, within each term is the potential for the creative discomfort of inconsistency and the relation through disconcert. Each time, and with each example, the very pairing in itself confounds the otherwise contradictory or incompatible meaning of each individual term to produce a startling telescopic concept. In other words, the content of the conjunction coincides with a wholly new function as an esoteric word.[220] The very term itself, 'intellectual property', must be deciphered in its conjunction of two seemingly different meanings, 'packed up into one word' combining the affect and presence of the intellectual with the rights, form and self-consciousness of property. In other words, it combines both the abstraction and community of the 'intellectual' with the possessiveness of the private in 'property'.

The common purpose of the intellectual, to which the discussion will turn in a moment, is then paired with the peculiarity and ownership of property (*proprietas*). It is thus 'proper' to maintain cultural outputs in connection with oneself, one's 'own' in a relationship of belonging. The disintegration of this link is thus not only 'improper' but also interferes with the property or quality of the thing itself. Property is intrinsically and inherently realized by virtue of its possession, its 'own' self. It would appear that by this logic interfering with possessive connections at the same time not only compromises the 'own' self but also denigrates the property itself.

Therefore, possession is not only understood in a relationship of 'selfhood' with respect to economic models, but also in respect of a mutually constitutive relationship with the person or thing in question.[221] Property (and indeed the 'properties' of intellectual property) in any relationship will be articulated on values of distinction, character and individuality. Property is in and of itself a principle of the genuine. To the common purpose of the intellectual is opposed the fiercely peculiar, restricted and private beast of property, which is not only of its own self but also of its 'own'.

Returning briefly to the untranslatability of proper nouns, the strict 'name' of property thus belongs to the notion of the 'real', the 'true', the 'complete' and the genuine. Property presents as in and of itself normal and exemplary, immediately displacing social and proliferative models of familiar production as not only uncivilized (improper) but also as abnormal (irregular and untrue). As exemplary, property is thus not only distinctive but also admirable, a kind of 'aesthetics' in ethics whereby to be property, to be in property, is to conform with social conduct and norms, to be respectable and respected, to be demanding and in demand.

However, property may also be a mere means to an end, a tool. And it is precisely this meaning that comes through in conjunction with intellectual and in the process of familiar production. The property in familiar production is a kind of

---

220   Deleuze, *Logic of Sense*, 45.

221   See 'Use'.

'share', as it were; a share which is possibly instrumental in that production but not in the object of that production, an investment in production but not distinguished in that production.

The conjunction of 'intellectual' and 'property' then reaffirms the notion of preference and taste, and that which is 'non-consumable' in the exchange of objects facilitated by this tool. The very notion of 'intellectual' is not merely necessary as intention; rather, it is significant in ushering in aesthetics and judgment.[222] The use of 'intellectual' (as distinct from intelligence) is relatively recent; applied as a noun in the nineteenth century to indicate (to name) a particular kind of person or work,[223] the term subsequently became associated with more negative connotations, with intellectualism posited as an alternative to rationalism.[224] It is in this sense that the contemporary notions of intellectual as objective and abstract are derived,[225] together with wider notions today of generalist thinking (as distinct from professional practice).[226] In this sense, 'intellectual' is associated with 'direct producers in the sphere of ideology and culture' as distinct from other examples of mental activity or effort (a kind of labour described by the latter, as distinct from the skill of the former).[227] The intellectual is therefore a producer, not in the sense of the labour power of the factory, but today in the sense of the social power of the intellectual.

Therefore, contrary to the 'own' self of property, the intellectual is the sphere of production, but nevertheless culturally and politically it has previously been somewhat counter to the 'business' of property.[228] Intellectual thus turns aside the meanings of 'property' and the personal with the notion of going beyond the private and into common circulation. In this way, the pairing of 'property' (and the resonances of self, own and private) with 'intellectual' ensures that, immediately, the term 'intellectual property' always already secures the credit of knowledge production and the debt of the user. In respect of the credit–debt relationship of knowledge: 'The argument about the relation of intellectuals to an established

---

222   See further the discussion of intellectual taste in Bourdieu, *Distinction*, 292–94.

223   Williams, *Keywords*, 169.

224   Williams, *Keywords*, 170.

225   Williams, *Keywords*, 170. Note, Williams also suggests that its unfavourable use and relegation as rationalism's alternative meant it also acquired implications of 'significantly, ineffectiveness'. In other words, there is the notion of a binary opposition between the material (practice) and intellectual (theory).

226   'Within universities the distinction is sometimes made between specialists or professionals, with limited interests, and intellectuals, with wider interests.' Williams, *Keywords*, 170.

227   Williams, *Keywords*, 170.

228   See the discussion in P Bourdieu, *Practical Reason: On the Theory of Action*, Polity P, Cambridge [1994]/1998: 24–30. Nevertheless, the university and other sites of intellectual production have more recently become somewhat factory-like, celebrating the qualities of consumerism in their students and services and products from academics.

social system, and therefore about their relative independence or incorporation in such a system, is crucially relevant.'[229] The expressed (as distinct from the expression) moves beyond the 'self' (the proper) and to society (the common). In other words, there is the tension between the proper and propertied 'self' and the objective and objectified 'intellectual'. However, the expression (the denomination of the intellectual production as patent, copyright, trade mark, design) renders it 'private' in the conventional structure of the labour market: 'exclusion, selection, hierarchy'.[230] In its more usual interpretation, then, the portmanteau 'intellectual property' thus ensures that the debt to the self is infinite.

In the digital, the relationship between the common production of the intellectual and the private self of property becomes more problematic.[231] Returning to the cliché of the 'democratization' of the digital environment, it is nevertheless notable the way in which the value of 'intellectual' circulates more prominently.[232] This is not something as simple as the rhetoric of 'democratization', but rather a notably ethical momentum in the digital space: 'It is an *ethical subject*.'[233] Rather than mere passive recipients of a debt which can never be paid, the propertyless 'peasants' are engaging directly with the productive labour of social power through the digital, social media and the like. They are addressing the Queen. This is not simply to celebrate uncritically the proliferation of opinion; rather, it is to understand a transformation in the mode of production itself. This conceptual and cultural transformation in the relationship to the intellectual, to expression and to production, presents a particular challenge to intellectual property in the digital, and a transformation which similarly unsettles conventional production relationships with areas of intellectual property, despite possibly being less disrupted by digital production on the face of things.[234]

Regardless of the potential of the digital, however, the environment itself does not present a simple crisis for intellectual property (or private property) in and of itself, because of the capacity of the mechanisms of capital to overwrite and resume this territory,[235] as it were: 'The conceptual crisis of private property does

---

229   Williams, *Keywords*, 171.

230   Negri, *Politics of Subversion*, 47.

231   Hardt and Negri explain: 'The concept of private property itself, understood as the exclusive right to use a good and dispose of all wealth that derives from the possession of it, becomes increasingly nonsensical in this new situation. There are ever fewer goods that can be possessed and used exclusively in this framework; it is the community that produces and that, while producing, is reproduced and redefined. The foundation of the classic modern conception of private property is thus to a certain extent dissolved in the postmodern mode of production.' Hardt & Negri, *Empire*, 302.

232   Negri, *Politics of Subversion*, 47–48.

233   Negri, *Politics of Subversion*, 48.

234   For instance, the pharmaceutical sector, the mechanical sector, trade marks.

235   For example, as discussed in more detail in earlier chapters, the conventional terms of use for various social media platforms frequently require relinquishing rights in

not become a crisis in practice, and instead the regime of private expropriation has tended to be applied universally.'[236] Nevertheless, 'Private property, despite its juridical power, cannot help becoming an ever more abstract and transcendental concept and thus ever more detached from reality.'[237] Thus, the 'private' of property is emerging as less meaningful within the way in which users access, use, acquire and distribute knowledge.[238] Instead, what is relevant is the 'private' in affect, in the relationship to the product, in a production in kinship, in the collective. Indeed, familiar production, in so doing, lays bare the producer, the observer, the user.

This ethic of community and kinship that is at work in familiar production is thus not necessarily a novel or strange concept to intellectual property. Indeed, arguably it is the principle motivating the concept of farmers' rights. This is the confrontation between the overwriting of intellectual production as property and the right to use that property by virtue of the labour towards the good itself (agricultural production): 'Traditional capitalist property law is based on labor: the one whose labor creates a good has the right to own it.'[239]

In this way, communication itself becomes a form of production. The size of the community, the wisdom of the crowd, becomes part of the authentication of quality and value in the production itself. The example of scientific research and publishing illustrates this very point, where work of large-scale collaborations is valued not in terms of traditional, originary, individualistic authorship in the conventional sense, but in the sense of corroboration through the collaboration network.[240] In this way, the scientific (research) community is itself the site (and

---

intellectual property in any content generated through that platform. The platform thus demands the interest in production in a strikingly similar way to that in the employer/employee relationship.

236   Hardt & Negri, *Empire*, 302.

237   Hardt & Negri, *Empire*, 302.

238   Hardt and Negri note: 'The mode of production of the multitude reappropriates wealth from capital and also constructs a new wealth, articulated with the powers of science and social knowledge through cooperation. Cooperation annuls the title of property. In modernity, private property was often legitimated by labor, but this equation, if it ever really made sense, today tends to be completely destroyed. Private property of the means of production today, in the era of the hegemony of cooperative and immaterial labor, is only a putrid and tyrannical obsolescence. The tools of production tend to be recomposed in collective subjectivity and in the collective intelligence and affect of the workers; entrepreneurship tends to be organized by the cooperation of subjects in general intellect. The organization of the multitude as political subject, as posse, thus begins to appear on the world scene. The multitude is biopolitical self-organization.' Hardt & Negri, *Empire*, 410–11.

239   Hardt & Negri, *Multitude*, 186–87.

240   MEJ Newman, 'The Structure of Scientific Collaboration Networks', *Proceedings of National Academic Sciences USA*, 16 January 2001, Vol 98.2: 404–409.

source) of collaborative production in the form of co-authorship networks.[241] Where work is of a lesser value, authors may 'opt out' of including their name on a publication.[242]

Production is now unspeakably social, an aggregation, as it were, with the social power of the digital leading to unanticipated, unpredictable, disorganized, heterogeneous, amorphous sites and horizons of innovation and change. Intellectual property frameworks and rights nevertheless persist in the digital, but such models are regularly criticized as opportunistic, and in some respects in contradistinction to the productive: 'Production is life itself. It is by virtue of this fact that everything that lives is part of the system of production. The forms of monetary exchange, the forms of command, the defense of property have, as a consequence, become more parasitic … Today, changes in the paradigm of production require their elimination.'[243]

However, as discussed in earlier chapters,[244] it is unclear that the cultures and models of production are appreciably different; at least, not yet. The digital is not a new economy, a new strategy, as such. It is a tool in conceptual change, but only if applied in that way. Indeed, the portmanteau 'digital economy' is itself a troubling confusion of the instrument with the strategy and one in concert with the way in which current legal and customary regimes charting the digital nevertheless internalize the architecture of traditional models of production, despite the 'user-led' rhetoric. All economy is digital. All digital is economicized. As a result, rather than the user deterritorializing modes of production, the risk is that life itself will become inextricably bound in the economic strategy of digital production.

The Lion and the Unicorn

I see nobody on the road

"I see nobody on the road," said Alice.

---

241   MEJ Newman, 'Coauthorship Networks and Patterns of Scientific Collaboration', *Proceedings of National Academic Sciences USA*, April 6 2004, Vol 101 (Suppl 1): 5200–205.

242   D Simone, 'Re-examining Authorship in Copyright Law: Lessons from Large Scientific Collaborations', Conference Paper, *Reinvigorating Legal Thought in Times of Change*, London, 6 June 2013.

243   Negri, *Negri on Negri*, 62–63. Indeed, Negri rejects the conjunction of the private and public when dealing with the production of knowledge (and the immaterial): 'Production cannot be founded on both the circulation of knowledge and the right to limit free access to it.' *Negri on Negri*, 62.

244   In particular, see the discussion of 'social power' and digital debt in Chapter 2.

"I only wish *I* had such eyes," the King remarked in a fretful tone. "To be able to see Nobody! And at that distance too! Why, it's as much as *I* can do to see real people, by this light!"[245]

And so the proper names of intellectual property (copyright, patent, trade mark, design) are taken to possess all the qualities of originality, patentability, registrability and so on. However, as seen wherever one looks in this inquiry, the nature of representation is such that the actual 'somebody' does not exist other than through the footprint, the imprint (the intellectual property object). This is the important paradoxical nature of ideas and information, manifest through intellectual property. But nobody, the user, is getting in the way.

In the course of this adventure it has become clear that the difficulty of well-worn models of use and production is the way in which the product itself is an effect of the market, of use, not only in the digital but also in many respects of contemporary innovation strategy. For the logic of innovation, this means that the product is just collateral or instrumental, but no longer an end in itself. Users are no longer to be understood as comprising a passive audience, they are orchestrating production. Exhausted versions of innovation as linear and calculable processes articulated upon notions of obsolescence, replacement, literally upon uselessness, are becoming less and less relevant and more and more damaging.

"His name is Haigha." (He pronounced it so as to rhyme with "mayor.")

"I love my love with an H," Alice couldn't help beginning, "because he is Happy. I hate him with an H, because he is Hideous. I fed him with – with – with Ham-sandwiches and Hay. His name is Haigha, and he lives –"

"He lives on the Hill," the King remarked simply, without the least idea that he was joining in the game, while Alice was still hesitating for the name of a town beginning with H. "The other Messenger's called Hatta. I must have *two*, you know – to come and go. One to come, and one to go."[246]

And so return, through sound, the March Hare (Haigha) and the Mad Hatter (Hatta), as messengers of the new. It was the instantaneity of the mad tea-party in which Alice experienced the always, already tea-time. And here through the Looking-glass she has been confronted by the speed of immobility in the smooth space of innovation. It simply 'makes sense' that the Hatter and the Hare should join her through the Looking-glass to confront a time of images without movement. This is the conjunction, as it were, of a world of production with a world of use, and the creation of new models of innovation.

Communication, the messenger itself, is thus always already in production, and the time of production is no longer identifiable as a logical, linear narrative.

245   *Through the Looking-Glass*, 165.
246   *Through the Looking-Glass*, 165.

It is unpredictable, it is instant, it is anytime and no time: 'An idea or an image comes to you not only in the office but also in the shower or in your dreams.'[247]

> "Who did you pass on the road?" the King went on, holding out his hand to the Messenger for some more hay.
>
> "Nobody," said the Messenger.
>
> "Quite right," said the King: "this young lady saw him too. So of course Nobody walks slower than you."
>
> "I do my best," the Messenger said in a sullen tone. "I'm sure nobody walks much faster than I do!"
>
> "He can't do that," said the King, "or else he'd have been here first."[248]

But indeed, Nobody was in fact there first. Or so it seemed. In the digital world of simultaneities, of the *now here* and the *here now*, the logic of the system is unravelling. In the digital environment, innovation is a series of simultaneities, a series of 'nows'. 'Invention', broadly speaking,[249] is instead an identity-effect. It appears as the relation between different referenced innovative events. Just as traditional knowledge and other 'ancient' forms of knowledge and incremental innovation are seemingly unthinkable within the conventional norms of innovation (as defined by intellectual property),[250] so too are many models of innovation that have emerged not merely through trading in digital environments and products, but through whole cultural and conceptual shifts in our relationship to knowledge and innovation in a civilization socializing in the digital.

In other words, the 'understanding' of the intellectual property system is slipping, and the environment is literally illogical. And this is the challenge for creators, producers and users. How does this interact with the intellectual property system, and how might the system look as a game without frontiers? It is not merely a question of changing the rules, but rather, an understanding of the relationship between knowledge creation and the tools for transacting in knowledge. Innovation is difference.

Consider for a moment the patent. The patent therefore represents that difference, but is not in itself innovation. The patent is the name by which we

---

247   Hardt & Negri, *Empire*, 111–12.

248   *Through the Looking-Glass*, 166–67.

249   To encompass not only the inventive activity in patent-based industry but also the 'invention' in other creative industries, including copyright, design and trade mark industries.

250   This is not the same thing as criticizing the intellectual property system for failing to adapt to traditional knowledge. Indeed, I have argued elsewhere that attempts to assimilate traditional knowledge within intellectual property risk destructive results, manipulating the knowledge to fit pre-conceived norms of intellectual property as well as assuming the intellectual property system is the only way in which to identify and exchange knowledge artefacts.

have come to call innovation as invention, but it is not innovation. In other words, we have a name rendering innovation events perspicuous within the intellectual property system; names, as the gnat pointed out earlier, are 'useful to the people that name them, I suppose'.[251] The patent is therefore evidence for innovation, and therefore a tool by which to facilitate the possibility of a market, as a sign-post for a product. However, the patent cannot point to innovation directly – there is 'nobody on the road'.

As seen all along, the product is no longer (nor perhaps has it ever been?) the repository of innovation. The product cannot capture innovation and in new and emerging markets the product is less and less relevant to the digital. Business models based on intellectual property can no longer simply assume the rights through a quantity, a physical good. Rather, they will instead succeed if conceived in relation to quality, that is, with respect to transactions in intellectual property and new economic strategies for exchanges in knowledge. And for this, the relevant quality is use. Innovation is not about static goods, but about the use of changes as identified and named in the system. Innovation is a field of creative transformation; it is not something to which one can point as co-existing with other 'objects' of innovation. As distinct from the intellectual property system, which renders innovation as discrete and sequential events, *fighting for the crown*,[252] innovation has become more and more meaningful through the Looking-glass as a simultaneous 'relation', a continuous kinship of familiar production, a relation between the objects of intellectual property, those moments of invention.

> "Who are at it again?" she ventured to ask.
> "Why, the Lion and the Unicorn, of course," said the King.
> "Fighting for the crown?"
> "Yes to be sure," said the King: "and the best of the joke is, that it's *my* crown all the while! Let's run and see them."[253]

In many ways, the Lion and the Unicorn consolidate the disjunctive synthesis experienced in the portmanteau, in the problem of 'intellectual property' itself: 'What is the relation between trying to solve it and solving it?'[254] It is almost intolerable to suggest this can be found: 'The very point of this discussion is to see the great difference.'[255] How is one to recognize a unicorn? And, in the same sense, it is safe to say, 'If a lion could talk, we could not understand him'.[256]

---

251 *Through the Looking-Glass*, 130.
252 *Through the Looking-Glass*, 167.
253 *Through the Looking-Glass*, 167.
254 Wittgenstein, *Lectures on the Foundations of Mathematics*, 64.
255 Wittgenstein, *Lectures on the Foundations of Mathematics*, 64.
256 Wittgenstein, *Philosophical Investigations*, 223e.

Is that a bargain?

> "Well, now that we *have* seen each other," said the Unicorn, "if you'll believe in me, I'll believe in you. Is that a bargain?"[257]

To an extent, therefore, there must be a leap of faith, a wager in the legitimacy bargain. In this way, the 'social contract' or 'balance' frequently proclaimed in intellectual property becomes one based on legitimacy with respect to a mutually constitutive belief and identity, as distinct from transferring all rights to a higher value, to the sovereign. At the same time, as in the dream itself, it can never be certain that one believes other than through language.

In other words, belief is not possible other than within the language-game: 'I've found a unicorn, not that a unicorn is analogous to this [picture] … The picture of the unicorn is used to model something after it and so is the picture of the pentagon. But the point is that the picture is in each case used in a completely different way.'[258] And as for the lion: 'What does "seeing the analogy" consist of? … If the lion had always been in the room it couldn't have been found. Suppose everyone had seen the white lion but hadn't realized it was a white lion. He suddenly realizes that this is the picture of that. But what does it come to, to say that he suddenly realizes this? He gives "white lion" a new meaning.'[259] Belief within the language-game is thus also 'The feeling of confidence'[260] such that one can play; that is, the trust and the conviction not only in the game, but also in the other players.[261]

Looking-glass cakes

> Alice had seated herself on the bank of a little brook, with the great dish on her knees, and was sawing away diligently with the knife. "It's very provoking!" she said in reply to the Lion (she was getting quite used to being called 'the Monster'). "I've cut off several slices already, but they will always join on again!"
>
> "You don't know how to manage Looking-glass cakes," the Unicorn remarked. "Hand it round first, and cut it afterwards."
>
> This sounded nonsense, but Alice very obediently got up, and carried the dish round, and the cake divided itself into three pieces as she did so. "*Now* cut it up," said the Lion, as she returned to her place with the empty dish.[262]

---

257 *Through the Looking-Glass*, 170.

258 Wittgenstein, *Lectures on the Foundations of Mathematics*, 65–66.

259 Wittgenstein, *Lectures on the Foundations of Mathematics*, 67.

260 Wittgenstein, *Philosophical Investigations*, §579.

261 Wittgenstein notes, 'Believing is a state of mind. It has duration; and that independently of the duration of its expression in a sentence, for example. So it is a kind of disposition of the believing person. This is shewn me in the case of someone else by his behaviour; and by his words.' Wittgenstein, *Philosophical Investigations*, 191e–92e.

262 *Through the Looking-Glass*, 171.

The Looking-glass cake is the quintessential manifestation of familiar production. The cake itself has all the qualities of the ways in which to conduct business in innovation in the digital cultural environment.

The fundamental ethic of the Looking-glass cake is to maximize consumption through the initial distribution and consumption of the cake, not through imposing artificial scarcity or other relevant aesthetic models of capital that have dominated the intellectual property framework. In other words, *hand it round first, and cut it afterwards.* Consumption is clearly not in terms of units but is instead motivated through the proliferation of experience, and thus affective value. Just as for the eggs, the more one uses, the more value one produces. Value is brought about by the proliferation of experience. Again, as for the eggs, the emphasis is on maximizing consumption and use, but in some respects the cake is a quantitative model as distinct from the qualitative model of the eggs.

In the Looking-glass cake, the goal is to distribute the content freely and widely, with value derived through other means subsequently to distribution. This principle is arguably at work in business models for digital distribution well-established in the media technologies.[263] However, notwithstanding input in modifying games, source code and other areas of innovation at work in these kinds of open models, a revolution on this model extends to the proliferation of experience as innovative in and of itself. Thus, what is of particular interest to the present discussion is the possibility of realizing these principles in a wider field of innovative and creative cultural output.

An example of this principle comes from the model of social media itself, where value is derived from the 'products' of life and participation in the communication itself. This includes not only the straightforward products (such as photographs) but also the indirect value derived from 'side-communications' of data on traffic, choices, user behaviour and so on. As noted in earlier discussions, participation in social media is becoming ubiquitous and obligatory, rendering the 'social' the archive and memory of all life, the memoranda of meaning and use, the memorandum-book of *your* feelings. Notwithstanding the concerns regarding the

---

263   For instance, the objective of dissemination in open source software is arguably accelerated and proliferative innovation upon the original source code. A similar principle is at work in the model of 'freemium' in casual gaming models in mobile technologies, as well as in software, media and web services. In the freemium model (itself a portmanteau word), a lite version of the game of software is distributed for free, creating the market and interest in the game, but additional levels or playing time must be purchased incrementally. Social media platforms like LinkedIn provide a basic free service, with a premium service attracting additional cost (aimed at a commercial customer base): A Levy & G Bensinger, 'LinkedIn joins ESPN, Skype in shifting from free to "freemium"', *Bloomberg*, 18 December 2009. This model has been extremely successful for the gaming industry: J Liu, 'Video games embrace China's freemium model to beat piracy', *BBC News*, 4 January 2013; J Brustein, 'How freemium products use our brains against us', *Bloomberg*, 11 July 2013. See further the discussion of freemium in C Anderson, *Free: The Future of a Radical Price*, Random House Business Books, London, 2009: 26–27.

reterritorialization of familiar production through the dominant business models and consistent terms and conditions of social media, these forms of communication have nevertheless thrived not only as a product, but also as a culture, because of the Looking-glass cake principle.

The challenge for industry is to examine and implement ways of maximizing the affective relationship to 'product'. Music has undoubtedly faced transformative challenges not only to its business models of distribution, but also to the very nature of the work itself (in a move away from the album to the single, through the advent of downloads)[264] and to the 'social' community of music. For instance, digital technologies of distribution have been linked to the closure of music stores,[265] once crucial sites for the generation of expertise, exclusivity and proximity to artists. The loss of this dimension of experiential, social life in music is paradoxically (given that these were sites originally for transactions in physical goods) an obstacle to transactions in affect. For the music industry, the variable here is the privity of user (fan) and artist, and the possibility of building upon the value of 'expertise' (as in the model of the eggs) together with the communion of music community and culture (as in the model of the Looking-glass cake). How might fans regenerate in ways more specific to the engine of music? Provide innovative production back to the industry? Become complicit in 'protecting' the work and art of the creators themselves?

> "I say, this isn't fair!" cried the Unicorn, as Alice sat with the knife in her hand, very much puzzled how to begin. "The Monster has given the Lion twice as much as me!"
>
> "She's kept none for herself, anyhow," said the Lion.[266]

In explaining the concepts of the just and justice in a lecture to children, Jean-Luc Nancy qualifies the distribution of cake not according to desert, as it were, but according to harm: 'We know, for example, that it is unfair to divide a cake into unequal parts … Yet you also understand that it can be entirely fair to give a very small piece of cake to someone or, indeed, not to give him or her any cake at all. If a child is diabetic, for instance, it is dangerous for him or her to eat too much cake.'[267] Is a little knowledge a dangerous thing?

---

264   An extensive consideration of this phenomenon is provided in the BBC Four Documentary, *When Albums Ruled the World*, broadcast 14 July 2013. This effect on the art of the album has even extended to associated industries as well, with the BBC's written Album Reviews cancelled in March 2013 due to budget cuts. The editor of Album Reviews is quoted as saying: 'The landscape is changing, not for the better.' See T Ingham, 'BBC Album Reviews to close this month', *Music Week*, 13 March 2013.

265   See the discussion in Chapter 4.

266   *Through the Looking-Glass*, 171.

267   Nancy, *God, Justice, Love, Beauty*, 42.

In other words, surplus value in the intellectual property system is not possible through mere exchange and equivalence of goods, even where increases in price might be possible, but rather through another (and additional) source of value. As seen, this source of value, once wholly dependent on labour-power, is now increasingly realized through social life. Thus, circulating the cake before it is divided (as in social media) increases productive output without changing constant capital, thereby ensuring an increase in surplus value. Social media builds intellectual property directly (through photographs and other outputs) as well as indirectly (through information on traffic, use and similar) in which the creator, however, will never hold the credit.

The concern remains, however, in the way in which such social media models may extend the user's debt, but nevertheless, the principles are crucial to a reconsideration of use as the nexus of value. Immediately, the problem of equivalence in the credit–debt relationship returns to be addressed. Whatever loss one feels at transformations in the way in which intellectual property might be disseminated, distributed, divulged in the digital, is transformed to gain by being part of that same exchange.

> "But it'll be quite easy to come, as soon as I've completed my new invention – for carrying one's-*self*, you know. It wants just a *little* more working out."
>
> "Won't that be very tiring, to carry *yourself*?" Sylvie enquired.
>
> "Well, no, my child. You see, whatever fatigue one incurs by *carrying*, one saves by *being carried*!"[268]

'It's My Own Invention'

I wasn't dreaming, after all

> There was no one to be seen, and her first thought was that she must have been dreaming about the Lion and the Unicorn and those queer Anglo-Saxon Messengers. However, there was the great dish still lying at her feet, on which she had tried to cut the plum-cake, "So I wasn't dreaming, after all," she said to herself, "unless – unless we're all part of the same dream. Only I do hope it's *my* dream, and not the Red King's! I don't like belonging to another person's dream."[269]

To account for the new – to believe in the Lion, the Unicorn, the Messengers, the great Looking-glass cake – that is, to facilitate new models of production, it is necessary to stop dreaming (the realm of judgment and moral certainty) and

---

268   *Sylvie and Bruno*, 140.
269   *Through the Looking-Glass*, 172.

succumb to intoxication (the realm of justice and ethics).[270] To bring new modes of existence into being one must awake from the dream, from judgment: 'Existence and judgment seem to be opposed on five points: *cruelty versus infinite torture, sleep or intoxication versus the dream, vitality versus organization, the will to power versus a will to dominate, combat versus war.*'[271] *So I wasn't dreaming after all.* Judgment must be renounced in order to bring new modes of existence into being, new modes of production, to allow the emergence of familiar production: 'It is not a question of judging other existing beings, but of sensing whether they agree or disagree with us, that is, whether they bring forces to us, or whether they return us to the miseries of war, to the poverty of the dream, to the rigors of organization.'[272]

The name of the song

> "You are sad," the Knight said in an anxious tone: "let me sing you a song to comfort you."
>
> "Is it very long?" Alice asked, for she had heard a good deal of poetry that day.
>
> "It's long," said the Knight, "but it's very, *very* beautiful. Everybody that hears me sing it – either it brings the *tears* into their eyes, or else –"
>
> "Or else what?" said Alice, for the Knight had made a sudden pause.
>
> "Or else it doesn't, you know. The name of the song is called 'Haddocks' Eyes.'"
>
> "Oh, that's the name of the song, is it?" Alice said, trying to feel interested.
>
> "No, you don't understand," the Knight said, looking a little vexed. "That's what the name is *called*. The name really *is* 'The Aged Aged Man.'"
>
> "Then I ought to have said, 'That's what the *song* is called'?" Alice corrected herself.

---

270   In *The Birth of Tragedy*, Nietzsche explains the 'separate art worlds of dream and intoxication, two physiological states which contrast similarly to the Apolline and the Dionysiac' (14) as opposites (visual and non-visual) akin to moral certainty and ethical judgment: 'on the one hand as the world of dream images, whose perfection is not at all dependent on the intellectual accomplishments or artistic culture of the individual; on the other as an ecstatic reality, which again pays no heed to the individual, but even seeks to destroy individuality and redeem it with a mystical sense of unity' (18). He explains that in Apollo 'the unshaken faith in that *principium* and the peaceful stillness of the man caught up in it have found their most sublime expression' (16); on the other hand, in Dionysiac intoxication 'subjectivity becomes a complete forgetting of the self' (17). F Nietzsche, *The Birth of Tragedy: Out of the Spirit of Music*, S Whiteside (trans), M Tanner (ed), Penguin, London, [1872]/1993.

271   Deleuze, *Essays Critical and Clinical*, 134.

272   Deleuze, *Essays Critical and Clinical*, 135.

"No, you oughtn't: that's another thing. The *song* is called 'Ways and Means': but that's only what it's *called*, you know!"

"Well, what *is* the song, then?" said Alice, who was by this time completely bewildered.

"I was coming to that," the Knight said. "The song really *is* 'A-sitting On a Gate': and the tune's my own invention."[273]

This passage recalls what are two fundamental concerns that have been revisited throughout this discussion;, namely, the deference of use to the proper nouns of intellectual property, and the problem of representation. Again, the interlocutory on names and naming conjures a series of nows, of simultaneous and yet disjunctive series: '"You can think now of *this* now of *this* as you look at it, can regard it now as *this* now as *this*, and then you will see it now *this* way, now *this*." – *What* way? There *is* no further qualification.'[274] The song is caught up in an infinite regress of representation and meaning: the name of the song is called, but the name is, but the song is called, but the song really is … and then finally the tune turns out not to be the Knight's invention at all. The genuine article, after all that time, is improper.

"But the tune *isn't* his own invention," she said to herself: "it's 'I give thee all, I can no more.'"[275]

The Eighth Square at last!

"The Eighth Square at last!" she cried, as she bounded over … and threw herself down to rest on a lawn as soft as moss, with little flower-beds dotted all about it here and there. "Oh, how glad I am to get here! And what *is* this on my head?" she exclaimed in a tone of dismay, as she put her hands up to something very heavy, that fitted tight round her head.

"But how *can* it have got there without my knowing it?" she said to herself, as she lifted it off, and set it on her lap to make out what it could possibly be. It was a golden crown.[276]

Alice, the user, the ultimate messenger of production and value, arrives at her destination. The message always arrives at its destination,[277] as its destination

---

273   *Through the Looking-Glass*, 180.

274   Wittgenstein, *Philosophical Investigations*, 200e.

275   *Through the Looking-Glass*, 181.

276   *Through the Looking-Glass*, 184.

277   Žižek notes that 'a letter always arrives at its destination' condenses a family of propositions (in Wittgenstein's sense), including the return of the sender's message from the receiver. S Žižek, *Enjoy Your Symptom! Jacques Lacan in Hollywood and Out*, Routledge, New York, 1992: 12. The user is indeed the original producer.

is always wherever it arrives.[278] The innovation always arrives at its solution or product, because its solution is always whatever is found in its use. Immediately incorporated into this logic is the riddle with no answer with which the Mad Hatter frustrates Alice. It is the joke, which the gnat wishes Alice had made. And indeed it is the Knight's invention, which is always already known. At last Alice is here, and yet she does not know where here is. Alice has arrived at her destination quite by chance, as it happens. But at last, she has arrived.

Queen Alice

A Queen so soon

> "Well, this *is* grand!" said Alice. "I never expected I should be a Queen so soon – and I'll tell you what it is, your Majesty," she went on in a severe tone (she was always rather fond of scolding herself), "it'll never do to loll about on the grass like that! Queens have to be dignified, you know!"[279]

The user becomes Queen. Alice has not only entered the game, but has prevailed.

Speak when you're spoken to!

> Everything was happening so oddly that she didn't feel a bit surprised at finding the Red Queen and the White Queen sitting close to her, one on each side: she would have liked very much to ask them how they came there, but she feared it would not be quite civil. However, there will be no harm, she thought, in asking if the game was over. "Please, would you tell me –" she began, looking timidly at the Red Queen.
> "Speak when you're spoken to!" the Red Queen sharply interrupted her.
> "But if everybody obeyed that rule," said Alice, who was always ready for a little argument, "and if you only spoke when you were spoken to, and the other person always waited for *you* to begin, you see nobody would ever say anything, so that –"
> "Ridiculous!" cried the Queen. "Why, don't you see, child –" here she broke off with a frown, and, after thinking for a minute, suddenly changed the subject of the conversation.[280]

Indeed, perhaps it is ridiculous, and yet the Queen has admonished not that one should speak *only* when one is spoken to, but that one *should* speak when one

---

278   Žižek acknowledges Barbara Johnson for the mantra: 'A letter always arrives at its destination since its destination is wherever it arrives.' Žižek, *Enjoy Your Symptom!*, 10.

279   *Through the Looking-Glass*, 184.

280   *Through the Looking-Glass*, 185.

is spoken to – that is, there must always be dialogue. Use, meaning and indeed production itself are dialogical. Thus, the 'manners' through the Looking-glass are those of dialogue, communication and innovation. There is nothing but speech, nothing but use. There is nothing but the language-game. Innovation, production and intellectual property as a 'remedy' for all the noise, are always already in dialogue.

Even a joke should have some meaning

"I'm sure I didn't mean –" Alice was beginning, but the Red Queen interrupted.

"That's just what I complain of! You *should* have meant! What do you suppose is the use of a child without any meaning? Even a joke should have some meaning – and a child's more important than a joke, I hope. You couldn't deny that, even if you tried with both hands."[281]

The joke is thus instrumental in the kinship of familiar production. The joke is free, an eruption, an intimacy in dialogue. The joke both increases the familiarity and at the same time offers the possibility of deference and custom. Even a joke should have some meaning: 'But has for instance a name which has *never* been used for a tool also got a meaning in the game? – Let us assume that "X" is such a sign and that A gives this sign to B – well, even such signs could be given a place in the language-game, and B might have, say, to answer them too with a shake of the head. (One could imagine this as a sort of joke between them.)'[282] The system of respect and manners (and class), which previously informed Alice's processes of recollection and recognition deployed at the outset of her journey in Wonderland, has now returned in all the glory of disrespect and custom (and kinship) through the Looking-glass and familiar production.

Manners are not taught in lessons

"I didn't know I was to have a party at all," said Alice, "but if there is to be one, I think *I* ought to invite the guests."

"We gave you the opportunity of doing it," the Red Queen remarked: "but I dare say you've not had many lessons in manners yet?"

"Manners are not taught in lessons," said Alice. "Lessons teach you to do sums, and things of that sort."[283]

Thus, the taste and aesthetics of familiar production, the ethics of intellectual property, are taught not through lessons (appealing to a higher value) but through practice, through experience, through experiment, through dialogue, through

---

281   *Through the Looking-Glass*, 185.

282   Wittgenstein, *Philosophical Investigations*, §42.

283   *Through the Looking-Glass*, 186.

use. That is, the distinction and aesthetics of markets in familiar production are generated by the 'affective' relationships to use.[284] The aesthetics of taste and judgment in the digital becomes a process not of selection but of dialogue and participation, of accumulation of expertise and 'manners'; that is, the kinship and coadaptation of familiar production. The 'own' self of familiar production is thus an aesthetic self, where taste is an inalienable quality of 'self', a property of self. In terms of the production of meaning and value through use, the dialogue between producer and consumer, the aesthetic conjunction and disjunction, as it were, will be fundamental.

Its temper would remain

> "Try another Subtraction sum. Take a bone from a dog. What remains?"
>
> Alice considered. "The bone wouldn't remain, of course, if I took it – and the dog wouldn't remain; it would come to bite me – and I'm sure *I* shouldn't remain!"
>
> "Then you think nothing would remain?" said the Red Queen.
>
> "I think that's the answer."
>
> "Wrong, as usual," said the Red Queen; "the dog's temper would remain."
>
> "But I don't see how –"
>
> "Why, look here!" the Red Queen cried. "The dog would lose its temper, wouldn't it?"
>
> "Perhaps it would," Alice replied cautiously.
>
> "Then if the dog went away, its temper would remain!" the Queen exclaimed.[285]

This is the problem of representation that has dogged the discussion every whit, as it were. What remains? The space of representation, of forgetting, of censure remains in order to suggest an infallible and untranslatable authenticity: 'It is through the subtraction of the remainder that reality is founded and gathers strength ... what else?'[286] This is the space of representation, the room for judging,[287] the necessary forgetting, the difference in repetition. This remainder is indeed the creative power of repetition, of difference. And what has been 'forgotten' in the representation of innovation within intellectual property? What is the remainder in copyright? Why, it is the user.

Words of one letter stick together

> "Of course you know your ABC?" said the Red Queen.

---

284 See the more detailed discussion of taste and affective labour in Chapter 4.

285 *Through the Looking-Glass*, 186–87.

286 Baudrillard, *Simulacra and Simulation*, 143.

287 Chapter 11.

"To be sure I do," said Alice.

"So do I," the White Queen whispered. "We'll often say it over together, dear. And I'll tell you a secret – I can read words of one letter! Isn't *that* grand? However, don't be discouraged. You'll come to it in time."[288]

Intellectual property and all its proper names (patent, copyright, trade mark, design) have to date been established in a profound relationship of designation such that there appears to be no difference between the word and the thing: 'Linguistics marks that moment when language takes itself as object.'[289] They have become seemingly untranslatable: 'The object derives its *thickness* from this speech. The word that designates it and that makes it *visible* is at the same time what strips it of its immediate meaning and deepens its meaning.'[290] And so the very dilemma of representation, the fundamental mystery of language, has been revealed through this present dialogue, this interlocutory with wonder. The crucial task is to question such words as unproblematic proper (and propertied) nouns, and to consider what is *not* read in the language of intellectual property. It is not reflection, but introsusception: 'It is not reflection, because no one needs philosophy to reflect on anything. It is thought that philosophy is being given a great deal by being turned into the art of reflection, but actually it loses everything.'[291] This inquiry is not about simply reflection and resemblance. It is about the passage straight through the Looking-glass.

Can you answer useful questions?

Here the Red Queen began again. "Can you answer useful questions?" she said. "How is bread made?"

"I know *that*!" Alice cried eagerly, "You take some flour –"

"Where do you pick the flower?" the White Queen asked. "In a garden, or in the hedges?"

"Well, it isn't *picked* at all," Alice explained: "it's *ground* –"

"How many acres of ground?" said the White Queen. "You mustn't leave out so many things."

"Fan her head!" the Red Queen anxiously interrupted. "She'll be feverish after so much thinking."[292]

The entire process has therefore been one of dialogue, of interlocutories – with the law, with philosophy, with literature, with Alice. The useful question is thus not, 'What is innovation?' It is the question of the inquiry itself.

---

288 *Through the Looking-Glass*, 187.
289 Lyotard, *Discourse, Figure*, 100.
290 Lyotard, *Discourse, Figure*, 82.
291 Deleuze & Guattari, *What is Philosophy?* 6.
292 *Through the Looking-Glass*, 187.

Do you know Languages?

> "Do you know Languages? What's the French for fiddle-de-dee?"
>
> "Fiddle-de-dee's not English," Alice replied gravely.
>
> "Who said it was?" said the Red Queen.
>
> Alice thought she saw a way out of the difficulty this time. "If you'll tell me what language 'fiddle-de-dee' is, I'll tell you the French for it!" she exclaimed triumphantly.
>
> But the Red Queen drew herself up rather stiffly, and said "Queens never make bargains."
>
> "I wish Queens never asked questions," Alice thought to herself.
>
> "Don't let us quarrel," the White Queen said in an anxious tone. "What is the cause of lightning?"
>
> "The cause of lightning," Alice said very decidedly, for she felt quite sure about this, "is the thunder – no, no!" she hastily corrected herself. "I meant the other way."
>
> "It's too late to correct it," said the Red Queen: "when you've once said a thing, that fixes it, and you must take the consequences."[293]

So finally, what emerges is the objective nature of the patent, copyright, trade mark, design, intellectual property – the very parameters of the language-game – as nonsense. That is, they have been sustained in the prevailing debate and discourse as untranslatable. However, 'We do not realize that we *calculate*, operate, with words, and in the course of time translate them sometimes into one picture, sometimes into another. – It is as if one were to believe that a written order for a cow which someone is to hand over to me always had to be accompanied by an image of a cow, if the order was not to lose its meaning.'[294]

A riddle with no answer

> Alice sighed and gave it up. "It's exactly like a riddle with no answer!" she thought.[295]

It is exactly like the cow, the white lion,[296] the unicorn.[297] The search for the answer has shown that the question itself is meaningless:[298] 'For in riddles one has no

293 *Through the Looking-Glass*, 188.
294 Wittgenstein, *Philosophical Investigations*, §449.
295 *Through the Looking-Glass*, 189.
296 Wittgenstein, *Lectures on the Foundations of Mathematics*, 64–65.
297 Wittgenstein, *Lectures on the Foundations of Mathematics*, 65–66.
298 Wittgenstein, *Lectures on the Foundations of Mathematics*, 64.

exact way of working out a solution. One can only say, "I shall know a good solution if I see it."[299]

Which bell? Which door?

> She was standing before an arched doorway, over which were the words QUEEN ALICE in large letters, and on each side of it there was a bell-handle; one marked "Visitors' Bell," and the other "Servants' Bell."
>
> "I'll wait till the song's over," thought Alice, "and then I'll ring the – the – *which* bell must I ring?" she went on, very much puzzled by the names. "I'm not a visitor, and I'm not a servant. There *ought* to be one marked 'Queen,' you know –"
>
> Just then the door opened a little way, and a creature with a long beak put its head out for a moment and said, "No admittance till the week after next!" and shut the door again with a bang.
>
> Alice knocked and rang in vain for a long time, but at last a very old Frog, who was sitting under a tree, got up and hobbled slowly towards her: he was dressed in bright yellow, and had enormous boots on.
>
> "What is it, now?" the Frog said in a deep, hoarse whisper.
>
> Alice turned round, ready to find fault with anybody. "Where's the servant whose business it is to answer the door?" she began.
>
> "Which door?" said the Frog.
>
> Alice almost stamped with irritation at the slow drawl in which he spoke. "*This* door, of course!"
>
> The Frog look at the door with his large dull eyes for a minute: then he went nearer and rubbed it with his thumb, as if he were trying whether the paint would come off; then he looked at Alice.
>
> "To answer the door?" he said. "What's it been asking of?"[300]

The user is now Queen, but where is her bell? Alice must make a decision, must take a chance. This time, Alice, the user, is not interpellated and positioned before the law,[301] but instead, the door is flung open itself. Alice, the user, is invited into play. Access to the user is now literally inscribed within the credit of the game. The ticket is presented. The letter has arrived. Neither a raucous noise[302] nor silence[303] greets Alice. Instead, 'a shrill voice was heard singing' and then 'hundreds of voices joined in the chorus'.[304]

---

299   Wittgenstein, *Lectures on the Foundations of Mathematics*, 84.
300   *Through the Looking-Glass*, 191.
301   See the discussion of the 'door' in Chapter 6.
302   *Wonderland*, 50.
303   Kafka, 'The Silence of the Sirens', 431.
304   *Through the Looking-Glass*, 192.

A lovely riddle

> "For it holds it like glue –
> Holds the lid to the dish, while it lies in the middle:
> Which is easiest to do,
> Un-dish-cover the fish, or dishcover the riddle?"[305]

The covers of language, the passwords of proper names, the deception of representation – all have served to obscure meaning. It is necessary to 'undiscover' the presumptions and assumed nature of the system itself, and to discover the riddle of the new, of the language-game: 'We remain unconscious of the prodigious diversity of all the everyday language-games because the clothing of our language makes everything alike.'[306] Rest assured, it may not be the easiest to do, but 'Something new (spontaneous, "specific") is always a language-game.'[307] In order to discover the motives of the system, one must uncover at the surface.

Just like pigs in a trough!

> "Take a minute to think about it, and then guess," said the Red Queen. "Meanwhile, we'll drink your health – Queen Alice's health!" she screamed at the top of her voice, and all the guests began drinking it directly, and very queerly they managed it: some of them put their glasses upon their heads like extinguishers, and drank all that trickled down their faces – others upset the decanters, and drank the wine as it ran off the edges of the table – and three of them (who looked like kangaroos) scrambled into the dish of roast mutton, and began to lap up the gravy, "just like pigs in a trough!" thought Alice.[308]

Take a minute to think, and then take a chance. Familiar production is all over the place, out of the way, all over the shop, directionless, instantaneous and gluttonous. Play the game right and we will all feast like pigs in a trough.

Something's going to happen!

> "Take care of yourself!" screamed the White Queen, seizing Alice's hair with both her hands. "Something's going to happen!"
>
> And then (as Alice afterwards described it) all sorts of thing happened in a moment. The candles all grew up to the ceiling, looking something like a bed of rushes with fireworks at the top. As to the bottles, they each took a pair of plates, which they hastily fitted on as wings, and so, with forks for legs, went fluttering

---

305   *Through the Looking-Glass*, 194.
306   Wittgenstein, *Philosophical Investigations*, 224e.
307   Wittgenstein, *Philosophical Investigations*, 224e.
308   *Through the Looking-Glass*, 194–95.

about: "and very like birds they look," Alice thought to herself, as well as she could in the dreadful confusion that was beginning.[309]

Something interesting is sure to happen, that is the only stability: up to the ceiling and all about so; rushing like a river; sudden flashes like fireworks; and winged lines of flight. The sweeping power of the affect guarantees that something interesting is sure to happen in the creative transformation of 'an endless succession of catatonic episodes or fainting spells, and flashes or rushes'[310] and the dreadful and fabulous confusion and intoxication of becoming.

As for you

> "And as for *you*," she went on, turning fiercely upon the Red Queen, whom she considered as the cause of all the mischief – but the Queen was no longer at her side – she had suddenly dwindled down to the size of a little doll, and was now on the table, merrily running round and round after her own shawl, which was trailing behind her.[311]

And so here we are, our tail at the beginning, once again chasing our trail. The problem has presented itself through the incidents of this story, the purpose itself deciphered through chance.

Shaking

> She took her off the table as she spoke, and shook her backwards and forwards with all her might.
> The Red Queen made no resistance whatever; only her face grew very small, and her eyes got large and green: and still, as Alice went on shaking her, she kept on growing shorter – and fatter – and softer – and rounder – and – [312]

The subject of familiar production is change, is becoming and that potential is unlimited: becoming bird, becoming kitten, all in an emergence of new assemblages of the power of the affect. Shaking the Red Queen backwards and forwards, Alice brings meaning to life, becoming as distinct from being, use as distinct from used.

---

309   *Through the Looking-Glass*, 195.
310   Deleuze & Guattari, *Thousand Plateaus*, 356.
311   *Through the Looking-Glass*, 196.
312   *Through the Looking-Glass*, 197.

Waking

    – and it really *was* a kitten, after all.[313]

Was it but a dream? Or was the adventure in Wonderland and the journey through the Looking-glass in fact an intoxication?[314] Has the journey of innovation been captured within the world of judgment, that is, within the dream?

    In dreams the doctrine of judgment is established,[315] and 'It is the dream that makes the lots turn (Ezekiel's wheel) and makes the lots pass in procession',[316] just as in the grand procession of the King and Queen of Hearts in Wonderland,[317] or the procession on the river through the Looking-glass.[318] Indeed, 'In the dream, judgments are hurled into the void', as in fact was Alice when she first went down the rabbit-hole, eager to seek the root of all things, attached to the logic and causality of a calculable and responsible future, 'without encountering the resistance of a milieu that would subject them to the exigencies of knowledge or experience'.[319] It is in the very assimilation of the dream, of the new, of an attempt to explain it and reassemble and resemble it, that something is lost, something is forgotten, something is necessarily censored, all in order to accommodate and perpetuate the world of representation and judgment: 'and if someone now shows me that this story was not the right story; that in reality quite a different one underlay it, so that I want to say, disenchanted, "Oh, *that's* how it was?", I have seemingly really been robbed of something.'[320]

    This is why the first question Alice must resolve is whether or not she was dreaming,[321] or whether this has been the intoxication of the eating and drinking of becoming: 'But once we leave the shores of judgment',[322] the shores to which Alice swam at the very start of her journey in Wonderland,[323] 'we also repudiate the dream in favor of an "intoxication," like a high tide sweeping over us.'[324] Upon waking, the retelling of the dream story renders the 'dream' (or the intoxication) 'an idea pregnant with possible implications'.[325]

---

313   *Through the Looking-Glass*, 197.

314   See Nietzsche, *The Birth of Tragedy*, particularly 1 and 2.

315   Nietzsche, *The Birth of Tragedy*, 14.

316   Deleuze, *Essays Critical and Clinical*, 129.

317   *Wonderland*, 66.

318   *Through the Looking-Glass*, 151–53.

319   Deleuze, *Essays Critical and Clinical*, 129.

320   Wittgenstein, *Culture and Value*, 78e–79e.

321   Deleuze explains: 'this is why the question of judgment is first of all knowing whether one is dreaming or not.' Deleuze, *Essays Critical and Clinical*, 129.

322   Deleuze, *Essays Critical and Clinical*, 130.

323   *Wonderland*, 17.

324   Deleuze, *Essays Critical and Clinical*, 130.

325   Wittgenstein, *Culture and Value*, 79e.

Which Dreamed It?

> "Now, Kitty, let's consider who it was that dreamed it all. This is a serious question, my dear, and you should *not* go on licking your paw like that – as if Dinah hadn't washed you this morning! You see, Kitty, it *must* have been either me or the Red King. He was part of my dream, if course – but then I was part of his dream, too! *Was* it the Red King, Kitty? You were his wife, my dear, so you ought to know – Oh, Kitty, *do* help to settle it! I'm sure your paw can wait!" But the provoking kitten only began on the other paw, and pretended it hadn't heard the question.[326]

Which do *you* think it was?

As with the Looking-glass cake, in order to account for the emergence of familiar production not merely as an irritation to conventional modes of production and the construction of value, but rather, as a new mode of existence and value itself, it is necessary to stop dreaming. It is necessary to renounce judgment, as it were, and seek to bring familiar production into meaningful existence and to recall the work *to come*. This is not to deprive the commercial potential of other modes of production, or indeed the commercial potential and value of familiar production itself. Rather, it is to understand how it is to bring familiar production into existence with the tools of intellectual property. Renouncing current business models focused on the war with familiar production is part of the emergence of new sites of value: 'What disturbed us was that in renouncing judgment we had the impression of depriving ourselves of any means of distinguishing between existing beings, between modes of existence, as if everything were now of equal value. But is it not rather judgment that presupposes preexisting criteria (higher values), criteria that preexist for all time (to the infinity of time), so that it can neither apprehend what is new in a existing being, nor even sense the creation of a mode of existence?'[327]

Alice might have been part of the King's dream, the user might be deferred as part of the doctrine of judgment in the dream of intellectual property, but the King was also part of Alice's dream, her intoxication, 'in the insomnia of sleep',[328] the emergence of familiar production as a new mode of existence. This is a sleep without judgment, a sleep of becoming: 'Judgment prevents the emergence of any new mode of existence. For the latter creates itself through its own forces, that is, through the forces it is able to harness, and is valid in and of itself inasmuch as it brings the new combination into existence. Herein, perhaps, lies the secret: to

---

326   *Wonderland*, 198.

327   Deleuze, *Essays Critical and Clinical*, 134–35.

328   Deleuze, *Essays Critical and Clinical*, 135. See further, Deleuze explains: 'The dream is rediscovered, no longer as a dream of sleep or a daydream, but as an insomniac dream. The new dream has become the guardian of insomnia.' Deleuze, *Essays Critical and Clinical*, 130.

bring into existence and not to judge ... What expert judgment, in art, could ever bear on the work to come?'[329]

Every utterance, stutter and stammer is always already in circulation, continuous, repeated and different. Memory is but a dream, but not as mere recollection; rather, to dream, just like Alice, is to undertake the task of difference. *There, now.*

Has that which has been proven impossible in fact been the very thing this inquiry was trying to do, after all? Or is *that* what this inquiry was trying to do?

Innovation.

What do *you* think it was?

---

329 Deleuze, *Essays Critical and Clinical*, 135.

After All

"So there's plenty of choice, after all, you see," said spokesman Hugh in conclusion.[1]

It is not possible to give the last word on these challenges that lie ahead. Indeed, I can but repeat it, throughout and all over.

The journey has been a 'tangled tale' of use and language: 'The concept of "seeing" makes a tangled impression. Well, it is tangled. – I look at the landscape, my gaze ranges over it, I see all sorts of distinct and indistinct movement; *this* impresses itself sharply on me, *that* is quite hazy. After all, how completely ragged what we see can appear!'[2]

This should not be taken lightly, for the journey of use is not mere fantasy and fairy tale, it is the chance and the provocation to consider the reform of the game: '"But the fairy tale only invents what is not the case: it does not talk *nonsense*." – It is not as simple as that. Is it false or nonsensical to say that a pot talks? Have we a clear picture of the circumstances in which we should say of a pot that it talked? (Even a nonsense-poem is not nonsense in the same way as the babbling of a child.)'[3]

It has been necessary to cast the discussion into wonder, as it is the very entanglement that is the question all along: 'This entanglement in our rules is what we want to understand (i.e. get a clear view of).'[4]

So this is a story with no beginning that refuses to conclude:

> "So you've got to the end of our racecourse?" said the Tortoise. "Even though it *does* consist of an infinite series of distances? I thought some wiseacre or other had proved that the thing couldn't be done?"
>
> "It *can* be done," said Achilles. "It *has* been done! *Solvitur ambulando.* You see the distances were constantly *diminishing*: and so –"
>
> "But if they had been constantly *increasing*?" the Tortoise interrupted. "How then?"
>
> "Then I shouldn't be *here*," Achilles modestly replied; "and *you* would have got several times round the world, by this time!"
>
> …
>
> "… Have you entered that in your notebook?"

---

1   *Tangled Tale*, 259.

2   Wittgenstein, *Philosophical Investigations*, 200e.

3   Wittgenstein, *Philosophical Investigations*, §282.

4   Wittgenstein, *Philosophical Investigations*, §125.

"I *have*!" Achilles joyfully exclaimed, as he ran the pencil into its sheath. "And at last we've got to the end of this ideal racecourse! Now you accept *A* and *B* and *C* and *D*, *of course* you accept *Z*."

"Do I?" said the Tortoise innocently. "Let's make that quite clear. I accept *A* and *B* and *C* and *D*. Suppose I *still* refused to accept *Z*?"

"Then Logic would take you by the throat, and *force* you to do it!" Achilles triumphantly replied. "Logic would tell you 'You can't help yourself. Now that you've accepted *A* and *B* and *C* and *D*, you *must* accept *Z*!' So you've no choice, you see."

"Whatever *Logic* is good enough to tell me is worth *writing down*," said the Tortoise. "So enter it in your book, please. We will call it:

(*E*) If *A* and *B* and *C* and *D* are true, *Z* must be true.

Until I've granted *that*, of course, I needn't grant *Z*. So it's quite a *necessary* step, you see?"

"I see," said Achilles; and there was a touch of sadness in his tone.

Here the narrator, having pressing business at the Bank, was obliged to leave the happy pair, and did not again pass the spot until some months afterwards. When he did so, Achilles was still seated on the back of the much-enduring Tortoise, and was writing in his note-book, which appeared to be nearly full. The tortoise was saying "Have you got that last step written down? Unless I've lost count, that makes a thousand and one. There are several millions more to come. And *would* you mind, as a personal favour – considering what a lot of instruction this colloquy of ours will provide for the Logicians of the Nineteenth century – would you mind adopting a pun that my cousin the Mock-Turtle will then make, and allowing yourself to be re-named Taught-Us?"

"As you please!" replied the weary warrior, in the hollow tones of despair, as he buried his face in his hands. "Provided that *you*, for *your* part, will adopt a pun the Mock-Turtle never made, and allow yourself to be re-named A Kill-Ease!"[5]

It seems, after all, 'In the end only chance has the possibility of openness.'[6] *So there's plenty of choice, after all, you see.*

The rest next time. It *is* next time![7]

---

5   *Tortoise to Achilles*, 455–56.
6   Bataille, *On Nietzsche*, 103.
7   *Wonderland*, 2.

What can be taken but never kept,
Gorged but never satisfied,
Consumed but never spent,
Makes one happy by accident,
Fortunate by misadventure, and
Successful without planning?

Bibliography

Scholarly Works and Reports

Abbeele G Van Den, 'Communism, the Proper Name', in Miami Theory Collective (ed), *Community at Loose Ends*, U of Minnesota P, Minneapolis, 1991: 30.

Abell P, 'Sociological Theory: What Has Gone Wrong and How to Put It Right, a View from Britain', in J Hage (ed), *Formal Theory in Sociology: Opportunity or Pitfall*, SUNY P, Albany NY, 1994: 105.

Adorno T, *Minima Moralia: Reflections from Damaged Life*, EFN Jephcott (trans), Verso, London, [1951]/1974.

Adorno T & Horkheimer M, *Towards a New Manifesto*, R Livingstone (trans), Verso, London, [1989]/2011.

Agamben G, *Homo Sacer: Sovereign Power and Bare Life*, D Heller-Roazen (trans), Stanford UP, Stanford, [1995]/1998.

Agamben G, *Profanations*, J Fort (trans), Zone, New York, [2005]/2007.

Anderson C, *Free: The Future of a Radical Price*, Random House Business Books, London, 2009.

Anderson C, *The Long Tail: How Endless Choice is Creating Unlimited Demand*, Random House Business Books, London, 2006.

Anderson ES, *Joseph A Schumpeter: A Theory of Social and Economic Evolution*, Palgrave Macmillan, Basingstoke, 2011.

Aristotle, *Physics*, Book VI, 9 in *The Complete Works of Aristotle*, J Barnes (ed), rev Oxford trans, Princeton UP, Princeton, 1984.

Aristotle, *Poetics*, in *The Complete Works of Aristotle: Volume II*, J Barnes (ed), rev Oxford trans, Princeton UP, Princeton, 1984.

Aust A, *Modern Treaty Law and Practice*, Cambridge UP, Cambridge, 2000.

Baird DG, Gertner RH & Picker RC, *Game Theory and the Law*, Harvard UP, Cambridge MA, 1994.

Bakshi H, Freeman A & Higgs P, *A Dynamic Mapping of the UK's Creative Industries*, NESTA, London, December 2012.

Barthes R, 'The Death of the Author', in R Barthes, *Image, Music, Text*, S Heath (trans), London, Fontana, 1977: 142.

Bataille G, *The Accursed Share: An Essay on General Economy, Volume I, Consumption*, Zone, New York, [1967]/1988.

Bataille G, *The Accursed Share: An Essay on General Economy, Volumes II and III*, R Hurley (trans), Zone, New York, [1976]/1991.

Bataille G, *Erotism: Death and Sensuality*, M Dalwood (trans), City Lights Books, San Francisco, [1957]/1986.

Bataille G, *Guilty*, B Boone (trans), Lapis P, Venice CA, [1961]/1988.Bataille G, *The Impossible*, R Hurley (trans), City Lights Books, San Francisco, [1962]/1991.

Bataille G, *On Nietzsche*, B Boone (trans), S Lotringer (intro), Paragon House, New York, [1945]/1992.

Baudrillard J, *Simulacra and Simulation*, SF Glaser (trans), U of Michigan P, Ann Arbor, [1981]/1994.

Baudrillard J, *Simulations*, P Foss et al. (trans), Semiotext(e), New York, 1983.

Baudrillard J, *Symbolic Exchange and Death*, IH Grant (trans), M Gane (intro), Sage, London, [1976]/1993.

Becker LC, *Property Rights: Philosophic Foundations*, Routledge & Kegan Paul, London, 1977.

Benjamin W, *The Arcades Project*, H Eiland & K McLaughlin (trans), Belknap-Harvard UP, Cambridge MA, [1988]/2002.

Benkler Y, 'Coase's Penguin, or Linux and the Nature of the Firm', 112, *Yale Law Journal* 2001: 369.

Benkler Y, *The Wealth of Networks*, Yale UP, New Haven, 2006.

Bergson H, *Creative Evolution*, A Mitchell (trans), The Modern Library, New York, [1911]/1944.

Blaug M, *Economic Theory in Retrospect*, Cambridge: Cambridge UP, 1983.

Bloch M, *Feudal Society 1: The Growth of Ties of Dependence*, LA Manyon (trans), 2nd ed, Routledge, London, 1962.

Bloch M, *The Historian's Craft*, Manchester UP, Manchester, 1992.

Borges JL, 'The Aleph', A Kerrigan (trans), in JL Borges, *A Personal Anthology*, A Kerrigan (ed), Grove P, New York, [1961]/1967: 138.

Borges JL, 'Kafka and His Precursors', JE Irby (trans), in *Labyrinths: Selected Stories and Other Writings*, DA Yates & JE Irby (eds), New Directions, New York, 1964: 199.

Borges JL, 'The Lottery in Babylon', JM Fein (trans), in JL Borges, *Labyrinths: Selected Stories and Other Writings*, DA Yates & JE Irby (eds), New Directions, New York, [1962]/1964: 30.

Borges JL, 'Of Exactitude in Science', in Borges JL, *A Universal History of Infamy*, NT di Giovanni (trans), Penguin, London, [1954]/1975: 131.

Borges JL, 'Pierre Menard, Author of the *Quixote*', JE Irby (trans), in JL Borges, *Labyrinths: Selected Stories and Other Writings*, DA Yates & JE Irby (eds), New Directions, New York, [1962]/1964: 36.

Bouquet M, 'Thinking and Doing Otherwise: Anthropological Theory in Exhibitionary Practice', in BM Carbonell (ed), *Museum Studies: An Anthology of Contexts*, Blackwell, Malden MA, 2012.Bourdieu P, *Distinction: A Social Critique of the Judgement of Taste*, R Nice (trans), Routledge, London, [1979]/1986.

Bourdieu P, *The Field of Cultural Production: Essays on Art and Literature*, R Johnson (ed), Polity P, Cambridge, 1993: 106.

Bourdieu P, *Language and Symbolic Power*, JB Thompson (ed), G Raymond & M Adamson (trans), Polity P, Cambridge, [1991]/1992.

Bourdieu P, *Practical Reason: On the Theory of Action*, Polity P, Cambridge [1994]/1998.

Bourdieu P, *Sociology in Question*, R Nice (trans), Sage, London, [1984]/1993. Bourdieu P & Darbel A, *The Love of Art: European Art Museums and their Public*, C Beattie & N Merriman (trans), Polity P, Cambridge, [1969]/1997.

Boyle A & Chinkin C, *The Making of International Law*, Oxford UP, Oxford, 2007.

Brace L, *The Politics of Property: Labour, Freedom and Belonging*, Edinburgh UP, Edinburgh, 2004.

Breward C, *Fashion*, Oxford History of Art, Oxford UP, Oxford, 2003.

Bronson P & Merryman A, *Top Dog: The Science of Winning and Losing*, Hatchette, New York, 2013.

Brouwer J & Mulder A, 'Information is Alive', in J Brouwer et al. (eds), *Information is Alive*, V2_/Nai, Rotterdam, 2003: 4.

Bryant WDA & Throsby D, 'Creativity and the Behavior of Artists', in VA Ginsburgh & D Throsby (eds), *Handbook of the Economics of Art and Culture*, North-Holland, Amsterdam, 2006, 507.

Buchanan M, *Forecast: What Physics, Meteorology and the Natural Sciences Can Teach Us About Economics*, Bloomsbury, London, 2013.

Burton A, *Vision & Accident: The Story of the Victoria and Albert Museum*, V&A Publications, London, 1999.

Calamar G & Gallo P, *Record Store Days: From Vinyl to Digital and Back Again*, Sterling, New York, 2012.

Canetti E, *Crowds and Power*, C Stewart (trans), Phoenix P, London, [1960]/2000.

Cerbone D, 'Don't Look but Think: Imaginary Scenarios in Wittgenstein's Later Philosophy', 37 *Inquiry* 1994: 159.

Certeau M de, *Heterologies: Discourse on the Other*, B Massumi (trans), W Godzich (foreword), U of Minnesota P, Minneapolis, [1986]/1993.

Conley VA, 'Of Rhizomes, Smooth Space, War Machines and New Media', in M Poster & D Savat (eds), *Deleuze and New Technology*, Edinburgh UP, Edinburgh, 2009: 32.

Cotter TF, 'Four Questionable Rationales for the Patent Misuse Doctrine', 12(2) *Minnesota Journal of Law, Science and Technology* 2011: 457.

Dasgupta P, 'The Economic Theory of Technology Policy: An Introduction', in P Dasgupta & P Stoneman (eds) *Economic Policy and Technological Performance*, Cambridge UP, Cambridge, 1987: 7.

Dasgupta P & Stiglitz J, 'Industrial Structure and the Nature of Innovative Activity', 90.358 *Economic Journal* 1980: 266.

Deane P & Cole WA, *British Economic Growth: 1688–1959*, 2nd ed, Cambridge UP, Cambridge, [1962]/1969.

Deazley R, *On the Origin of the Right to Copy: Charting the Movement of Copyright Law in Eighteenth-Century Britain (1695–1775)*, Hart, Oxford, 2004.

Deleuze G, *Difference and Repetition*, P Patton (trans), Columbia UP, New York, [1968]/1994.

Deleuze G, *Empiricism and Subjectivity: An Essay on Hume's Theory on Human Nature*, CV Boundas (trans), Columbia UP, New York, [1953]/1991.

Deleuze G, *Essays Critical and Clinical*, DW Smith & MA Greco (trans), Verso, London, [1993]/1998.

Deleuze G, *Foucault*, S Hand (trans and ed), U of Minnesota P, Minneapolis, [1986]/1988.

Deleuze G, *The Logic of Sense*, M Lester & C Stivale (trans), CV Boundas (ed), Columbia UP, New York, [1969]/1990.

Deleuze G, *Negotiations: 1972–1990*, M Joughin (trans), Columbia UP, New York, [1990]/1995.

Deleuze G, *Proust and Signs: The Complete Text*, R Howard (trans), Athlone P, London, [1964]/2000.

Deleuze G, *Spinoza: Practical Philosophy*, R Hurley (trans), City Lights Books, San Francisco, [1981]/1988.Deleuze G, *Two Regimes of Madness: Texts and Interview, 1975–1995*, D Lapoujade (ed), A Hodges & M Taormina (trans), Semiotext(e), New York, 2006.

Deleuze G & Guattari F, *Anti-Oedipus: Capitalism and Schizophrenia*, R Hurley et al. (trans), M Foucault (preface), U of Minnesota P, Minneapolis, [1972]/1983.

Deleuze G & Guattari F, *On the Line*, J Johnston (trans), Semiotext(e), New York, 1983.

Deleuze G & Guattari F, *A Thousand Plateaus: Capitalism and Schizophrenia*, B Massumi (trans), U of Minnesota P, Minneapolis, [1980]/1987.

Deleuze G & Guattari F, *What is Philosophy?* H Tomlinson & G Burchill (trans), Verso, London, [1991]/1994.

Deleuze G & Parnet C, *Dialogues*, H Tomlinson & B Habberjam (trans), Columbia UP, New York, [1977]/1987.

Denny M, 'Framing the Victorians: Photography, Fashion, and Identity', in J Potvin (ed), *The Places and Spaces of Fashion 1800–2007*, Routledge, New York, 2009: 34.

Derrida J, *Acts of Literature*, D Attridge (ed), Routledge, New York, 1992.

Derrida J, *Archive Fever: A Freudian Impression*, E Prenowitz (trans), U of Chicago P, Chicago, [1995]/1996.

Derrida J, 'Force of Law: The "Mystical Foundation of Authority"', in D Cornell et al. (eds), *Deconstruction and the Possibility of Justice*, Routledge, New York, [1990]/1992: 3.

Derrida J, *Given Time: I. Counterfeit Money*, P Kamuf (trans), U of Chicago P, Chicago, 1992.

Derrida J, *Negotiations: Interventions and Interview, 1971–2001*, E Rottenberg (ed, trans), Stanford UP, Stanford, 2002.

Derrida J, *Spectres of Marx: The State of the Debt, The Work of Mourning, and the New International*, P Kamuf (trans), Routledge, New York, [1993]/1994.

Derrida J, *The Truth in Painting*, G Bennington & I MacLeod (trans), U of Chicago P, Chicago, [1978]/1987.

Dickensen D, *Property in the Body: Feminist Perspectives*, Cambridge UP, Cambridge, 2007.

Eechoud M, Hugenholtz PB, Gompel S van, Guibault L & Helberger N, *Harmonizing European Copyright Law: The Challenges of Better Law Making*, Information Law Series 19, Kluwer Law International, Alphen aan den Rijn, 2009.

English B, *A Cultural History of Fashion in the 20th Century*, Berg, Oxford, 2007.

Eno B, in interview with Hans Ulrich Obrist, in HU Obrist, *Interviews: Volume 1*, Charta, Milan/Fonazione Pitti Immagine Discovery, Florence, 2003: 214.

Foucault M, 'Photogenic Painting', in G Deleuze, M Foucault & G Fromanger, *Le Peinture Photogénique*, A Rifkin (intro), S Wilson (ed), Black Dog, London, 1999: 83.

Frankel S & Gervais D, 'Plain Packaging and the Interpretation of the TRIPs Agreement', *Vanderbilt Journal of Transnational Law*, forthcoming.

Freeman A, 'London's Creative Sector: 2004 Update', Greater London Authority, London, 2004.

Freeman C & Soete L, *The Economics of Industrial Innovation*, 3rd ed, MIT P, Cambridge MA, 1997.

Ghosh S, 'The Fable of the Commons: Exclusivity and the Construction of Intellectual Property Markets', 40 *UC Davis Law Review* 2007: 855.

Gibson J, *Community Resources: Intellectual Property, International Trade and Protection of Traditional Knowledge*, Ashgate, Aldershot, 2005.

Gibson J, *Creating Selves: Intellectual Property and the Narration of Culture*, Ashgate, Aldershot, 2006.

Gibson J, *Intellectual Property, Medicine and Health: Current Debates*, Ashgate, Farnham, 2009.

Gibson J, 'The Lay of the Land: The Geography of Traditional Cultural Expression', in CB Graber & M Burri-Nenova (eds), *Intellectual Property and Traditional Cultural Expressions in a Digital Environment*, Edward Elgar, Cheltenham, 2008: 182.

Gide A, *The Journals of André Gide*, J O'Brien (trans), Alfred A Knopf, New York, 1955.

Godelier M, *Rationality and Irrationality in Economics*, B Pearce (trans), Verso, London, [1966]/2012.

Gorz A, *Capitalism, Socialism, Ecology*, C Turner (trans), Verso, London, [1991]/1994.

Gorz A, *Critique of Economic Reason*, G Handyside & C Turner (trans), Verso, London, [1988]/1989.

Habermas J, *Legitimation Crisis*, T McCarthy (trans), Beacon P, Boston, 1976.

Habermas J, *Theory of Communicative Action Volume II: Lifeworld and System, A Critique of Functionalist Reason*, T McCarthy (trans), Polity P, Cambridge, [1981]/1989.

Hantelman D von, 'Affluence and Choice: The Social Significance of the Curatorial', in B von Bismarck et al. (eds), *Cultures of the Curatorial*, Sternberg P, Berlin, 2012: 41.

Hardin G, 'The Tragedy of the Commons', 162 *Science* 1968: 1243.

Hardt M & Negri A, *Empire*, Harvard UP, Cambridge MA, [2000]/2001.

Hardt M & Negri A, *Multitude: War and Democracy in the Age of Empire*, Hamish Hamilton-Penguin, London, 2004.

Hargreaves I, *Digital Opportunity: A Review of Intellectual Property and Growth*, May 2011.

Harris JW, *Property and Justice*, Oxford UP, Oxford, [1996]/2001.

Hobsbawm EJ, *Primitive Rebels: Studies in Archaic Forms of Social Movement in the Nineteenth and Twentieth Centuries*, Manchester UP, Manchester, 1959.

Hooper R & Lynch R, *Copyright Works: Streamlining Copyright Licensing for the Digital Age*, Intellectual Property Office, Cardiff, July 2012.

International Necronautical Society (INS), 'Interim Report on Recessional Aesthetics', in T McCarthy, S Critchley et al., *The Mattering of Matter Documents from the Archive of the International Necronautical Society*, Sternberg P, Berlin, 2012: 238.

Kafka F, 'Advocates', T Stern & J Stern (trans), F Kafka, *The Complete Short Stories*, NN Glatzer (ed), Minerva, London, 1992: 449.

Kafka F, 'Before the Law', W Muir & E Muir (trans), in F Kafka, *The Complete Short Stories*, NH Glatzer (ed), Minerva, London, 1992: 3.

Kafka F, 'The Silence of the Sirens', W Muir & E Muir (trans), in F Kafka, *The Complete Short Stories*, NH Glatzer (ed), Minerva, London, 1992: 430.

Kant I, *The Critique of Judgement*, JC Meredith (trans), Oxford, Clarendon P, [1790]/1952.

Kelley TJ with Littman J, *The Art of Innovation: Lessons in Creativity from IDEO, America's Leading Design Firm*, Currency, New York, 2001.

Kojève A, *Introduction to the Reading of Hegel: Lectures on the Phenomenology of Spirit*, JH Nichols Jr (trans), Cornell UP, Ithaca, [1947]/1980.

Krauss R, *The Optical Unconscious*, MIT P, Cambridge MA, 1993.

Landes M & Posner R, *The Economic Structure of Intellectual Property Law*, Belknap-Harvard UP, Cambridge MA, 2003.

Latour B, *Reassembling the Social: An Introduction to Actor-Network Theory*, Oxford UP, Oxford, 2005.

Law J & Hassard J (eds), *Actor Network Theory and After*, Blackwell, Oxford, [1999]/2006.

Lazzarato M, *The Making of the Indebted Man*, JD Jordan (trans), Semiotext(e), Los Angeles, [2011]/2012.

Leach ER, *Rethinking Anthropology*, Berg, Oxford, [1961]/2004.

Lindegaard S, *The Open Innovation Revolution*, Wiley, Hoboken NJ, 2010.

Lipovetsky G & Manlow V, 'The "Artialization" of Luxury Stores', in J Brand et al. (eds), *Fashion and Imagination: About Clothes and Art*, ArtEZ Press, Arnhem, 2009: 154.

Lyotard J-F, *The Differend: Phrases in Dispute*, G Van Den Abbeele (trans), U of Minnesota P, Minneapolis, [1983]/1988.

Lyotard J-F, *Discourse, Figure*, A Hudek & M Lydon (trans), J Mowitt (intro), U of Minnesota P, Minneapolis, [1971]/2011.

Lyotard J-F, *The Inhuman: Reflections on Time*, G Bennington & R Bowlby (trans), Stanford UP, Stanford, [1988]/1991.

Lyotard J-F, *Peregrinations: Law, Form, Event*, Columbia UP, New York, 1988.

Lyotard J-F, *Political Writings*, B Readings & K Paul (trans), U of Minnesota P, Minneapolis, 1993.Lyotard J-F, *The Postmodern Condition: A Report on Knowledge*, G Bennington & B Massumi (trans), F Jameson (foreword), U of Minnesota P, Minneapolis, [1979]/1984.

Lyotard J-F, 'Representation, Presentation, Unrepresentable', in J-F Lyotard, *The Inhuman: Reflections on Time*, G Bennington & R Bowlby (trans), Stanford UP, Stanford, [1988]/1991: 119.

Lyotard J-F & Thébaud J-L, *Just Gaming*, W Godzich (trans), U of Minnesota P, Minneapolis, [1979]/1985.

Madden M, Lenhart A, Cortesi S, Gasser U, Duggan M, Smith A & Beaton M, *Teens, Social Media, and Privacy*, Pew Research Center, Washington DC, 21 May 2013.

Malinauskaite J, 'Harmonisation of Competition Law in the Context of Globalisation', 21(3) *European Business Law Review* 2010: 369.

Manco T, *Stencil Graffiti*, Thames & Hudson, London, 2002.

Marazzi C, *The Violence of Financial Capitalism*, K Lebedeva & JF McGimsey (trans), new ed, Semiotext(e), Los Angeles, 2011.

Marshall L, *Understanding Copyright Law*, 5th ed, LexisNexis, New Providence NJ, 2010.Marx K, *Capital: A Critique of Political Economy, Volume I*, E Mandel (intro), B Fowkes (trans), Penguin, London, 1976.

Marx K, *Grundrisse: Foundations of the Critique of Political Economy*, M Nicolaus (trans), Penguin, London, 1973.

Marx K, *Writings of the Young Marx on Philosophy and Society*, Hackett, Indianapolis, [1967]/1997.

Marx K & Engels F, *The Communist Manifesto: A Modern Edition*, E Hobsbawm (intro), Verso, London, [1848]/2012.

Massumi B, 'The Archive of Experience', in J Brouwer et al. (eds), *Information is Alive*, V2_/NAI, Rotterdam, 2003: 142.

Massumi B, *A User's Guide to Capitalism and Schizophrenia: Deviations from Deleuze and Guattari*, MIT P, Cambridge MA, 1992.

Miller C & Osborn RN, 'Innovation as a Contested Terrain: Planned Creativity and Innovation Versus Emergent Creativity and Innovation', in MD Mumford et al. (eds), *Multi-Level Issues in Creativity and Innovation*, JAI P, Bingley, 2008: 169.

Moore GE, 'Wittgenstein's Lectures in 1930–33', in L Wittgenstein, *Philosophical Occasions 1912–1951*, JC Klagge & A Nordmann (eds), Hackett, Indianapolis, 1993: 46.

Munzer SR, *A Theory of Property*, Cambridge UP, Cambridge, 1990.

Nancy J-L, 'From the Imperative to Law', in BC Hutchens (ed), *Jean-Luc Nancy: Justice, Legality and World*, Continuum, London, 2012: 11.

Nancy J-L, *God, Justice, Love, Beauty: Four Little Dialogues*, S Clift (trans), Fordham UP, New York, [2009]/2011.

Negri A, *Art & Multitude, Nine Letters on Art, Followed by Metamorphoses: Art, and Immaterial Labour*, E Emery (trans), Polity P, Cambridge, [2009]/2011.

Negri A with Dufourmantelle A, *Negri on Negri*, MB De Bevoise (trans), Routledge, New York, 2004.

Negri A, *The Politics of Subversion: A Manifesto for the Twenty-First Century*, J Newell (trans), Polity P, Cambridge, [1989]/2005.

Negri A, *Reflections on Empire*, E Emery (trans), Polity P, Cambridge, [2003]/2008.

Neumann J von & Morgenstern O, *Theory of Games and Economic Behaviour*, Princeton UP, Princeton, 1983.

Newman MEJ, 'Coauthorship Networks and Patterns of Scientific Collaboration', *Proceedings of National Academic Sciences USA*, April 6 2004, Vol 101 (Suppl 1): 5200.

Newman MEJ, 'The Structure of Scientific Collaboration Networks', *Proceedings of National Academic Sciences USA*, 16 January 2001, Vol 98.2: 404. Nietzsche F, *The Birth of Tragedy: Out of the Spirit of Music*, S Whiteside (trans), M Tanner (ed), Penguin, London, [1872]/1993.

Nietzsche F, *The Gay Science*, W Kaufmann (trans), Vintage-Random House, New York, [1887]/1974.

Nietzsche F, *On the Genealogy of Morals*, W Kaufmann & RJ Hollingdale (trans), [1887], and *Ecce Homo*, W Kaufmann (trans and ed), [1908], Vintage-Random House, New York, [1969]/1989.

Obrist HU, *Interviews: Volume 1*, Charta, Milan/Fonazione Pitti Immagine Discovery, Florence, 2003.

Oddi AS, 'Un-Unified Economic Theories of Patents: The Not-Quite-Holy Grail', 71 *Notre Dame Law Review*, 1996: 267.

OECD *Frascati Manual*, 6th ed, OECD, Paris, 2002.

Orwell G, *Nineteen Eighty-Four*, Secker & Warburg, London, 1949.

Owens N & Cannon Jones A, *A Comparative Study of the British and Italian Textile and Clothing Industries*, DTI Economics Paper No 2, DTI, April 2003.

Patton P, *Deleuze and the Political*, Routledge, Abingdon, 2000.

Peters B, *Innovation and Firm Performance*, Physica-Verlag, Heidelberg, 2008.

Plant A, 'The Economic Theory Concerning Patents for Inventions', 1(1) *Economica* 1934: 30.

Raqs Media Collective, 'To Culture: Curation as an Active Verb', in B von Bismarck et al. (eds), *Cultures of the Curatorial*, Sternberg P, Berlin, 2012.

Ricoeur P, *Memory, History, Forgetting*, K Blamey & D Pellauer (trans), U of Chicago P, Chicago, [2004]/2006.

Ricoeur P, *Time and Narrative: Volume III*, K Blamey & D Pellauer (trans), U of Chicago P, Chicago, [1985]/1988.

Robertson B, 'The South Kensington Museum in Context: An Alternative History', 2(1) *Museum and Society* March 2004: 1.

Rocard M & Meledo C, 'Capitalism, Crisis, and Ethics: An Interview with Michel Rocard', 12 *BC Journal* Spring 2009: 5.

Rothwell R, 'Towards the Fifth-Generation Innovation Process', 11(1) *International Marketing Review* 1994: 7.

Roughton A, Cook T, Spence M & Johnson P, *The Modern Law of Patents*, 2nd ed, LexisNexis, London, 2010.

Ruskin J, *'Unto this Last': Four Essays on the First Principles of Political Economy*, John Wiley & Son, New York, 1879.

Sahlins M, *Stone Age Economics*, Routledge, London, [1972]/2004.

Schroeder J, 'Education: Introduction', in ME Leighton & L Surridge (eds), *The Broadview Anthology of Victorian Prose, 1832–1902*, Broadview P, Ontario, 2012: 149.

Schumpter JA, *Business Cycles: A Theoretical, Historical and Statistical Analysis of the Capitalist Process*, McGraw-Hill, New York, 1939.

Schumpeter JA, *Capitalism, Socialism and Democracy*, R Swedberg (intro), Routledge, London, [1943]/1994.Sherman B & Bently L, *The Making of Modern Intellectual Property Law*, Cambridge UP, Cambridge, 1999.

Simone D, 'Re-examining Authorship in Copyright Law: Lessons from Large Scientific Collaborations', Conference Paper, *Reinvigorating Legal Thought in Times of Change*, London, 6 June 2013.

Simpson MD, *Making Representations: Museums in the Post-Colonial Era*, Routledge, London, 2011.

Smith DW, '"A Life of Pure Immanence": Deleuze's "Critique et Clinique" Project', Introduction to G Deleuze, *Essays Critical and Clinical*, DW Smith and MA Greco (trans), Verso, London, [1993]/1998: xi.

Soros G, *Open Society: Reforming Global Capitalism*, Little, Brown, London, 2000.

Spinoza, B de, *A Theological-Political Treatise, and a Political Treatise*, Cosimo Books, New York, [1883]/2007.

Stiegler B, 'The Indexing of Things', in U Ekman (ed), *Throughout: Art and Culture Emerging with Ubiquitous Computing*, MIT P, Cambridge MA, 2013: 493.

Stiglitz JE, *Making Globalization Work*, Penguin, London, 2006.

Stoneman P, *The Handbook of Economics of Innovation and Technological Change*, Blackwell, Oxford, 1995.

Stoneman P, *Soft Innovation: Economics, Product Aesthetics and the Creative Industries*, Oxford UP, Oxford, 2010.Swedberg R, *Joseph A Schumpeter: His Life and Work*, Polity P, Cambridge, 1991.

Swedberg R, *Principles of Economic Sociology*, Princeton UP, Princeton, 2009.

Swedberg R, *Schumpeter: A Biography*, Princeton UP, Princeton, 1995.

Thomas G & Caulton T, 'Communication Strategies in Interactive Space', in SM Pearce (ed), *Exploring Science in Museums*, Athlone P, London, 1996: 107.

Thompson EP, 'The Moral Economy of the English Crowd in the Eighteenth Century', 50 *Past and Present* February 1971: 76.

Tournier M, *The Mirror of Ideas*, JF Krell (trans), U of Nebraska P, Lincoln, [1994]/1998.

Troy NJ, 'Poiret's Modernism and the Logic of Fashion', in G Riello & P McNeil (eds), *The Fashion History Reader: Global Perspectives*, Routledge, London, 2010: 455.

Tuomi I, *Networks of Innovation: Change and Meaning in the Age of the Internet*, Oxford UP, Oxford, 2002.

Ulnwick A, *What Customers Want: Using Outcome-Driven Innovation to Create Breakthrough Products and Services*, McGraw-Hill, New York, 2005.

Ursin LØ, Hoeyer K & Skolbekken J-A, 'The Informed Consenters: Governing Biobanks in Scandinavia', in H Gottweis & A Petersen (eds), *Biobanks: Governance in Comparative Perspective*, Routledge, Abingdon, 2008: 177.

Virilio P, *The Information Bomb*, C Turner (trans), London, Verso, [1998]/2000.

Virilio P. 'Virilio: Cyberesistance Fighter: An Interview with Paul Virilio', D Dufresne (interview), J Houis (trans), *Aprés-Coup Psychoanalytic Association*, 1999. Available at www.apres-coup.org/mt/archives/title/2005/01/cyberesistance.html.

Voss CA, 'Technology Push and Need Pull: A New Perspective', 14(3) *R&D Management* 1984: 147.

Wallace AFC, *The Social Context of Innovation*, U of Nebraska P, Lincoln, [1982]/2003.

Weber S, 'Literature – Just Making It', B Massumi (trans), Afterword in J-F Lyotard & J-L Thébaud, *Just Gaming*, W Godzich (trans), U of Minnesota P, Minneapolis, [1979]/1985: 101.

Wellbon OG, 'Demeanor', 76 *Cornell Law Review* 1990–91: 1075.

Williams R & Edge D, 'The Social Shaping of Technology', 25 *Research Policy* 1996: 856.

Williams R, *Keywords: A Vocabulary of Culture and Society*, Fontana, London, [1976]/1988.

Wittgenstein L, *The Big Typescript TS 213*, CG Luckhardt & MAE Aue (eds & trans), Blackwell, Malden MA, 2005.

Wittgenstein L, *The Blue and Brown Books*, 2nd ed, Harper & Row, New York, [1958]/1965.

Wittgenstein L, *Culture and Value: A Selection from the Posthumous Remains*, GH von Wright & H Nyman (eds), A Pichler (rev ed), P Winch (trans), rev ed, Blackwell, Malden MA, [1977]/1998.

Wittgenstein L, *Last Writings on the Philosophy of Psychology: The Inner and the Outer 1949–1951, Volume 2*, GH von Wright & H Nyman (eds), CG Luckhardt & MAE Aue (trans), Blackwell, Malden MA, 1993.

Wittgenstein L, *Notebooks 1914–1916*, GH von Wright & GEM Anscombe (eds), GEM Anscombe (trans), 2nd ed, Blackwell, Malden MA, [1961]/1998.

Wittgenstein L, *On Certainty*, GEM Anscombe & GH von Wright (eds), D Paul and GEM Anscombe (trans), Blackwell, Malden MA, [1969]/1975.

Wittgenstein L, *Philosophical Grammar*, R Rhees (ed), A Kenny (trans), Blackwell, Malden MA, [1974]/1980.

Wittgenstein L, *Philosophical Investigations*, GEM Anscombe (trans), 3rd ed, Blackwell, Oxford, [1953]/1995.

Wittgenstein L, *Philosophical Occasions: 1912–1951*, JC Klagge & A Nordmann (eds), Hackett, Indianapolis, 1993.

Wittgenstein L, *Remarks on a Philosophy of Psychology: Volume 1*, GEM Anscombe et al. (eds), Blackwell, Oxford, 1991.

Wittgenstein L, *Remarks on the Foundations of Mathematics*, GH von Wright et al. (eds), GEM Anscombe (trans), 3rd ed, Macmillan, New York, 1956.

Wittgenstein L, *Tractatus-Logico-Philosophicus*, DF Pears & BF McGuinness (trans), B Russell (intro), Routledge, London, [1921]/1974.

Wittgenstein L, *Wittgenstein's Lectures on the Foundations of Mathematics, Cambridge, 1939*, C Diamond (ed), U of Chicago P, Chicago, 1976.

Wittgenstein L, *Zettel*, GEM Anscombe & GH von Wright (eds), GEM Anscombe (trans), U of California P, Berkeley, 1967.

Woods L, 'The Fall', in P Virilio, *Unknown Quantity*, Thames & Hudson, London, 2002: 150.

Žižek S, *Enjoy Your Symptom! Jacques Lacan in Hollywood and Out*, Routledge, New York, 1992.

News Comment, Film, Music

Adams R, 'Boston "witchhunt" on social media sites – and a bad week for the old guard', *The Guardian*, 22 April 2013.

'Banksy's Slave Labour mural auctioned in London', *BBC News*, 3 June 2013.

Belgrave K, 'Why I'm done with Facebook (and moving on to Twitter)', *Tarrytown-SleepyHolly*, 21 May 2013.

Brustein J, 'How freemium products use our brains against us', *Bloomberg*, 11 July 2013.

Carroll R, 'Facebook gives way to campaign against hate speech on its pages', *The Guardian*, 29 May 2013.

Cellan-Jones R, 'Sexism campaign: Facebook learns a lesson', *BBC News*, 29 May 2013.

Clark L, 'Don't Copyright Me launches to protect US schoolkids' homework', *Wired*, 12 February 2013.

Collen J, 'Maryland Board wants copying homework to be a federal offense', *Forbes*, 22 February 2013.

Combes R, 'The copyright hub', *ALCS News*, 20 May 2013.

'Companies pull Facebook ads over violent content', *CBS News*, 29 May 2013.

Dalrymple J, 'Delivering innovation vs delivering products', *The Loop Magazine*, 8 January 2013.

De Castella T & McClatchey C, 'UK riots: what turns people into looters?' *BBC News*, 9 August 2011.

Fagenson Z, 'Banksy street murals pulled from Miami auction after controversy', *Reuters*, 23 February 2013.

'Fatboy Slim DJs at House of Commons', *BBC News*, 7 March 2013.

Finch D, McIntyre S & Sundberg K, 'The Vancouver Riot hangover: crowdsourced justice and public shaming', *Centennial Reader*, June 2012.

Fox I, 'Paul Smith: a designer in his own fashion', *The Guardian*, 10 June 2013.

Games Without Frontiers, Lyrics P Gabriel, Music P Gabriel, Charisma-Universal, UK, 1980.

'Give us our Banksy back: Haringey urges Miami auctioneers to halt sale of mural', *Evening Standard*, 23 February 2013.

Glover A, 'Independent record shops say they are open for business', *BBC News*, 16 January 2013.

Guarini D, 'Record Store Day: saving independent music stores since 2008', *Huffington Post*, 20 April 2012.

Hall K, 'Jurors found guilty of Facebook and Google contempts', *Law Society Gazette*, 29 July 2013.

Halliday J, 'Facebook juror and defendant guilty of contempt', *The Guardian*, 14 June 2011.

Heisler Y, 'Tim Cook on the state of IP protection: our product cycles move much quicker than the court system', *The Unofficial Apple Weblog (TUAW)*, 21 May 2013.

Ingham T, 'BBC Album Reviews to close this month', *Music Week*, 13 March 2013.

Ingham T, 'Digital Copyright Exchange: UK Music, PPL, PRS applaud Hooper', *Music Week*, 31 July 2012.

Ingham T, 'Digital Copyright Exchange: who's going to pay for it?' *Music Week*, 31 July 2012.

'Juror admits contempt of court over Facebook contact', *BBC News*, 14 June 2011.

'Juror denies contempt of court over Facebook paedophile post', *The Guardian*, 23 July 2013.

'Jurors jailed for contempt of court over internet use', *BBC News*, 29 July 2013.

Kane GC, 'What can managers learn about social media from the Boston Marathon bombing?' *MIT Sloan Management Review*, 25 April 2013.

Kemp N, 'Consumers urge brands to boycott Facebook over domestic violence', *Marketing Magazine*, 22 May 2013.

Kotz D, 'Injury toll from Marathon bombs reduced to 264', *Boston Globe*, 24 April 2013.

Koulopoulos T, 'Innovation isn't about new products, it's about changing behavior', *Fast Company*, 31 July 2012.

Kurjata A, 'Social media, crowd-sourced justice, and the Vancouver Riots', 18 June 2011, www.andrewkurjata.ca/blog/2011/06/18/crowd-sourced-justice/.

Lee D, 'Facebook bows to campaign groups over "hate speech"', *BBC News*, 29 May 2013.

Lessig L, 'Free, as in Beer', *Wired*, September 2006, 94.

Levy A & Bensinger G, 'LinkedIn joins ESPN, Skype in shifting from free to "freemium"', *Bloomberg*, 18 December 2009.

Liu J, 'Video games embrace China's freemium model to beat piracy', *BBC News*, 4 January 2013.

Luscombe R, 'Banksy mural: I'm being scapegoated, says Miami art dealer', *The Guardian*, 22 February 2013.

Magid L, 'Apple's "next big thing" will be an innovation, not an invention', *Forbes*, 26 March 2013.

'Open letter to Facebook', *Huffington Post*, 21 May 2013.

Orlowski A, 'Cameron's "Google Review" sparked by killer quote that never was', *The Register*, 21 March 2012.

Orlowski A, 'Hargreaves' Digital Copyright Exchange will never happen', *The Register*, 23 May 2011.

Popescu I, 'Counterfeit goods exhibition at Bucharest Palace of Justice marks World Intellectual Property Day', *Romania Insider*, 26 April 2013.

Reid R, 'What to do when attacked by pirates', *Wall Street Journal*, 1 June 2012.

Robinette R, 'Twitter outlasts Facebook as a "cool" site for young people', *Trinity Tripod*, 22 April 2013.

'Social media vigilantes cloud Boston bombing investigation', *NPR* broadcast, 22 April 2013.

Steadman I, 'Reddit users are hosting a witch-hunt for the Boston Marathon bomber', *Wired*, 17 April 2013.

'Stingrays slaughtered in QLD', *ABC AM*, broadcast 13 September 2006.

Walker P, 'Boston bombing identification attempts on social media end in farce', *The Guardian*, 19 April 2013.

Wanderlust (2012). Dir: D Wain; SP: D Wain and K Marino.

Warner J, 'Intellectual Property reform cannot be dictated by Google', *The Telegraph*, 3 March 2011.

Warzel C, 'Teens explain why they don't care about Facebook anymore', *BuzzFeed*, 21 May 2013.

When Albums Ruled the World, BBC 4 Documentary broadcast 14 July 2013.

White M, 'Fatboy Slim – review', *The Guardian*, 7 March 2013.

Index